AF796486

Metallothionein

Proceedings of the »First International Meeting on Metallothionein and Other Low Molecular Weight Metal-binding Proteins«
Zürich, July 17–22, 1978

Edited by Jeremias H. R. Kägi and Monica Nordberg

1979 Birkhäuser Verlag
Basel, Boston, Stuttgart

FEBS Symposia Series 59

QP
552
.M47
I55
1978

Library of Congress Cataloging in Publication Data
International Meeting on Metallothionein and Other
Low Molecular Weight Metal-binding Proteins, lst,
Zürich, 1978.
Metallothionein.
(Experientia · Supplementum; 34)
Sponsored by the FEBS Committee for Advanced
Courses and others.
1. Metallothionein – Congresses. I. Kägi, Jeremias,
1930. II. Nordberg, Monica. III. FEBS Committee for
Advanced Courses. IV. Series.
QP552.M47I55 1978 591.1'9245 79-21385
ISBN 3–7643–1036–7

CIP-Kurztitelaufnahme der Deutschen Bibliothek
Metallothionein: proceedings of the 1. Internat.
Meeting on Metallothionein and Other Low Molecular
Weight Metal-binding Proteins, Zürich, July 17–22,
1978 / ed. by Jeremias H. R. Kägi and Monica
Nordberg. – Basel, Boston, Stuttgart: Birkhäuser,
1979.
(Experientia: Suppl.; 34) (FEBS symposia series; 59)
ISBN 3–7643–1036–7
NE: Kägi, Jeremias H. R. [Hrsg.]); International
Meeting on Metallothionein and Other Low Molecular
Weight Metal-binding Proteins ⟨01, 1978, Zürich⟩

All rights reserved. No part of this publication may be
reproduced, stored in a retrieval system, or transmitted,
in any form or by any means, electronic, mechanical,
photocopying, recording or otherwise, without the prior
permission of the copyright owner.
© Birkhäuser Verlag Basel, 1979
ISBN 3–7643–1036–7

Printed in Switzerland

P R E F A C E

Metallothionein plays a central role in trace metal metabolism. Since its first recognition by Margoshes and Vallee in 1957, a large amount of research effort has been devoted to its study in many branches of the life sciences. These efforts revealed that, upon entering the living cell, essential trace elements such as zinc and copper and toxic elements such as cadmium and mercury are passed onto this ubiquitous, low molecular weight protein. Thus, metallothionein has emerged as a key compound for the understanding of the physiological mechanism regulating the flow of metals through the organism in health and disease, of the molecular pathway of the metals in biosynthesis and degradation of metalloenzymes, and of metal toxicity and detoxification.

This book comprises the proceedings from the "First International Meeting on Metallothionein and Other Low Molecular Weight Metal-binding Proteins", held in Zürich, Switzerland, July 17 - 22, 1978. Besides highlighting the present frontiers of the title subject, it presents an up-to-date review of the established body of knowledge on the topic and thus addresses itself to workers in the field as well as to newcomers and other interested scientists.

The meeting was organized on the initiative of Dr. M. Piscator as an interdisciplinary endeavour by the Department of Biochemistry of the University of Zürich, Switzerland, and the Department of Environmental Hygiene, Karolinska Institute, Stockholm, Sweden. It was held under the joint sponsorship of the American Society of Biological Chemists, the Division of Biological Chemistry of the American Chemical Society, the FEBS Committee for Advanced Courses, the Scientific Committee on the Toxicology of Metals under the Permanent Commission and International Association on Occupational Health, and the Swiss Biochemical Society. It was attended by 38 scientists from the fields of biochemistry, protein chemistry, bioinorganic chemistry, physiology, nutrition, toxicology, and pathology. Dr. B.L. Vallee accepted an invitation to serve as honorary chairman of the meeting. Drs. J.H.R. Kägi and M. Piscator served as chairmen, Dr. M. Nordberg as secretary. The meeting was opened by welcoming addresses of Dr. A. Gilgen, Minister of Education, Dr. W. Siegenthaler, Dean of the Faculty of Medicine, and Dr. G.F. Nordberg, Secretary of the Scientific Committee on the Toxicology of Metals under the Permanent Commission and International Association on Occupational Health. A plenary introductory lecture was given by Dr. B.L. Vallee on the topic "Metallothionein: Historical Review and Perspectives". The conference was conducted mainly by holding discussions in groups and in the plenum on the working papers of the participants and on a draft report which summarized present knowledge on the conference subject and which, together with the working papers, was distributed prior to the meeting. This collective effort resulted in the completion of the report "Metallothionein and Other Low Molecular Weight Metal-binding Proteins" included in this volume. In addition to these activities, there were plenary lectures by Drs. C.-I. Brändén on the "Organization of Metal-chelating Centers in Proteins" and H.A.O. Hill on "Spectroscopic Approaches to the Study of Metalloproteins" as well as a general discussion on the topic "Metallothionein: Queries and Conjectures" that was chaired by Dr. B.L. Vallee.

The arrangement of the contributions to this volume reflects the emphasis placed on the collective effort of the participants during the meeting. The first part comprises the introductory lecture, the report and the general discussion. The second part contains the working papers submitted which were reviewed before the meeting by a preeditorial group consisting of Drs. J.H.R. Kägi, G.F. Nordberg, M. Nordberg, and M. Piscator at a meeting in Odense, Denmark, on May 8th and 9th, 1978.

On behalf of the organizers and the participants, we acknowledge the generous financial support received from the Federation of European Biochemical Societies (FEBS), the Schweizerische Naturforschende Gesellschaft, the Department of Education of the Kanton Zürich, the Permanent Commission and International Association on Occupational Health as well as from the Swedish companies Draco AB and NIFE Jungner AB. Our sincere thanks are also due to the University of Zürich for providing the facilities for the meeting.

Our special gratitude goes to Dr. F. Leuthardt, emeritus Professor and Chairman of the Department of Biochemistry of the University of Zürich, who with his continued advice and encouragement as well as with financial assistance greatly helped to make this meeting successful. Thanks are also due to Drs. P.N. Campbell, O. Hayaishi, G. Semenza, and W.J. Whelan of the International Union of Biochemistry (IUB), to Drs. M. Gruber, G. Bernardi and S.P. Datta of the Federation of European Biochemical Societies (FEBS), to Dr. B.L. Horecker of the Division of Biological Chemistry of the American Chemical Society, and to Dr. L.T. Friberg of the Scientific Committee on the Toxicology of Metals for their kind support.

The editors are greatly indebted to all contributors and especially to Drs. M. Piscator, G.F. Nordberg and Y. Kojima who took part in the planning of this volume. We also acknowledge the assistance of the editorial staff of Birkhäuser Verlag, Basel/Boston/Stuttgart and offer our thanks to the publishers who have permitted reproduction of Figures and Tables from their publications.

Finally, it is a privilege to thank our secretaries Mss. Genia de Vallier and Helena Hasselgren for their devoted effort in the preparation and organization of the meeting and in the production of this volume.

May 1979

Jeremias H.R. Kägi
Monica Nordberg

TABLE OF CONTENTS

PART II

LIST OF PARTICIPANTS

Robert D. Andersen	Laboratory of Nuclear Medicine University of California at Los Angeles 900 Veteran Avenue Los Angeles, Calif. 90024 USA
Arne Bakka	Institute for Surgical Research Division of Clinical Pharmacology University of Oslo Rikshospitalet Langesgate Oslo 1 Norway
Joseph V. Bannister	Department of Physiology and Biochemistry The University of Malta Msida Malta
Christine Berger	Biochemisches Institut der Universität Zürich Zürichbergstrasse 4 CH-8028 Zürich Switzerland
J. Lemuel Bethune	Biophysics Research Laboratory Department of Biological Chemistry Harvard Medical School Boston, Mass. 02115 USA
Frank O. Brady	Division of Biochemistry, Physiology and Pharmacology The University of South Dakota Vermillion, S.D. 57069 USA
Carl-Ivar Brändén	Department of Chemistry Agricultural College of Sweden S-750 07 Uppsala 7 Sweden
Ian Bremner	Nutritional Biochemistry Department The Rowett Research Institute Bucksburn Aberdeen AB2 9SB U.K.

M. George Cherian	Department of Pathology University of Western Ontario Health Sciences Centre London, Ontario N6A 5C1 Canada
Robert J. Cousins	Department of Nutrition Rutgers University Thompson Hall New Brunswick, N.J. 08903 USA
Gary W. Evans	USDA Agricultural Research Service Human Nutrition Laboratory 2420 Second Avenue North Grand Forks, N.D. 58201 USA
H. Allen O. Hill	Department of Inorganic Chemistry University of Oxford South Parks Road Oxford OX1 3QR U.K.
Jeremias H.R. Kägi	Biochemisches Institut der Universität Zürich Zürichbergstrasse 4 CH-8028 Zürich Switzerland
Martin D. Kamen	Department of Chemistry University of California at San Diego Box A-002 La Jolla, Calif. 92093 USA
Masami Kimura	Department of Experimental Toxicology National Institute of Industrial Health 21-1, Nagao 6-chome, Tama-ku Kawasaki 213 Japan
Margrit M. Kissling	Biochemisches Institut der Universität Zürich Zürichbergstrasse 4 CH-8028 Zürich Switzerland
Yutaka Kojima	Biochemisches Institut der Universität Zürich Zürichbergstrasse 4 CH-8028 Zürich Switzerland

Konrad Lerch	Biochemisches Institut der Universität Zürich Zürichbergstrasse 4 CH-8028 Zürich Switzerland
Alton Meister	Department of Biochemistry Cornell University Medical College 1300 York Avenue New York, N.Y. 10021 USA
Gunnar F. Nordberg	Department of Community Health and Environmental Medicine Odense University J.B. Winsløvsvej 19 DK-5000 Odense Denmark
Monica Nordberg	Department of Environmental Hygiene Karolinska Institute S-104 01 Stockholm Sweden
Robert W. Olafson	Australian Institute of Marine Science Cape Ferguson Townsville, Qld. 4810 Australia
Noriko Otaki	Department of Experimental Toxicology National Institute of Industrial Health 21-1, Nagao 6-chome, Tama-ku Kawasaki 213 Japan
James D. Otvos	Department of Molecular Biophysics and Biochemistry Yale University Medical School 333 Cedar Street New Haven, Conn. 06510 USA
Julian Overnell	Institute of Marine Biochemistry St. Fittick's Road Aberdeen AB13RA U.K.
Jiri Pařízek	Environmental Health Criteria and Standards Division of Environmental Health World Health Organization CH-1211 Geneva 27 Switzerland

Jerzy K. Piotrowski	Medical Academy of Lodz Institute of Environmental Research and Bioanalysis Narutowicza 120 A 90-145 Lodz Poland
Magnus Piscator	Department of Environmental Hygiene Karolinska Institute S-104 01 Stockholm Sweden
Pablo Pulido	Centro Médico Docente La Trinidad Apartado 80.474 Caracas 108 Venezuela
John R. Riordan	Department of Clinical Biochemistry The Hospital for Sick Children University of Toronto 555 University Avenue Toronto, Ontario M5G 1X8 Canada
Heinz Rupp	Anorganische Biochemie Physiologisch-chemisches Institut der Universität Tübingen Hoppe-Seyler-Strasse 1 D-7400 Tübingen 1 Federal Republic of Germany
Zahir A. Shaikh	Department of Pharmacology and Toxicology University of Rochester School of Medicine and Dentistry 601 Elmwood Avenue Rochester, N.Y. 14642 USA
William R.D. Smith	The Courtauld Institute of Biochemistry The Middlesex Hospital Medical School Mortimer Street London W1P 7PN U.K.
Bert L. Vallee	Biophysics Research Laboratory Department of Biological Chemistry Harvard Medical School Boston, Mass. 02115 USA

Milan Vašák — Biochemisches Institut der Universität Zürich
Zürichbergstrasse 4
CH-8028 Zürich
Switzerland

Michael Webb — MRC Toxicology Unit
Woodmansterne Road
Carshalton, Surrey SM5 4EF
U.K.

Ulrich Weser — Anorganische Biochemie
Physiologisch-chemisches Institut
der Universität Tübingen
Hoppe-Seyler-Strasse 1
D-7400 Tübingen 1
Federal Republic of Germany

Akira Yoshida — Department of Biochemical Genetics
City of Hope National Medical Center
1500 East Duarte Road
Duarte, Calif. 91010
USA

Dr. Philip D. Whanger, Department of Agricultural Chemistry, Oregon State University, Corvallis, Oregon 97331, USA, was invited to participate and contributed a working paper which was discussed during the meeting and is published in this volume. Unfortunately, he could not attend the meeting.

PART I

Metallothionein
Jeremias H. R. Kägi and Monica Nordberg, eds.
Birkhäuser Verlag, Basel/Boston/Stuttgart, 1979

METALLOTHIONEIN: HISTORICAL REVIEW AND PERSPECTIVES

Bert L. Vallee

Biophysics Research Laboratory
Department of Biological Chemistry
Harvard Medical School
and the
Division of Medical Biology
Peter Bent Brigham Hospital
Boston, Massachusetts

EARLIEST HISTORY

The detection, isolation and characterization of metallothionein is interwoven inextricably with the search for a biological role and function of cadmium: The discovery of metallothionein is the direct consequence of an inductive search for a biological role of that element. In 1941, D.P. Maliuga was the first to determine cadmium quantitatively in natural products. His publication carried the title: "Cadmium and Organisms", and appeared in Doklady Akad. Nauk. U.S.S.R., (the official Journal of the Academy of Sciences of the U.S.S.R.) 31, 145-147, No. 2, 1941; it was abstracted briefly in Chemical Abstracts 37.1, 894, 1943. He performed cadmium analyses polarographically and validated the procedure colorimetrically by means of the dithizone procedure. Maliuga analyzed a wide variety of biological materials, among them the bark of the Aspen tree (1.6×10^{-4}% dry weight) and a number of algae and certain aquatic species (Table I). In the latter a constant Cd/Zn = 1/20 ratio was found.

The last, very short paragraph of this manuscript reported the observation of this study which proved to be particularly significant and, hence, is cited in full for historical reasons:

"The latest data which are at our disposal with regard to the cadmium content in the live matter refer to its content in human organs. Thus, we have found that the kidneys of a sick person, who had died after a serious poisoning with volatile organic lead components, contained 3.3×10^{-3}% (= 33 μgm/gm wet

TABLE I

Organism	Origin	Quantity of cadmium in the dry matter of the organism, %
Sea algae		
Fam. Fucaceae, Fucus serratus	Dalnie Zelentzy (Murmansk Biolog. Station. 1938).....	3.4×10^{-5}
Fam. Fucaceae, Ascophyllum nodosum	Do........................	2.9×10^{-5}
Fam. Fucaceae, Fucus inflatus	Ekaterininsky harbour, Barents Sea. 1929..........	1.6×10^{-5}
Fam. Laminariaceae, Allaria esculenta	Dalnie Zelentzy............	1.0×10^{-4}
Fam. Laminariaceae, Phyllaria dermatodea	Do........................	1.0×10^{-5}
Fam. Laminariaceae, Laminaria saccharina	Ekaterininsky harbour......	1.4×10^{-5}
Sea flowering plants		
Fam. Potamogetaceae, Lostera marina	Do........................	2.3×10^{-5}
Sea animals		
Fam. Gastropoda, Pecten islandicus	Do........................	Traces
Fam. Asteriidae, Asterias rubens	Dalnie Zelentzy............	1.1×10^{-4}
Fam. Gadidae, Gadus collarius	Do........................	Traces

Maliuga, 1941

weight) cadmium in the raw matter. In the kidneys of another sick person who had died from a heart disease, the Cd content was equal to $3.8 \times 10^{-4}\%$, respectively. The quantity of zinc was 5-6 times that of cadmium. Along with this, cadmium was found in some other organs in the spleen, in the lungs, etc."

The cadmium content of the human kidneys seemed exceptionally large to me.

Two features of this manuscript, while implicit, deserve emphasis. First, it was published in the Mineralogy section of these Proceedings. Second, it appeared in the year 1941, when World War II had broken out. As a consequence, a decade was to elapse before the full manuscript became accessible in the West -- and yet additional time passed before an English translation appeared. A copy became available to me in the early to midfifties as a result of translation, Number R.J. 1037 of the Associated Technical Services, Inc., POB 271, East Orange, New Jersey, U.S.A., an organization which had then become engaged in making Russian scientific literature of the war time period accessible to Western scientists.

In fact, it was soon thereafter that through this channel I came upon another relevant manuscript by A.O. Voinar entitled "The Occurrence and Role of Cadmium in the Organisms of Animals and Man" in Mikroelementy y Zhizni Rastenii i Zhivotnykh, Akad. Nauk. S.S.S.R., Trudy Konf. Mikroelement, 1950, 580-583 (1952) constituting translation RJ-296 of the above services.

Although this report bears a publication date about a decade later than that referred to above, owing to the circumstances, it came into my hands at nearly the same time as Maliuga's. The precise year in which this occurred is not apparent from any of my records but -- judging by circumstantial evidence -- it must have been somewhere between 1953 and 1955. Voinar's manuscript, apparently presented initially at a conference in 1950, does not give references but reports colorimetric (dithizone) and spectrographic procedures for cadmium determinations developed by the author and carried out by one E.G. Lipskaya in his laboratory. These showed that Cd is present uniformly in the organs of representative specimens of different classes of vertebrates: fish, amphibians, reptiles and mammals, including man. The cadmium content of the liver and kidneys of the human, other mammals and -- to a lesser extent -- amphibians, were found to be highest, those of reptiles and birds were negligible.

N.B. Nasel'skii, also working in the same author's laboratory, was reported to have investigated changes in the cadmium content of human organs as a function of embryonic and postnatal development. Voinar suggested that his data established "definite trends in the distribution of cadmium in relation to age". The data in Table II were interpreted to show that the cadmium content of human embryos is high during the early stages of embryonic development, decreases at about 5 months, rises to a maximum at 6-7 months and then falls until parturition. Table III was thought to demonstrate a significant increase in the Cd content of liver, kidney and spleen with increasing age.

It should be appreciated that the data in Tables II and III are in terms of µg or mg Cd/100 g of ash. No information is given regarding the mode of ashing[1], the dry or wet weight of tissue, analytical methodology, its precision and accuracy, numbers of organs or replicate samples studied, their physiological or pathological state, nor is any statistical analysis presented. The last column of Table III entitled "Average cadmium content of organs of embryos" is quite misleading, of course. It simply averages the data of Table II at different times of development for each organ analyzed. These are scattered so widely, that they were thought to constitute variations reflecting the developmental process (see above) -- but in the absence of data to the contrary could well represent methodological error. In any event, the presentation of their average would seem to have a purpose diametrically opposite to the physiological interpretation given.

I have gone to some length to point out that it was not possible then to establish meaningful relationships between Voinar's and Maliuga's data. Be that as it may, taken in conjunction, the two publications left little doubt that cadmium is present uniformly in human tissues, and that its possible role was deserving of close inspection.

The balance of Voinar's manuscript, moreover, purported to demonstrate a hypoglycemic effect of $CdCl_2$ when injected into rabbits, thus raising the level of curiosity of the reader,

---

[1] The temperature at which samples are ashed, the weight of the total ash and its equivalent wet or dry weight are missing, but as is apparent even from some of the earliest quantitative analytical work on the element's occurrence in biological matter, Cd is lost by volatilization on dry ashing unless special precautions are taken (Cholak and Hubbard, 1944; Klein and Wichmann, 1945; Shively et al., 1949). When sulfate is added, $CdSO_4$ forms (MP 1000°C) and serves as a "fixative" allowing dry ashing at lower temperatures e.g. 450-550°C.

To avoid volatilization during dry ashing, many authors have employed wet ashing procedures, but these generally introduce contamination, as is apparent from high "blank" values. However, dry ashing may also result in low recoveries of Cd due to difficulty of leaching the element from large amounts of acid-insoluble ash of biological materials, particularly plants.

Table II

Average Cadmium Content of Various Organs in Relation to the Age of the Embryo

Age of embryo, months	Liver	Kidney	Heart	Muscle	Lungs	Skin
	Cadmium content, micrograms per 100 g. of ash					
4	1176	470	1282	1247	277	471
5	289	42	695	219	74	62
6	1648	641	199	74	58	476
7	619	134	687	334	268	354
8	468	583	-	413	264	-
9	295	344	-	141	65	79

Table III

Cadmium Content of Human Organs and Tissues at Various Ages as Determined by Emission Spectroscopy

(In milligrams per 100 grams of ash)

Organs and tissues	Age of man: 4 yrs.	18 yrs.	40 yrs.	64 yrs.	Average cadmium content of organs of embryos according to the chemical colorometric method (dithiazone)
Liver	3.5	10	16	16	0.748
Kidney	0.5	8	20	20	0.369
Testicles	0.3	0.5	1.0	0.5	-
Thyroid gland	0.5	1.0	1.0	1.0	-
Pancreas	-	1.5	1.5	-	0.287
Spleen	0.3	0.3	1.0	1.5	0.303
Muscles	1.0	3.0	3.0	3.0	0.405
Bladder	1.0	1.0	1.0	1.0	-
Stomach (muscular layer)	1.0	1.0	1.0	1.0	-
Heart	2.0	3.0	3.0	3.0	0.623
Skin	-	0.3	0.5	-	0.304
Brain	0.2	0.3	0.3	-	from traces to 0.031
Lungs	-	0.2	0.2	-	0.168

though in retrospect, to little avail. Efforts in our laboratory to reproduce those data were not successful.

Had it not been for the fact that zinc metabolism and zinc metalloenzymes had meanwhile attracted my intense interest, there would have been little incentive for my search of the cadmium literature. I had been struck by the abundance of zinc in biological matter and had developed long-range plans to pursue its occurrence and function in enzymes, areas of study virtually non-existent at that time. This decision was to form the basis for most of the work of my scientific career. Save for the essentiality of zinc to the growth and development of rodents (Todd et al.,1934; Bertrand and Bhattacherjee, 1934) and its presence in the enzyme carbonic anhydrase (Keilin and Mann, 1940) very little was known about its role at that time (Vallee and Altschule, 1949). But in short order, we were able to identify a series of zinc metalloenzymes containing stoichiometric quantities of functionally essential zinc atoms, detected both by quantitative emission spectrography and colorimetry (Vallee, 1955).

While pursuing these problems, the remarkable similarity both of the atomic structures and chemical reactivities of zinc and cadmium became of interest to me. The conjoint presence of cadmium and zinc in zinc ores, the difficulties encountered in separating them and ascertaining their identities by chemical means implied that biological matter might discriminate less between their absorption, utilization and perhaps even function than might be thought. Hence, I decided to search for cadmium proteins and enzymes, their biological roles, normal and abnormal metabolism, much as I had proceeded to study the biological role of zinc.

The operational concept of "metalloproteins" had already suggested a "transperiodic" approach to the mode of association and function of metals in biology (Vallee, 1954, 1957). This viewpoint subsequently proved valid for other metals, e.g. Co, Ni, and Mn, whose participation in biolobical processes was then also either unknown or dubious and which were certainly then not known to be constituents of metalloenzymes.

ISOLATION OF A CADMIUM PROTEIN FROM HORSE KIDNEY

I began work on the isolation of a hypothetical cadmium protein or enzyme in 1955, a year after the construction of the Biophysics Research Laboratory whose facilities had then become available. We revised the existent colorimetric dithizone method (Saltzman, 1953) and established a quantitative emission spectrographic procedure utilizing a porous-cup graphite spark excitation source for zinc and other metals (Vallee, 1955).

My first experiments simply reexamined the data referred to above, both to verify or reject their validity and to find the species and organs which would be most suitable for the work intended, based on the highest cadmium content per gram of tissue.

Both methods rapidly confirmed the presence of cadmium in most tissues and the kidney proved to be the organ containing the highest concentrations (Table IV). A search for the

TABLE IV

DISTRIBUTION OF CADMIUM IN THE HUMAN BODY (μg/g wet wt)

Kidney	36
Liver	2.6
Pancreas	1.4
Lung	0.8
Spleen	0.7
Heart	0.6
Testis	0.6
Muscle (skeletal)	0.1
Bone	0.1
Brain	0.1

species whose kidneys contained the greatest accumulations of Cd was an obvious extension of this work followed rapidly by the finding that the cadmium content of the cortex far exceeded that of the medulla; this ensured a substantial purification step simply by dissection of these two gross anatomical components of the kidney.

Table V shows the average values obtained by dithizone analyses of multiple kidney samples from a number of domestic animals. We elected to examine species containing organs of sufficient mass to facilitate the isolation of a cadmium protein in adequate amounts and yields to permit its characterization. It must be recalled that the then state of the art of protein purification made organ size an important consideration in the choice of the starting material, both for ease of separation and physico-chemical characterization. Clearly, from these points of view the equine and human kidney presented the best starting material among those examined (Table V). Solely anatomical dis-

TABLE V

CADMIUM IN MAMMALIAN KIDNEY (μg/g wet weight)

Species	Cortex	Medulla
Horse	94	1.7
Human	36	12
Cattle	0.6	0.2
Lamb	3.1	0.6
Dog	0.4	0.1

section of the equine kidney ensured nearly 100-fold accumulation of cadmium. This high cadmium/protein ratio of equine kidney cortex determined our selection of that species which presented us with a "specific metal content" optimal for further purification. This advantage seemed analogous to high "specific enzymatic activities" of tissues sought when undertaking purification of enzymes.

The successful isolation of a cadmium protein from equine kidney cortex, the first of its kind, was reported by Marvin Margoshes and Bert L. Vallee in a communication to the Editor of the Journal of the American Chemical Society 79, 4813 (1957) entitled "A Cadmium Protein from Equine Kidney Cortex", now 21 years ago. It is here reprinted with permission of the Journal of the American Chemical Society (Table VI).

The results of these earliest metal analyses revealed that in addition to cadmium, the protein which had been the objective of my search, clearly contained substantial concentrations of zinc and -- in the case of the equine renal material -- lesser but significant amounts of copper and iron.

Soon after the publication of this manuscript, fractions containing 43 mg of cadmium/gm of protein were obtained. Ultracentrifugation in a synthetic boundary cell showed the final products of seven successive fractionations to be monodisperse, with an uncorrected sedimentation value varying from 1.10 to 1.98. Electrophoresis of the final material of a later preparation in acetate, tris and veronal buffers, μ 0.05 revealed one component at pH 4.5, two at pH 5.5 and three at pH 6.5 and 8.6. The main component comprising 70% of the total material at pH 6.5 and 8.6, moved towards the anode, the two minor ones moved towards the cathode. At lower pH all components migrated toward the cathode. Cd was bound to the main and one of the minor components, both

Table VI[2]

[Reprinted from the Journal of the American Chemical Society, 79, 4813 (1957).]
Copyright 1957 by the American Chemical Society and reprinted by permission of the copyright owner.

A CADMIUM PROTEIN FROM EQUINE KIDNEY CORTEX

Sir:

Cadmium never has been demonstrated to be an integral part of a natural product although present in various species.[1,2,3] Physiological function has not been shown.

Colorimetric analyses[4] of human, horse, cow, hog, and sheep kidney for cadmium led to the choice of horse kidney cortex for fractionation.

Fractionation of horse kidney cortex with ethanol and ammonium sulfate gave a product containing 20 to 25 *milligrams* of cadmium per gram dry weight of trichloroacetic acid precipitable material in successive fractionations. Ultracentrifugation[5] in a synthetic boundary cell showed the final products of four successive fractionations to be monodisperse with a sedimentation constant (uncorrected for viscosity and diffusion) varying from 0.94 to 1.22×10^{-13}. Paper electrophoresis at *p*H 8.5 of the product of one fractionation showed three components moving toward the cathode, the slowest comprising about 70% of the total material.

The fractions were analyzed colorimetrically[4] and by emission spectrography[6] (Table I). The cadmium content rises throughout the fractionation, a 30-fold increase from the first extract to the product. Cadmium is not removed by dialysis at *p*H 7, but is by treatment with hot trichloroacetic acid. With the exception of zinc, the other metals present initially and introduced during fractionation are removed and are low in concentration in the final material. Isomorphism cannot be excluded as an explanation of the substantial, although lesser, increase of zinc content as fractionation proceeds, nor can it be ruled out that cadmium is associated with one and zinc with another of the three electrophoretic fractions.

The product contains 14% nitrogen, measured on the material precipitated by trichloroacetic acid. It reacts positively to the biuret and ninhydrin tests. Hydrolysis and paper chromatography showed serine, glycine, aspartic, and glutamic acids, among other amino acids not identified. The last fraction (Table I, Fraction VII) contains about 1% of hexoseamine.[7] A carbazole test for uronides[8] was negative.

There is no ultraviolet absorption maximum near 280 mμ at *p*H 7 or *p*H 12, indicating a low content of aromatic groups. Absorption bands have not been found in the visible region. The infrared spectrum of a potassium bromide pellet of the lyophilized product closely resembles those obtained for several proteins.[9]

The low sedimentation constant and high metal content of this material are indicative of a low molecular weight protein, probably containing a small number of cadmium atoms per molecule.

TABLE I

EMISSION SPECTROGRAPHIC AND COLORIMETRIC ANALYSES OF HORSE KIDNEY CORTEX FRACTIONS

Cadmium determined by spectrography[6] and by colorimetry[4]; all other metals determined spectrographically. Protein measured by dry weight of material precipitated by trichloroacetic acid. Data expressed as μg./g. wet weight of cortex, and as μg./g. protein for the fractions.

	Cortex	Fraction I	Fraction II	Fraction III	Fraction IV	Fraction V	Fraction VI	Fraction VII
Preparation E: colorimetric cadmium	82.7	1130	3050	17,100	14,300	*a*	*b*	24,500
Preparation C: colorimetric cadmium	83.7	754	3440	8,900	9,990	17,100	21,600	22,400
Spectrographic cadmium	137	687	2630	9,730	*c*	*c*	20,700	24,200
Spectrographic zinc	91	340	1440	3,600	3,330	*c*	4,910	5,880
Other metals (spectrographic)[e]	829	2410	2200	4,910	3,190	45,200[d]	15,400[d]	2,470

[a] Protein concentration too small to measure by trichloroacetic acid precipitation. [b] Sample lost. [c] Sample size inadequate. [d] Contamination with Mg and Ca introduced with ammonium sulfate. [e] Mg, Ca, Ba, Sr, Al, Fe, Mn, Cr, Pb; Na, K, Cu not determined.

Characterization of this unusual natural product is in progress.

This work was supported by the National Institutes of Health and the Howard Hughes Medical Institute.

(1) D. P. Maliuga, *Compt. rend. Acad. Sci. U.S.S.R.*, **31**, 145 (1941).
(2) A. K. Klein and H. J. Wichmann, *J. Assoc. Off. Agric. Chem.*, **28** 257 (1945).
(3) A. O. Voinar, *Trudy Konf. Mikroelement* 1950, 580 (1952); *Akad. Nauk U.S.S.R.*, Translation R-J-296, of Associated Technical Services, East Orange, New Jersey.
(4) B. E. Saltzman, *Anal. Chem.*, **25**, 493 (1953).
(5) The ultracentrifugations were done by Mr. Paul M. Reilly of the Biophysics Research Laboratory.
(6) B. L. Vallee, in "Advances in Protein Chemistry," **10**, 317 (1955).
(7) R. J. Winzler, in "Methods of Biochemical Analysis," D. Glick, ed., Vol. 2, Interscience Publishers, New York, N. Y., 1955, pp. 292–293.
(8) Z. Dische, *ibid.*, p. 343.
(9) L. J. Bellamy, "The Infrared Spectra of Complex Molecules," John Wiley and Sons, Inc., New York, N. Y., 1954, pp. 192–196.
(10) Fellow of the Kettering Foundation, Dayton, Ohio.

BIOPHYSICS RESEARCH LABORATORY OF THE DEPARTMENT OF MEDICINE
HARVARD MEDICAL SCHOOL AND PETER BENT BRIGHAM HOSPITAL
BOSTON 15, MASSACHUSETTS
MARVIN MARGOSHES[10]
BERT L. VALLEE

RECEIVED AUGUST 1, 1957

2 This whole article including its Table will be referred to as Table VI in the present manuscript.

giving a positive ninhydrin reaction on paper. Zinc specific staining with dithizone further indicated that zinc was also bound to the main component (Vallee, 1958).

A summary of this work, one year subsequent to the first publication, concludes: "Thus far no biological activity has been ascribed to this protein. The remarkable aggregation of cadmium would lead to the inference that it has a significant biological function. But at present, it remains a metalloprotein in search of biological activity", an epitaph which remains valid some 20 years and many studies and publications later.

METALLOTHIONEIN

Meanwhile, Dr. Margoshes had left the laboratory to join the National Bureau of Standards and Dr. Kägi decided to pursue the characterization of this material to whose knowledge he has contributed so importantly ever since.

We were able to isolate it, demonstrate its protein character containing 5.9% cadmium, 2.2% zinc, 0.45% iron, 0.18% copper, 14.9% nitrogen and 8.5% sulfur, the latter two equivalent to 16.4% and 9.3%, respectively, for the metal-free protein (Kägi and Vallee, 1961). In view of the high metal and sulfur contents, we named the metal containing protein metallothionein and its apoprotein thionein. This nomenclature was deliberate, as the data suggested the possibility that the same apoprotein might well turn out to complex either any one of these metals -- and yet others -- or different combinations of those found. In the light of subsequent events, this conjecture proved valid and the choice of terminology prudent.

Metallothionein was purified and characterized to homogeneity as reported in two publications by Kägi and Vallee in 1960 and 1961. Table VII summarizes the essential properties of vertebrate metallothioneins as now known (Kägi et al., 1974).

At that time, the preparative and analytical methods were laborious and exceedingly time-consuming, requiring nearly 3 months from initiation to completion with very poor yields (Kägi and Vallee 1960, 1961) compared to about one week by later procedures with analogous results (Kägi et al., 1974).

It is important to recall the reasons for the technical difficulties that were then experienced, both due to the state of the art of metal analyses and of protein purification. Determinations for Cd (or Zn) had to be performed chemically by means of diphenylthiocarbazone, limiting the total number of analyses that could be completed to about 12 per man day (i.e. 6 in duplicate).

TABLE VII

GENERAL PROPERTIES OF VERTEBRATE METALLOTHIONEINS

Sources:	Kidney, Liver
Molecular Weight:	6500-6900 (Thionein: 6100)
Elemental Composition:	16.5% N 10.9% S 6-11% Metal (Cd + Zn)
Amino Acid Composition:	30-35% Cys 1 g Atom Metal/3 Cys 1 Met/Chain no Trp, Phe, Tyr, His no Disulfide Bridges
Structure:	Primary - single chain Secondary/tertiary - tightly folded chain Axial ratio - 6 (Prolate ellipsoid)

An equivalent number of quantitative spectrographic analyses to determine all metals present in each sample i.e. a total of 120 analyses, required about 3 weeks from start to finish. This should be compared with 30 determinations per element per hour(!) now feasible.

The laborious metal determinations of the early work, albeit sensitive and accurate, imposed severe limitations on the rate of progress, since they constituted the critical monitoring step: in the absence of any known biological activity, purification depended entirely on the metal content of fractions. For a considerable period of time, work was confined largely to dithizone analyses in search for the presence of cadmium, the element whose presence and possible role had induced the search. Other metallothioneins in this or other tissues complexed with different metals would not have been detected by monitoring their isolation through cadmium measurements, of course. We found the Zn, Cu and Fe content through spectrographic analyses. However, spectral analyses of all samples obtained by fractionation were clearly not feasible, for reasons of time and economy.

AMINO ACID COMPOSITION, CADMIUM CONTENT AND ABSORPTIVITY

An unusual problem, intrinsic to the amino acid composition of the protein further complicated the problem of isolation in a manner which was not apparent at the outset: metallothionein shows no absorption band between 260 to 300 nm, a region where proteins containing the aromatic amino acids phenylalanine, tyrosine and tryptophan characteristically absorb radiation. Remarkably, these amino acids are completely absent from metallothionein, as are histidine and cystine -- while cysteine is very abundant (see below), a most unusual amino acid composition. Hence, the specific absorbance at 278 nm, which usually serves to guide the isolation of a given protein, could not be employed in this instance.

Actually cadmium metallothionein turned out to exhibit a characteristic absorption spectrum: a cadmium-dependent, broad shoulder at 250 nm ultimately proved to be of critical value and importance for its isolation and characterization. Had it not been for the absence of the aromatic amino acids, this absorption near 250 nm due to cadmium mercaptide complexes might not have been detected at this early stage. Fortuitously, the lack of the absorption at 278 nm provided a "window" making the cadmium-mercaptide chromophore visible.

This absorption -- indirectly --reveals the most striking and unusual feature of the protein: cysteinyl residues account for 30-35% of the total amino acid composition, the highest cysteine and, hence, -SH content of any protein known. There are no S-S bridges. Instead, each three cysteines bind one atom of Cd (or other metal) (Kägi and Vallee, 1960, 1961). The cadmium mercaptide chromophore accounts for the absorbance at $\sim$ 250 nm which is abolished by removal of Cd from metallothionein and restored by readdition of Cd to thionein.

In thionein, the presence and number of cysteinyl residues was established by titration with Ag^+, pCMB, N-ethylmaleimide, and, in metallothionein, by the former two methods. In contrast to cadmium mercaptides, zinc mercaptides absorb at lower wavelengths, i.e. 215 nm, consistent with studies of Zn and Cd complexes of mercaptoethanol (Kägi and Vallee, 1960, 1961). The stability constants of the Cd- and Zn-thionein species have been measured.

Considerable confusion has arisen from the non-integral values of amino acid analyses of seemingly homogeneous material, obtained from the same organ of the same animal of a given species. As has long since been known, the existence of isometallothioneins satisfactorily accounts for these seeming discrepancies. The recognition of multiple isometallothioneins due to mutational changes have, in fact, greatly advanced insight regarding the possible evolution of this unusual protein.

Much as the search for a Cd protein motivated the investigations leading to the isolation of this protein, it will be recalled that the earliest results of spectrographic data established its association with other metals e.g. Zn, Cu and Fe in the native state (Margoshes and Vallee, 1957) and called for searches of metallothioneins containing different metals in other organs and species. Such studies were facilitated greatly by the subsequent introduction of atomic absorption spectrometry, which seemingly -- all at once -- altered the outlook for success by removing earlier analytical obstacles.

From 1957 to 1970, the protein was studied almost exclusively in my laboratory and I shall not recapitulate the extension of the work to the proteins in the liver of both the horse, human and rabbit -- the latter by Piscator, 1964. Subsequent to 1970 work was extended to the rat, calf, dog, seal, mouse, chicken and fish, and most recently to mollusks, crustaceans, yeast and Neurospora crassa. Establishment of primary sequences began quite some time ago and Professor Kägi and his collaborators, who have pioneered in this field, will cover these subjects in considerable detail.

I will limit myself here to a summary of the essential general criteria which have emerged for the identification of metallothioneins:

<u>Criteria for the Identification of Metallothioneins</u>

1. Their metal content is very high, usually comprising Cd, Zn or Cu.
2. Their cysteine -- and hence potential -SH content is similarly remarkable, constituting about 30-35% of the total amino acid composition.
3. The ratio of metal ions bound to -SH groups is 1/3.
4. Generally, there are 7 g atom of Me^{2+} per 20 to 21 -SH groups of metallothionein.
5. Aromatic amino acids, histidine and disulfides are usually absent. The mammalian proteins contain only a single methionine.
6. The metallothioneins exhibit no protein absorption band near 280 nm.
7. Cd-thionein absorbs radiation near 250 nm. Other metallothioneins absorb radiation at the maximum of their particularly predominant metal mercaptides.
8. The molecular weights of these proteins range from 6000-7000.
9. Gel exclusion chromatography and the position of an excluded peak corresponding to that of a <u>known</u> metallothionein is insufficient to identify a <u>new, unknown</u> metallothionein.

In view of their importance, a few comments on some spectroscopic properties seem indicated, since they have served as the foundation for much of the current work.

ABSORPTION SPECTRA AND OPTICAL ROTATORY DISPERSION OF ZINC AND CADMIUM METALLOTHIONEINS.

Zinc and cadmium proteins do not absorb radiation in the visible region of the spectrum. However, native metallothioneins containing both zinc and cadmium, isolated from equine and human kidney, absorb radiation, maximally near 250 nm. The molar difference extinction coefficient for cadmium-thionein is approximately 1.45×10^{4} $M^{-1}cm^{-1}$ (Kägi and Vallee, 1961; Pulido et al., 1966). This absorption maximum reflects a charge transfer band of cadmium mercaptide complexes such as that observed in the presence of excess of mercaptoethanol which has a molar difference extinction coefficient of approximately 1.5×10^{4} $M^{-1}cm^{-1}$ at 25°. While metallothionein is the only naturally occurring cadmium protein known, cadmium displaces the zinc of liver alcohol dehydrogenase, generating an absorption maximum near 250 nm which long ago signalled the involvement of -SH groups in binding the metal to that enzyme (Druyan and Vallee, 1964), as confirmed later on by X-ray diffraction structure analysis of that enzyme (Brändén et al., 1975). Complete replacement of cadmium by zinc in metallothionein results in a zinc mercaptide charge transfer band at 215 nm (Kägi and Vallee, 1961).

The cadmium-mercaptide chromophore is optically active (Ulmer et al., 1962). The rotatory dispersion of the protein is plain at wavelengths longer than 300 nm; the rotatory dispersion of thionein, the metal-free protein, remaining, plain to wavelengths as low as 244 nm.

The binding of lead to thionein generates a broad, low absorption band near 340 nm and a marked but diffuse increased absorbance at wavelengths shorter than 300 nm. These absorption bands result in complex anomalous rotatory dispersion of the lead protein characterized by a trough at 350 nm, and a sharp peak at 220 nm. Two or more Cotton effects appear to be responsible, but it has not yet been possible to isolate their individual contributions to the total dispersion.

The protein from human kidney cortex (Pulido et al., 1966) exhibits large positive Cotton effects centered at 254 nm, the region of the metal mercaptide chromophore. The effect is reversibly abolished and restored by removal and addition of Cd^{2+}.

The magnitude of the 254 nm Cotton effect of human and equine metallothionein is a function of the cadmium content. The sum of the difference rotations, $\Delta[\alpha]$, of thionein and of metallothionein at 268 and 245 nm indicate a total amplitude of the Cotton effect of 1570° for human and 2250° for the equine protein, in direct proportion to their cadmium contents. Hence, based on cadmium, in both proteins the Cotton effect has a total molar amplitude, $[M]^T$, of 42,500°, reflecting the cadmium content. The almost identical rotatory dispersion characteristics of the two metallothioneins support the postulate of a unique, sterically limited, asymmetric orientation of the multidentate metal ligand sites, examples of extrinsic Cotton effects due to the interaction of a chromogenic group, not part of the peptide chain, with asymmetric loci of these proteins (Ulmer and Vallee, 1965).

More recent circular dichroism studies have shown that this cadmium Cotton effect is the result of several optically active transitions. The difference ellipticity spectrum of cadmium-thionein versus thionein shows positive maxima at 257 nm ($\Delta[\Theta]_{Cd}$ = +36,000) and at 224 nm ($\Delta[\Theta]_{Cd}$ = +70,000) and a negative one at 240 nm ($\Delta[\Theta]_{Cd}$= -12,000). Zinc-thionein exhibits analogous circular dichroism features associated with the zinc-mercaptide chromophore at lower wavelengths. The difference ellipticity spectrum of zinc-thionein versus thionein displays positive maxima at 239 nm ($\Delta[\Theta]_{Zn}$ = +15,000) and at 222 nm ($\Delta[\Theta]_{Zn}$ = +29,000). A negative maximum occurs at 2o7 nm ($\Delta[\Theta]_{Zn}$ = -7,400).

While the Cotton effects of metallothionein are detectable only by measuring rotation below 300 nm, the rotatory power of the optically active cadmium-mercaptide absorption band renders the specific rotation more positive at longer wavelengths. In the equine protein, e.g., this is reflected at wavelengths as far removed from the chromophore as the sodium D line where measurements reveal a shift from -59° to -20° on addition of cadmium to thionein. If the rotational contribution of the metal chromophore were to remain unrecognized, the change in specific rotation consequent to cadmium removal or restoration might be attributed, erroneously, to an alteration in protein conformation.

FUNCTION OF METALLOTHIONEIN

The function of metallothioneins has been and remains unknown, but there has been no dearth of speculation regarding their role. Our first discussion of its properties and possible biological function indicated that its physical properties are consistent with a wide range of potential homeostatic mechanisms including e.g. transport, catalysis, storage, immune phenomena and detoxification (Kägi and Vallee, 1960). Clearly, these speculations constituted an enumeration of the most obvious possibilities, but proof or refutation of each one of these could engage the life's work of many, perhaps without palpable demonstration of the validity of any one of them.

There has been a surfeit of explanations regarding the role(s) of metallothioneins but knowledge has not increased proportionately, confirming a favorite aphorism: "What we do not know, we explain to each other". Moreover, given the human condition "when experiment fails to confirm a favorite speculation -- this rarely results in its abandonment". Detoxification of metals early became a favorite in the "horse race of hypothetical functions"; the subsequent social preoccupations with pollution and environmental contamination provided further impetus for this view.

Quite some time ago and prior to the isolation of metallothionein, Schroeder (1956) classified metal ions in biological systems as either "normal" or "abnormal", though the basis of this categorization was as bizarre as the classification was simplistic. According to Schroeder, cadmium was "abnormal". Even though it takes but one manuscript to place a hypothesis on record, it requires decades and dozens of publications to change the resultant perceptions. Schroeder greatly reinforced the impact of these views through his subsequent writings (Schroeder and Balassa, 1961; Schroeder, 1965) in which he stated that, in fact, cadmium causes essential hypertension. It was not long until it became an environmental enemy par excellence.

The description of Itai-itai disease in Japan and nephrotoxic disorders due to e.g. cadmium or mercury poisoning generated both a suitable climate for studies disposed to detect specific biological mechanisms for the defense against "bad" metals.

Aided by the fact that metal ions can induce formation of the protein, metallothionein, many workers interested in these related problems became aware of its properties which seemed to make it an ideal candidate for metal detoxification. As a consequence, the numbers of publications on metallothionein, 9 between 1957 to 1967, rose exponentially to more than 120 during the succeeding 10 years and publications asserting a role for metallothionein in detoxification have mushroomed. For a while at least, that particular speculation was accepted so unhesitantly that it was no longer felt necessary to prove its valditiy.

Having received a fair proportion of the manuscripts of that period for review, I hesitate to be explicit regarding the basis of these developments. Leaving aside the always real possibility that this protein does serve some function in detoxification, it was a most difficult period in which to obtain research support, and there may well have been a justified and understandable interest in scientific survival, encouraging the orientations, direction and objectives both of work and publications. However, the exclusive preoccupation of many workers with a detoxifying role of metallothionein may have been excessive and accounts, at least in part, for the fact that different hypotheses regarding the function of that protein were not examined with similar avidity.

Factually, the detoxification hypothesis was inferred by some to have received early support from the finding in my laboratory that certain preparations of metallothionein isolated from human kidneys obtained at post-mortem examination contained mercury; the patients whose kidneys were employed had received mercurial diuretics for the therapy of cardiac failure, and this was presumed to be the source of the metal (Pulido et al., 1966). Subsequently, mercury has been found in metallothionein of liver and kidney of rats by other workers.

In vitro, metallothionein binds many heavy metals but in vivo, thionein appears to bind only cadmium, zinc, copper and mercury. Significant amounts of metallothionein have been detected in fetal tissue, suggesting a pre- and postnatal role.

The discovery of forms of metallothionein in human liver which contains solely zinc (Bühler and Kägi, 1974) suggests that, at least in this tissue, the primary biological role of the protein might be related to zinc. An involvement in zinc transport, storage or detoxification is equally plausible, since no physiological system performing such functions for Zn has been identified yet. The discovery that rat liver metallothionein is synthesized in response to parenteral administration of zinc (Webb, 1972; Piotrowski et al., 1974; and Davies et al., 1973) also gives credence to a specific role in zinc metabolism. But on the other hand, there is no direct evidence that metallothionein does have such a role.

An involvement of metallothionein in copper metabolism has also been suggested, since a copper protein, thought to be metallothionein, has been isolated from liver (Rydén and Deutsch, 1978). Alterations in copper homeostasis in Wilson's disease have been thought to result from synthesis of an abnormal "metallothionein" with increased affinity for copper (Evans et al., 1973).

A role of metallothionein in amino acid transport analogous to glutathione (Tate et al., 1973; Meister, 1973) has also been proposed (Kägi et al., 1974).

Taking a completely contrary point of view, there is the possibility that the apoprotein, thionein, rather than the metal-containing protein plays a cardinal metabolic role. If this were so, then Zn, Cd, Cu and Hg might inhibit an essential biological function. The reported increase in production of the protein in instances of heavy metal poisoning could then be interpreted to be an attempt to overcome this inhibition.

I am unaware of a specific, biological role that can be attributed to metallothionein so far, though recent studies of J.L. Bethune (unpublished data) suggest an unexpected potential: The protein selectively binds GTP. This could signal a role of metallothionein in mediating transfer of this nucleotide and, hence, its involvement in nucleic acid synthesis, transcription, translation or formation of coenzymes. While this, too, is a speculative hypothesis, it points up new avenues of thought on the capacity of this protein which deserve future exploration.

Having now been involved with this protein for some twenty years, I can't resist joining the fun by adding some new speculations, just so that the old ones don't become orthodox and established, albeit without cause.

Ever since we found metallothionein, we have performed experiments too numerous to recapitulate to identify its possible biological role. We have examined its potential enzymatic activities, since cadmium is known to activate many enzymes (Table VIII) toward numerous and different substrates. We have studied its potential hormonal actions, its possible participation in systemic and cellular transport, oxido-reduction of -SH groups, maintenance of cellular redox potentials, and detoxification, of course, particularly regarding lead intoxication (Vallee and Ulmer, 1972). In our hands, at least, it has not proven possible to reach conclusions as to the function of this protein.

The remarkable amino acid composition, metal binding capacity and size are the outstanding features of the molecule. I do not know of any molecule like it! In what manner -- if at all -- could its composition serve as a guide to its function? Perhaps it could serve in aspects of biochemistry which have yet to be explored and where the active molecular species are presently quite as unknown as the underlying biochemistry.

In addition to those mentioned earlier, embryonic development, mitogens, and cellular differentiation come to mind as areas just beginning to be explored by biochemical means. It is conceivable that molecules of unusual composition serve these objectives much as DNA and RNA function in the transmission of genetic information, histones in cell division and fibrous proteins such as collagen and elastin in connective tissue stabilization and muscle in contraction and motility.

TABLE VIII

ENZYMATIC ACTIVITIES ENHANCED BY CADMIUM

Enzyme	Source	Reference
Acid phosphatase	rat liver	1
Adenosine triphosphatase	fowl red blood cells	2
Alanyl leucine dipeptidase	bacterial	3
Alkaline phosphatase	rabbit and rat liver; rat lung	1,4
δ-Aminolevulinic acid dehydratase	fowl liver	2
Amylase	malt	5
Arginase	yeast	6,7
Carboxypeptidase A (esterase)	bovine pancreas	8
Carboxypeptidase B (esterase)	swine pancreas	9
Carnosinase	swine kidney	10
Cholinesterase	rat brain, spleen, heart, kidney	11
Cytochrome oxidase	pigeon brain (B_{12} deficient)	12
Glucose oxidase	bacterial	13
Glucose 6-phosphate dehydrogenase	mouse liver and heart	14
Glutamate formimino transferase	bacterial	15
Histidine ammonia lyase	bacterial; rat and guinea pig liver	16,17,18
Malic dehydrogenase	mouse liver and heart	14
Oxaloacetate decarboxylase	bacterial	19,20
Phosphopyruvate carboxylase	plant	21
Phosphopyruvate hydratase	yeast	22,23
Phosphorylase	rat liver	24
Pyruvate dehydrogenase	bacterial	25
Pyruvate decarboxylase	wheat germ; yeast	26,27
Prolidase	swine kidney	28
Succinic dehydrogenase	pigeon liver; heart and skeletal (B_{12} deficient)	12

References to Table VIII

1. Nowak, B. 1969. Med. Pracy 20:333
2. Kench, J.E., Gubb, P.J.D. 1970. Biochem. J. 120:27P
3. Berger, J., Johnson, M.J., Peterson, W.H. 1938. J.Biol. Chem. 124:385
4. Kosmider, S., Zacharewicz, M., Zajusz, K. 1968. Arh. Hig. Rada Toksikol. 19:191
5. Hultin, E. 1956. Ark. Kemi 2:135
6. Edlbacher, S., Zeller, A. 1937. Z. Physiol. Chem. 245:65

7. Voinar, A.O. 1952. Mikroelem. Zhizni Rast. Zhivotn. Tr. Konf., 1st, Moscow, 1950, 580
8. Vallee, B.L. 1964. Fed. Proc. 23:8
9. Folk, J.E., Gladner, J.A. 1961. Biochim. Biophys. Acta 48:139
10. Rosenberg, A. 1960. Biochim. Biophys. Acta 45:297
11. Pham-Huu-Chanh, Plancade, Y. 1971. Biochem. Pharmacol. 20:729
12. Nepesova, Z.N. 1969. Zdravookhr. Turkm. 2:23
13. Dalby, A., Blackwood, A.C. 1955. Can. J. Microbiol. 1:733
14. Weber, C.W., Reid, B.L. 1969. Toxicol. Appl. Pharmacol. 14:420
15. Kato, A., Nakaya, A., Shinomura, Y., Suda, M. 1954. J. Biochem. Tokyo 41:771
16. Kato, A., Yoshioka, Y., Watanabe, M., Suda, M. 1955. J. Biochem. Tokyo 42:305
17. Suda, M., Tominata, K., Nakaya, A., Kato, A. 1953 J. Biochem. Tokyo 40:257
18. Koizumi, T., Uchida, M., Ichihara, K. 1956. J. Biochem. Tokyo 43:345
19. Plaut, G.W.E., Lardy, H.A. 1949. J. Biol. Chem. 180:13
20. Herbert, D. 1951. Symp. Soc. Exp. Biol. 5:52
21. Walker, D.A. 1957. Biochem. J. 67:73
22. Malmstrom, B.G. 1959. The Enzymes, ed. P.D. Boyer, H. Lardy, K. Myrbäck, V: 471. New York: Academic
23. Wold, F., Ballou, C.E. 1957. J. Biol. Chem. 227:313
24. Sporn, A., Dinu, I., Stoenescu, L. 1970. Rev. Roum. Biochim. 7:299
25. Yamamoto, M. 1955. Kekkaku 30:473
26. Green, D.E., Herbert, D., Subrahmanyan, V. 1941. J. Biol. Chem. 138:327
27. Singer, T.P., Pensky, J. 1952. J. Biol. Chem. 196:375
28. Neuman, R.E., Smith, E.L. 1951. J. Biol. Chem. 193:97

Mitogens and messenger molecules serving the induction of organ systems, e.g. nerve, and blood vessel growth, relate to biological problems which are poorly understood but highly specific. Perhaps such processes depend on molecules whose composition and structure reflect the tasks to which they are devoted, much as in the examples cited above.

I am hardly buoyed by having to conclude the discussion of my twenty years with metallothionein on some more speculations. While provocative perhaps, they are lacking totally in experimental support -- at best they are the fiction of my imagination. But, who knows? -- The truth may be stranger than fiction!

ACKNOWLEDGMENT

This work was supported by Grant-in-Aid GM-15003 from the National Institutes of Health of the Department of Health, Education and Welfare.

REFERENCES

Bethune, J. (unpublished data).

Bertrand, J. and Bhatacherjee, R.D. (1934) Comp. Rend. Acad. Sci. 198, 1923.

Brändén, C.I., Jörnvall, H., Eklund, H. and Furugren, B. (1975) in The Enzymes, Vol. XI, P.D. Boyer, ed., 3rd Ed., 104

Bühler, R.H.O., and Kägi, J.H.R. (1974) FEBS Lett. 39, 229.

Davies, N.T., Bremmer, I., Mills, C.F. (1973) Bioch. Soc. Trans. 1, 985.

Druyan, R. and Vallee, B.L. (1964) Biochem. 3, 944.

Evans, G.W., Dubois, R.S. and Hambidge, J.M. (1973) Science 181, 1175.

Kägi, J.H.R. (1970) 8th Int. Congr. Biochem., Switzerland, Sept. 1970, abstracts p. 130-131.

Kägi, J.H.R., Himmelhoch, S.R., Whanger, P.D., Bethune, J.L., and Vallee, B.L. (1974) J. Biol. Chem. 249, 3537.

Kägi, J.H.R. and Vallee, B.L. (1960) J. Biol. Chem. 235, 3460.

Kägi, J.H.R. and Vallee, B.L. (1961) J. Biol. Chem. 236, 2435.

Kägi, J.H.R., Koyima, Y., and Berger, C. (1978) Fed. Proc. 37, 1286.

Keilin, D. and Mann, T. (1940) Biochem. J. 34, 1163.

Maliuga, D.P. (1941) Doklady Akad. Nauk, U.S.S.R. 31, No. 2, 145; Translation RJ-296 of the Associated Technical Services, Inc., POB 271, East Orange, New Jersey, U.S.A.; Chem. Abstr. 37.1 (1943), 894.

Margoshes, M. and Vallee, B.L. (1957) J. Am. Chem. Soc. 79, 4813.

Meister, A. (1972) Science 80, 33.

Piotrowski, J.K., Trojanowska, B., Wisniewska-Knypl, J.M., Bolanowska, W. (1974) Toxicol. Appl. Pharmacol. 21, 11.

Piscator, M. (1964) Nord. Hyg. Tidskr. 45, 76.

Pulido, P., Kägi, J.H.R., and Vallee, B.L. (1966) Biochem. 5, 1768.

Rydén, L. and Deutsch, H.F. (1978) J. Biol. Chem. 253, 519.

Saltzman, B.E. (1953) Anal. Chem. 25, 493.

Schroeder, H.A. (1956) Adv. Int. Med. 8, 259.

Schroeder, H.A. (1965) J. Chron. Dis. 18, 647.

Schroeder, H.A. and Balassa, J.J. (1961) J. Chron. Dis. 14, 236.

Tate, S.S., Ross, L.L., Meister, A. (1973) Proc. Natl. Acad. Sci. USA 70, 1447.

Todd, W.R., Elvehjem, C.A. and Hart, E.B. (1934) Am. J. Physiol. 107, 146.

Ulmer, D.D., Kägi, J.H.R. and Vallee, B.L. (1962) Biochem. Biophys. Res. Commun. 8, 327.

Ulmer, D.D. and Vallee, B.L. (1965) Adv. Enzymol. Vol. 27, 37.

Vallee, B.L. (1955) Adv. Prot. Chem., Vol. 10, 317.

Vallee, B.L. (1954) in Principles of Internal Medicine, T.R. Harrison, ed., 1st Ed., 437.

Vallee, B.L. (1957) in Principles of Internal Medicine, T.R. Harrison, ed., 2nd Ed., 465.

Vallee, B.L. (1960) Proc. IVth Int. Congr. Biochem., Symp. 8, Vienna, 1958. Permagon Press Ltd., London, 138.

Vallee, B.L. and Altschule, M.D. (1949) Physiol. Rev. 29, 370.

Vallee, B.L. and Ulmer, D.D. (1972) Ann. Rev. Biochem. 41, 91.

Voinar, A.O. (1962) in Mikroelementy v Zhizni Rastenii i. Zhivotnykh, Akad. Nauk, S.S.S.R., Trudy Konf. Mikroelement, 1950, 580-583; Translation RJ-296 of the Associated Technical Services, Ind., POB 271, East Orange, New Jersey, U.S.A.

Webb, M. (1972) Biochem. Pharmacol. 21, 2751.

METALLOTHIONEIN AND OTHER LOW MOLECULAR WEIGHT METAL-BINDING PROTEINS

Report from the "First International Meeting on Metallothionein and Other Low Molecular Weight Metal-binding Proteins", organized by the Department of Biochemistry, University of Zürich, Zürich, Switzerland, the Department of Environmental Hygiene, Karolinska Institute, Stockholm, Sweden, and the Scientific Committee on the Toxicology of Metals under the Permanent Commission and International Association on Occupational Health, held in Zürich, Switzerland, July 17 - 22, 1978.

Andersen, R.D., Bakka, A., Bannister, J.V., Berger, C., Bethune, J.L., Brady, F.O., Brändén, C.-I., Bremner, I., Cherian, M.G., Cousins, R.J., Evans, G.W., Hill, H.A.O., Kägi, J.H.R., Kamen, M.D., Kimura, M., Kissling, M.M., Kojima, Y., Lerch, K., Meister, A., Nordberg, G.F., Nordberg, M., Olafson, R.W., Otaki, N., Otvos, J.D., Overnell, J., Pařízek, J., Piotrowski, J.K., Piscator, M., Pulido, P., Riordan, J.R., Rupp, H., Shaikh, Z.A., Smith, W.R.D., Vallee, B.L., Vašák, M., Webb, M., Weser, U., and Yoshida, A.

Edited by Monica Nordberg and Yutaka Kojima

PROLOGUE

This report constitutes the outcome of a group effort of 38 scientists from different fields who participated in the "First International Meeting on Metallothionein and Other Low Molecular Weight Metal-binding Proteins", held in Zürich, Switzerland, July 17 - 22, 1978. The purpose of this meeting was to summarize and appraise present knowledge on metallothionein and other low molecular weight metal-binding proteins in order to facilitate and stimulate further research in this field. Practically, this collaborative effort was accomplished by centering discussions on a draft report which was prepared by M. Nordberg and Y. Kojima from published data and from working papers submitted by the participants and which was distributed to the participants in advance of the conference. The scientific discussions and the accompanying rewriting of the draft report during the meeting took place mainly in four subgroups which dealt with the

1) Occurrence and chemical characterization of metallothionein and other low molecular weight metal-binding proteins (Sections 3,4 and 8).
   Participants: Kissling (chairman), Andersen, Bannister, Berger, Brady, Bremner, Kamen, Kimura, Kojima, Lerch, Meister, Olafson, Otaki, Overnell, Pařízek, Piotrowski, Pulido, Riordan, Vallee, Weser, Yoshida.

2) Physicochemical properties of metallothionein (Section 5).
   Participants: Rupp (chairman), Bannister, Bethune, Brändén, Hill, Kägi, Kamen, Kimura, Otvos, Vallee, Vašák, Weser, Yoshida.

3) Biosynthesis and degradation of metallothionein (Section 6)
   Participants: Webb (chairman), Andersen, Brady, Bremner, Cousins, Evans, M. Nordberg, Shaikh.

4) Role of metallothionein in metal metabolism and toxicity (Section 7).
Participants: G. Nordberg (chairman), Bakka, Bremner, Cherian, Evans, Kimura, Meister, M. Nordberg, Overnell, Pařízek, Piotrowski, Piscator, Riordan, Shaikh, Smith.

The entire report was also discussed in plenary sessions (chairmen: J.H.R. Kägi and M. Piscator, secretary: M. Nordberg). Recommendations were worked out on questions concerning the definition and the nomenclature of the title subject. After the meeting, the report was reviewed and edited by an Editorial Committee consisting of I. Bremner, J.H.R. Kägi, Y. Kojima, G.F. Nordberg, M. Nordberg, and M. Piscator. The edited manuscript was resubmitted to all participants for corrections and improvements.

Monica Nordberg

Yutaka Kojima

Monica Nordberg: Department of Environmental Hygiene, Karolinska Institute, S-104 01 Stockholm, Sweden

Yutaka Kojima: Biochemisches Institut der Universität Zürich, Zürichbergstrasse 4, CH-8028 Zürich, Switzerland

TABLE OF CONTENTS

1. INTRODUCTION

The important role of metals in living organisms is increasingly being recognized. Many metalloenzymes and metalloproteins are now known to form part of the catalytic machinery of the cell. In addition, in recent years a number of widely occurring non-enzymic metalloproteins have been identified and are of growing interest in many fields of biological science.

In particular, investigators of nutrition and of metabolic functions of essential trace elements such as iron, zinc and copper have become aware of the special significance of carriers and regulators in the homeostatic mechanisms for these metals. Similarly, prompted by the recognition of occupational exposure to metals and by the growing problems of environmental metal pollution, especially by the occurrence of, for example, Minamata and Itai-itai disease, there has been an impressive increase in research activities in toxicology and in occupational and environmental hygiene regarding the handling of metals such as mercury and cadmium by the human body.

These metals have in common that, upon entering the organism, they combine with specific proteins. Interest has been focused in particular on a low molecular weight cysteine-rich protein, metallothionein. This protein is now thought to occupy a central position in metal metabolism and is therefore attracting the attention of many scientists in biochemistry, physiology, toxicology, nutrition, and medicine. This interest is reflected by the greatly increased number of publications on this topic in recent years. Thus, it seems timely to summarize the existing knowledge on metallothionein and related proteins on the occasion of the First International Meeting on this subject.

2. HISTORY OF METALLOTHIONEIN RESEARCH

Metallothionein was discovered by Margoshes and Vallee in 1957. They isolated the cadmium- and zinc-containing protein from equine kidney, an organ known to accumulate cadmium (Vallee, 1979).

Kägi and Vallee (1960,1961a) further purified and characterized this material which they found to be a low molecular weight, metal-binding, cysteine-rich protein. In view of this unique composition, Kägi and Vallee (1960) named it "metallothionein".

In 1964, Piscator found that metallothionein was present in increased quantity in livers of rabbits exposed to cadmium. He postulated that the biosynthesis of this protein is induced by this metal and that metallothionein plays a role in metal detoxification.

In 1966, Pulido et al. reported that metallothionein also occurs in the human. Metal analyses of the protein isolated from human kidney showed that besides cadmium, zinc and copper, it also contained mercury in certain circumstances.

In the decade following the first report, only nine publications and abstracts (Margoshes and Vallee, 1957; Vallee, 1960; Kägi, 1960; Kägi and Vallee, 1960,1961a,b; Ulmer et al., 1962; Piscator, 1964; Pulido et al., 1966) concerning metallothionein appeared. However, since the late sixties, an exponentially increasing number of researchers have contributed to the knowledge of this protein.

3. DEFINITION, NOMENCLATURE AND OCCURRENCE

3.1 Definition

Historically, the term "metallothionein" has been used to designate the cadmium-, zinc- and copper-containing sulfur-rich protein from equine renal cortex (Kägi and Vallee, 1960).

This protein was characterized as follows (Kägi et al., 1974; Kojima et al., 1976):

- molecular weight: 6000 - 7000[1)]
- high metal content
- characteristic amino acid composition (high cysteine content, no aromatic amino acid nor histidine)
- optical features characteristic of metal thiolates (mercaptides)
- unique amino acid sequence (fixed distribution of cysteinyl residues)

Proteins resembling equine renal metallothionein in several of these features can be designated as "metallothionein".

3.2 Nomenclature

The terminology employed for different metallothioneins and their derivatives has long been confusing (Kägi and Kojima, 1979). Based on the large body of information

---

1) On gel filtration, native metal-containing metallothionein moves like a spherical protein of molecular weight 10,000.

now available on the composition and structure of these proteins, the plenum of the First International Meeting on Metallothionein and Other Low Molecular Weight Metal-binding Proteins has now adopted the following recommendations.

3.2.1 Nomenclature of metallothionein

- The term "metallothionein" should be used for all proteins fitting the definition given in Section 3.1.[2] This designation should be employed in publication titles, running titles and as a key word for computer filing.
- The term "thionein" should be used only to specify the apoprotein of metallothionein. The term should be avoided in publication titles and running titles and should not be used as a key word for computer filing. The designation "apometallothionein" may also be used to specify the apoprotein.
- More specific terms such as "cadmium-metallothionein" and "cadmium-thionein" can be employed when metallothionein contains only cadmium. For metallothionein containing another metal, an analogous nomenclature should be adopted. The metal prefixed to the word "thionein" should not be used to designate the metal employed for metallothionein induction.
- When metallothionein contains more than one metal, for example, 5.3 gram atoms of cadmium

---

2) For unambiguous identification of a protein as metallothionein, especially when isolated from a new source, it is recommended that its resemblance to equine renal metallothionein be documented by an analysis of its amino acid composition and/or of its amino acid sequence.

and 1.7 gram atoms of zinc, terms such as "(cadmium, zinc)-metallothionein", "(cadmium, zinc)-thionein" and "($Cd_{5.3}$, $Zn_{1.7}$)- thionein" can be used. The sequence in which the metals are named should indicate the order of their relative abundance.

- In contexts where specification of the metal composition is not available or is of no special interest, the term "metallothionein" should be used.

3.2.2 Nomenclature of multiple forms of metallothionein

The nomenclature adopted is based on the recommendation of the "Nomenclature of Multiple Forms of Enzymes" IUPAC-IUB Commission on Biochemical Nomenclature (CBN) (1976).

- The term "multiple forms of metallothionein" or "isometallothioneins" should be used as a broad term covering all metallothioneins occurring naturally in a single species. These terms should apply only to forms of metallothioneins arising from genetically determined differences in primary structure and not to forms derived by modification of the same primary sequence nor to forms differing only in metal composition.
- For isometallothioneins, the term "metallothionein" should be used, followed by a number.
- Where complex isometallothionein patterns occur, the numbers may be used to designate the major groups with subscript lower case letters applied consecutively for the individual subforms separated by additional procedures (1_a, 1_b, 1_c; 2_a, 2_b, etc.).
- In cases where classification can be based on amino acid sequence type, a Greek letter code is recommended for chain designation.

3.3 Occurrence

Besides the proteins designated as metallothionein, there are a number of proteins that are similar to metallothionein. It is likely that many of them may turn out to be metallothionein if more data become available. These proteins are also described in this section.

3.3.1 Occurrence in vertebrates

The widespread occurrence of metallothionein in vertebrate tissues is documented in Table 1.

In mammals, most work has been concentrated on the proteins from kidney, liver and intestines although metallothioneins have also been reported to occur in other organs (Table 1).

The amount of metallothionein in different species and tissues is variable. On induction of the biosynthesis by certain metals such as cadmium or zinc, the concentration of metallothionein has been reported to increase up to 40-fold (Piotrowski et al., 1974a; Trojanowska and Piotrowski, 1974). Large concentrations of metallothionein occur in fetal liver (Bremner et al., 1977; Hartmann and Weser, 1977; Rydén and Deutsch, 1978; Webb, 1979).

3.3.2 Occurrence in invertebrates

There is now good evidence that metallothionein occurs in several phyla of the animal kingdom. Metallothioneins from the crustacean Scylla serrata have recently been characterized (Olafson et al., 1979).

Cadmium-binding proteins have also been isolated from the phylum mollusca (Howard and Nickless,

TABLE 1 Occurrence of Metallothionein in Vertebrates

Human	liver	M	Kägi,	1970
		M	Wiśniewska-Knypl et al.,	1971
			Shaikh and Lucis,	1972
		M A	Bühler and Kägi,	1974
			Norheim and Steinnes,	1976
		M S	Kissling and Kägi,	1977
		M S	Kissling and Kägi,	1979
	fetal liver	M S	Rydén and Deutsch,*	1978
		M	Webb,	1979
	kidney	M	Pulido et al.,	1966
		M	Wiśniewska-Knypl et al.,	1971
		M	Shaikh and Lucis,	1972
	heart	M	Wiśniewska-Knypl et al.,	1971
	testis	M	Wiśniewska-Knypl et al.,	1971
	epithelial skin cells		Rugstad and Norseth,	1975
	lymphoblasts	A	Riordan and Madapallimattam,	1979
	fibroblasts	A	Riordan and Madapallimattam,	1979
	fetal fibroblasts		Lucis et al.,	1970
	HeLa cells		Lucis et al.,	1970
Monkey	kidney		Shaikh and Lucis,	1972
		M A	Kimura et al.,	1979b
	epithelial kidney cells		Lucis et al.,	1970
	kidney cells	M A	Kimura et al.,	1979b
Horse	liver	M A	Kägi,	1970
		M A	Kägi et al.,	1974
		M S	Kojima et al.,	1979
	kidney		Margoshes and Vallee,	1957
		M	Kägi and Vallee,	1960
		M	Kägi and Vallee,	1961a
		M A	Kägi,	1970
			Roosemont,	1972
		M A	Kägi et al.,	1974
		M S	Kojima et al.,	1976
		M S	Kojima et al.,	1979
Bovine	liver	M	Evans et al.,	1970
	fetal liver	M A	Hartmann and Weser,	1977
	duodenum	M	Evans et al.,	1970
Calf	liver		Bremner and Marshall,	1974a
		M A	Bremner and Marshall,	1974b
Sheep	liver	M A	Bremner et al.,	1977
	fetal liver	M A	Bremner et al.,	1977
	kidney	M A	Bremner and Young,	1977
Pig	liver	M A	Bremner and Young,	1976a
	kidney		Cousins et al.,	1973
	kidney cortex cells	M	Webb and Daniel,	1975

M: Protein classified as metallothionein by the author(s)
A: Amino acid composition reported in reference
S: Amino acid sequence reported in reference

TABLE 1 Occurrence of Metallothionein in Vertebrates (continued)

Dog	liver	M	Vostal and Cherian,	1974
	kidney	M	Vostal and Cherian,	1974
	spleen		Amacher and Ewing,	1975
Rabbit	liver	M	Piscator,	1964
		M A	Nordberg et al.,	1972
		M	Kimura and Otaki,	1975
		M S	Kimura et al.,	1979a
	fetal liver	M	Webb,	1979
	kidney	M	Nordberg et al.,	1974
		M A	Suda et al.,	1974
		M	Kimura and Otaki,	1975
Hamster	fetal liver	M	Webb,	1979
	macrophages	M	Waters et al.,	1975
Rat	liver		Shaikh and Lucis,	1969
			Jakubowski et al.,	1970
			Piotrowski and Bolanowska,	1970
			Lucis et al.,	1972
		M A	Winge and Rajagopalan,	1972
		M A	Weser et al.,	1973a
		M A	Weser et al.,	1973b
		M	Cherian,	1974
		M	Piotrowski et al.,	1974a
		M	Piotrowski et al.,	1974b
		M	Trojanowska and Piotrowski,	1974
		M A	Bremner and Davies,	1975
		M	Chen and Ganther,	1975
		M	Richards and Cousins,	1975a
		M	Richards and Cousins,	1975b
		A	Riordan and Gower,	1975a
		A	Riordan and Gower,	1975b
			Sabbioni and Marafante,	1975a
			Sabbioni and Marafante,	1975b
		M A	Sokolowski and Weser,	1975
		M A	Winge et al.,	1975a
		M A	Bremner and Young,	1976b
			Bryan and Hidalgo,	1976
		M	Feldman and Cousins,	1976
		M	Lafont et al.,	1976
		M	Squibb et al.,	1977
		M	Andersen and Weser,	1978
		M A	Andersen et al.,	1978
		M	Winge et al.,	1978
		M	Cherian,	1978
		M	Cherian et al.,	1978
		M	Andersen and Weser	1979
		M S	Kissling et al.,	1979
		M A	Shaikh,	1979
	fetal liver		Kaszper et al.,	1976
		M	Webb,	1979
	liver cell line	M	Rudd and Herschman,	1978
	liver parenchymal cells	M	Failla and Cousins,	1978a

M: Protein classified as metallothionein by the author(s)
A: Amino acid composition reported in reference
S: Amino acid sequence reported in reference

TABLE 1 Occurrence of Metallothionein in Vertebrates (continued)

Rat	kidney		Shaikh and Lucis,	1969
		M	Wiśniewska et al.,	1970
		M A	Kimura et al.,	1974
		M	Piotrowski et al.,	1974a
		M	Piotrowski et al.,	1974b
		M	Trojanowska and Piotrowski,	1974
		M	Chen and Ganther,	1975
		M	Bremner et al.,	1978
			Sabbioni et al.,	1978
	intestine	M	Chen and Ganther,	1975
		M A	Richards and Cousins,	1977a
		M	Evans and Johnson,	1978
			Sabbioni et al.,	1978
		M A	Evans,	1979
	spleen		Shaikh and Lucis,	1971a
			Amacher and Ewing,	1975
	pancreas		Shaikh and Lucis,	1971a
			Sabbioni et al.,	1978
	placenta		Lucis et al.,	1972
		M	Lafont et al.,	1976
	testis	M	Chen and Ganther,	1975
	brain	M	Chen and Ganther,	1975
Mouse	liver	M	Nordberg et al.,	1971a
			Shaikh and Lucis,	1972
		M A	Nordberg et al.,	1975b
		M S	Huang et al.,	1977
		M	Tsunoo et al.,	1978
		M S	Huang et al.,	1979
	kidney		Shaikh and Lucis,	1972
		M	Nordberg and Nordberg,	1973
	pancreas	M	Nordberg and Nordberg,	1973
	placenta	M	Sonawane et al.,	1975
	testis	M	Nordberg,	1971
	blood plasma	M	Nordberg et al.,	1971b
	erythrocytes	M	Nordberg et al.,	1971b
	fibroblasts		Rugstad and Norseth,	1978
Seal	liver		Olafson and Thompson,	1974
Chicken	liver	M A	Weser et al.,	1973a
		M A	Weser et al.,	1973b
Teleost fish	liver		Olafson and Thompson,	1974
		A	Bouquegneau et al.,	1975
			Marafante,	1976
			Overnell et al.,	1977
	kidney		Bouquegneau et al.,	1975
			Marafante,	1976
	gill		Bouquegneau et al.,	1975

M: Protein classified as metallothionein by the author(s)
A: Amino acid composition reported in reference
S: Amino acid sequence reported in reference

1975; Noël-Lambot, 1976; Olafson et al., 1979) indicating that this group of invertebrates may also contain metallothioneins.

3.3.3 Occurrence in microorganisms

Prinz and Weser (1975) purified a copper-containing protein with a molecular weight of 9500 from Saccharomyces cerevisiae which was cultivated in a medium containing a final concentration of 0.2 mM copper salt. The properties of this exclusively copper-containing protein were similar to those of mammalian metallothionein.

Another copper-binding protein was isolated from Neurospora crassa grown on a medium supplemented with 0.5 mM $CuSO_4$ (Ammer et al., 1978; Lerch, 1979). Although its molecular weight (2600) is much smaller than that of other metallothioneins, its amino acid sequence shows total agreement in the positions of all cysteinyl residues and of some seryl residues with those of the amino-terminal half (Residue No. 3-27) of the mammalian metallothioneins.

A cadmium- and zinc-binding macromolecule of molecular weight of 10,000 to 12,000 has been reported in a blue-green alga, Anacystus nidulans, grown in a medium containing $CdCl_2$ and $ZnCl_2$ (Maclean et al., 1972).

4. BIOCHEMICAL PROPERTIES

4.1 Isolation and quantification

Isolation procedures have included fractionation by salt and/or organic solvents, heat treatment, gel filtration, ion exchange chromatography and isoelectric focusing. However, gel filtration and ion exchange chromatography have become the most popular methods. To prevent loss of the metal, all chromatographic separations should be carried out above pH 7. Isoelectric focusing yields reproducible results when prestarted gradients are employed (Nordberg et al., 1972; Nordberg, 1978). Care has to be taken to avoid autoxidation in the case of copper-containing proteins (Bremner and Young, 1977; Hartmann and Weser, 1977; Rydén and Deutsch, 1978). Mercaptoethanol has been used as an additive during the preparation.

Separations based on charge properties (Kägi and Vallee, 1961a; Pulido et al., 1966; Shaikh and Lucis, 1971a,b; Nordberg et al., 1972) resulted in resolution into different "forms" of metallothionein. These forms were shown to differ in amino acid composition (Nordberg et al., 1972). All mammalian tissues examined and a crustacean contain at least two distinct forms of the protein (Kojima and Kägi, 1978; Olafson et al., 1979). In some cases, variant forms of metallothionein could not be separated from each other, although their existence could be demonstrated by sequence analyses (Huang et al., 1979; Kimura et al., 1979a; Kissling and Kägi, 1979; Kojima et al., 1979).

The metallothionein content in animal tissues can be determined by in vitro binding with radioactive mercury (Piotrowski et al., 1973a; Żelazowski and Piotrowski, 1977a) or cadmium (Chen and Ganther, 1975) and both procedures give comparable results (Kotsonis and Klaassen, 1977). Using the mercury-binding assay, Żelazowski and

Piotrowski (1977b) determined the metallothionein concentration in mammalian tissues. The concentration of the protein referring to the wet weight of the tissues ranged from 7 µg/g in rabbit muscle to about 500 µg/g in dog liver. There are reports on the generation of antibodies against metallothionein (Madapallimattam and Riordan, 1977; Tohyama and Shaikh, 1978; Vander Mallie and Garvey, 1978) and on their use in metallothionein quantification in tissues (Brady et al., 1979).

4.2 Chemical composition: metal, cysteine, amino acid, cysteine/metal

While metallothionein was first recognized by the presence of cadmium (Margoshes and Vallee, 1957), it was realized even in the earliest studies that its metal content was variable and that the balance was made up of zinc and copper, suggesting that the three elements may replace one another (Margoshes and Vallee, 1957; Kägi and Vallee, 1960,1961a). In fact, metallothionein judged to be pure by chromatographic criteria may often be heterogeneous with respect to metal ion composition, as suggested by the frequent observation of nonintegral zinc, cadmium, and copper contents. However, the total metal content of metallothioneins which contain the 2b metals, cadmium and/or zinc, is constant, reaching 6 to 7 gram atoms per mole in the most purified preparations (Kägi, 1970; Nordberg et al., 1972; Bühler and Kägi, 1974; Kägi et al., 1974; Nordberg et al., 1975b). Copper-containing metallothionein may have a larger total metal content (Bremner and Young, 1976a,b; Weser et al., 1977; Rupp and Weser, 1978).

The composition of metallothionein preparations with respect to the different metals varies with the tissue origin of the protein and with the pretreatment of the animal with metals. Especially noteworthy is the difference in metal composition between metallothionein from the liver and the kidney of individuals not subjected to excessive exposure to metals. Thus, in equine kidney, cadmium tends to be the predominant metallic component, while in the liver of the same organism zinc prevails by far over cadmium (Kägi et al., 1974).

In metallothionein isolated from fetal liver, copper and zinc were found to be the major constituents (Bremner et al., 1977; Hartmann and Weser, 1977; Rydén and Deutsch, 1978).

The most unique aspect of the mammalian metallothioneins is the high content of cysteine (approximately 33 residue %) which exceeds even that of the high sulfur proteins of wool. There is also a relatively large proportion of serine (approximately 14 residue %) and of basic amino acids (approximately 13 residue % of lysine plus arginine). All contain a single methionine. The Neurospora protein shows a similar abundance of cysteine (28.0%) and serine (28.0%). Also typical of all metallothioneins is the complete lack of aromatic amino acids and histidine (Table 2).

The protein contains neither disulfide nor free sulfhydryl (-SH) groups. All cysteinyl side chains are bound to the metal ions in thiolate (mercaptide) complexes. For metallothionein containing mainly cadmium and/or zinc, the ratio of cysteinyl residues to the sum of the metals is close to 3 (Kägi and Vallee, 1960), suggesting the existence of 6 to 7 trithiolate (trimercaptide) metal-binding sites (Kojima et al., 1976). For yeast copper-metallothionein and vertebrate metallothionein fully loaded with Cu(I), the ratio is close to 2 (Rupp and Weser, 1974; Bremner and Young, 1976a,b; Weser et al., 1977). For copper-metallothionein from Neurospora crassa, it is close to 1 (Lerch, 1979).

TABLE 2 Amino Acid Composition of Metallothioneins (MT) whose Complete Sequences are known

Amino acid	Human MT-2	Equine MT-1A	Equine MT-1A Var.L-54	Equine MT-1A Var.R-39	Equine MT-1B	Mouse MT-I	Mouse MT-II	Neurospora MT
Ala	7	5	5	5	7	5	6	1
Arg	-	2	2	3	1	-	-	-
Asn	1	1	1	1	1	1	1	2
Asp	3	2	2	2	2	3	3	1
Cys	20	20	20	20	20	20	20	7
Gln	1	1	1	1	2	1	3	-
Glu	1	1	1	1	1	-	-	-
Gly	5	7	7	6	5	5	4	6
Ile	1	-	-	-	-	-	1	-
Leu	-	-	1	-	-	-	-	-
Lys	8	6	6	6	7	7	8	1
Met	1	1	1	1	1	1	1	-
Pro	2	3	3	3	2	2	2	-
Ser	8	8	7	8	8	9	10	7
Thr	2	3	3	3	1	5	1	-
Val	1	1	1	1	3	2	1	-
Total residues	61	61	61	61	61	61	61	25

For references, see Fig. 1.

Fig. 1 Amino Acid Sequences of Metallothioneins (MT)

	Sequence (positions 1, 10, 20, 30, 40, 50, 60)
Human MT-1	Ac-M D P N C S C A $\overset{A}{\underset{T}{P}}$ $\overset{G}{V}$ G $\overset{S}{V}$ C $\overset{T}{A}$ C A G S C K C K E C K C T S C K K S C C S C C P V G C A K C A Q G C I C K G($\overset{T}{A}$,S)$\overset{D}{E}$ K C C S C A-OH
Human MT-2	Ac-M D P N C S C A A G D S C T C A G S C K C K E C K C T S C K K S C C S C C P V G C A K C A Q G C I C K G A S D K C C S C A-OH
Equine MT-1A	Ac-M D P N C S C P T G G S C T C A G S C K C K E C R C T S C K K S C C S C C P G G C A R C A Q G C V C K G A S D K C C S C A-OH
Variant L-54	Ac-M D P N C S C P T G G S C T C A G S C K C K E C R C T S C K K S C C S C C P G G C A R C A Q G C V C K G A L D K C C S C A-OH
Variant R-39	Ac-M D P N C S C P T G G S C T C A G S C K C K E C R C T S C K K S C C S C C P R G C A R C A Q G C V C K G A S D K C C S C A-OH
Equine MT-1B	Ac-M D P N C S C V A G E S C T C A G S C K C K Q C R C A S C K K S C C S C C P V G C A K C A Q G C V C K G A S D K C C S C A-OH
Rabbit MT-B	X-M D P N C S C A A D G(S,C,T,C,A,T,S,C)K C K $\overset{A}{E}$ C K C T S C K K S C C S C C P S G C A K C A Q G C I C K G A S D K C S C C A-OH
Mouse MT-I	Ac-M D P N C S C S T G G S C T C T S S C A C K D C K C T S C K K S C C S C C P V G C S K C A Q G C V C K G A A D K C T C C A-OH
Mouse MT-II	Ac-M D P N C S C A S D G S C S C A G A C K C K Q C K C T S C K K S C C S C C P V G C A K C S Q G C I C K Q A S D K C S C C A-OH
Neurospora MT	H-G D C G C S G A S S C N C G S G C S C S N C G S K-OH
Rat MT-2	Ac-M D P N C S C A T D G S C S C A G S C K C K Q C K C T S C K K S C C S C C P . . . (Partial sequence)
Human fetal MT	X-M D P N X S X A . . . (Partial sequence)

One-letter symbols:

A = Alanine	M = Methionine
C = Cysteine	N = Asparagine
D = Aspartic acid	P = Proline
E = Glutamic acid	Q = Glutamine
G = Glycine	R = Arginine
I = Isoleucine	S = Serine
K = Lysine	T = Threonine
L = Leucine	V = Valine

X = Undetermined

Other symbols:

Ac = Acetyl
H = Free amino terminus
OH = Free carboxyl terminus

References:

Human MT-1	Kissling and Kägi, 1979
Human MT-2	Kissling and Kägi, 1977, 1979
Equine MT-1A	Kojima et al., 1979
Equine MT-1B	Kojima et al., 1976, 1979
Rabbit MT-B	Kimura et al., 1979a
Mouse MT-I	Huang et al., 1977, 1979
Mouse MT-II	Huang et al., 1979
Neurospora MT	Ammer et al., 1978; Lerch, 1979
Rat MT-2	Kissling et al., 1979
Human fetal MT	Rydén and Deutsch, 1978

4.3 Amino acid sequence

The amino acid sequences are now known for metallothioneins from horse (Kojima et al., 1976,1979), man (Kissling and Kägi, 1977,1979), rabbit (Kimura et al., 1979a), mouse (Huang et al., 1977,1979), and Neurospora crassa (Ammer et al., 1978; Lerch, 1979). Partial sequences have also been reported for metallothioneins from human fetal liver (Rydén and Deutsch, 1978) and from rat liver (Kissling et al., 1979). In all mammalian metallothioneins, the polypeptide chain contains 61 residues with N-acetylmethionine and alanine at the amino and carboxyl terminus, respectively (Fig. 1).

The amino acid sequence reveals a fairly uniform distribution of the cysteinyl residues along the polypeptide chain (Fig. 2). The most striking feature is the predominance of -Cys-x-Cys- sequences (x stands for an amino acid other than cysteine). The basic amino acids and serine tend to be located in juxtaposition to the cysteinyl residues.

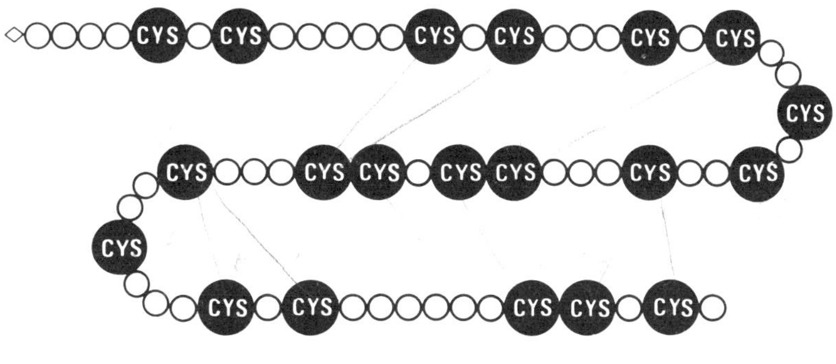

Fig. 2 Mode of Distribution of Cysteinyl Residues

4.4 Multiple forms

A comparison of the sequences of the two major variants of equine metallothioneins, MT-1A and MT-1B, reveals remarkable similarities (Fig. 1). The positions of the cysteines as well as of the adjacent serines and the basic amino acids are identical. Amino acid replacements which give rise to the variant forms of metallothionein are generally located outside of these regions and are explicable mainly by single base changes.

The two principal forms of equine metallothionein differ in 7 amino acid positions and the two principal forms of mouse liver metallothionein, MT-I and MT-II, differ in 15 positions. This extent of substitution exceeds that to be expected for allelic proteins and implies that they are coded by different cistrons. It is therefore proposed that the two isometallothioneins be representative of two principal sequence types.

In addition to these two principal forms, some allelic polymorphic variants occur in man, horse, and rabbit which differ only by a single amino acid substitution and might thus be attributed to allelic cistrons arising from point mutations. The joint occurrence of variant forms and subforms of metallothionein seems to be the major cause of difficulties in the purification of these proteins and in arriving at preparations with an integral amino acid composition.

4.5 Evolutional aspects

Comparison of the amino acid sequences of all the mammalian metallothioneins (Fig. 1) reveals remarkable structural similarity (Table 3). Virtually, all positions of cysteine are preserved. Furthermore, although the metallothionein from Neurospora crassa has a molecular weight signif-

TABLE 3 Amino Acid Sequence Similarity of Metallothioneins (MT)

The figures show percentage identity or, in brackets, percentage homology judged from conservative homologous replacements in pairs of proteins. Homology is scored on the basis of Lys (K) = Arg (R), Ala (A) = Gly (G), Ser (S) = Thr (T), Asp (D) = Glu (E) and Val (V) = Ile (I).

	Human MT-2	Equine MT-1A	Equine MT-1B	Mouse MT-I	Mouse MT-II	Neurospora MT
Human MT-2	100 (100)					
Equine MT-1A	89 (93)	100 (100)				
Equine MT-1B	90 (95)	89 (90)	100 (100)			
Mouse MT-I	80 (84)	80 (85)	79 (80)	100 (100)		
Mouse MT-II	84 (85)	77 (85)	79 (84)	75 (82)	100 (100)	
Neurospora MT	32 (44)	32 (40)	32 (44)	40 (44)	32 (40)	100 (100)

icantly smaller than that of the mammalian proteins (consisting of 25 amino acid residues), the primary structure is remarkably similar to the amino-terminal part of these proteins and shows 32% to 40% overall identity in aligned residues (Table 3). These data indicate that the gene coding for the Neurospora copper-metallothionein is evolutionally related to the primordial gene of the vertebrate metallothioneins. It is also significant that two isometallothioneins have been found in the crustacean Scylla serrata (Olafson et al., 1979).

On species-species comparison of vertebrate isometallothioneins, without consideration of any consistent isolation parameters such as electrophoretic mobility or ion-exchange elution position, it is remarkable that there has been no apparent selective functional divergent evolution working on the isoproteins to enable a best-fit alignment of either isometallothionein on comparison of two species. This observation is strongly in support of the functional identity of the two isoproteins since the duplicated genes have considerable antiquity.

5. PHYSICOCHEMICAL PROPERTIES

5.1 Molecular weight

Native equine metallothionein is a single chain protein with a molecular weight ranging from 6500 to 7000 depending on the metal composition (Kägi et al., 1974). This value is consistent with the chain weight of approximately 6000 deduced from sequence data (Kojima et al., 1976). When determined by gel filtration, the molecular weight of equine and all other mammalian metallothioneins is considerably higher, i.e., about 10,000. This discrepancy is attributed to the nonglobular shape of the protein. The molecular mass and the measured Stokes radius (16.1 Å) are consistent with the measures of a prolate ellipsoid that has an axial ratio of 6 (Kägi et al., 1974). Gel filtration of oxidized apometallothionein in the presence of guanidine HCl or urea (random coil conformation) yields chain weights in agreement with sequence data (Bühler and Kägi, 1974; Tsunoo et al., 1978). The similar behaviour on gel filtration of native mammalian metallothioneins from different sources implies that their molecular size and shape are identical.

5.2 Ionic and optical properties

All metallothioneins characterized thus far are negatively charged at neutral pH (Bühler and Kägi, 1974). Isoelectric points have been reported to be close to 4 (Nordberg et al., 1972; Cherian, 1974; Nordberg et al., 1975b). At neutral pH, free boundary electrophoresis of native equine renal metallothionein yields a charge of -1.89 per molecule (Bethune et al., 1979).

On lowering the pH, the metal ions bound to the protein are released yielding the apometallothionein (thionein). The pH at which 50% of the metal is dissociated depends on the protein concentration and on the metal ion concerned. It ranges between pH 3.5 and 4.5 for zinc, pH 2.5 and 3.5 for Cd, and below pH 1 for Cu (Kägi and Vallee, 1960, 1961a; Rupp and Weser, 1978; Weser and Rupp, 1979). The apoprotein is stable at low pH but polymerizes rapidly by disulfide formation when adjusted to neutral pH in the absence of metal (Weser et al., 1973b).

The optical features of the protein reflect both the metal and amino acid composition. In accordance with the absence of aromatic amino acids, there is no protein absorbance at 280 nm. There are, however, broad absorption bands of the charge-transfer type which are typical of the

TABLE 4 Absorption and Circular Dichroism of Metal Thiolate (Mercaptide) Bands of Metallothionein

Metal	Wavelength nm	Molar absorption coefficient [a] $\Delta\varepsilon$	Molar ellipticity [a] $\Delta[\Theta]$
Cd [c),d)]	257 [b)]		+36'000
	245	16'500	
	242		0
	240		-12'000
	235		0
	224		+70'000
Zn [c),e)]	239		+15'000
	222		+29'000
	212.5	17'000	0
	207		- 7'400
	199		0
Cu [f)]	359		+ 2'300
	302		-10'300
	283		-14'000
	280 [g)]	3'650	
	280 [g)]	2'870 [h)]	
	245		+23'000

a) $\Delta\varepsilon$ and $\Delta[\Theta]$ refer to the molarity of the metal.
b) Values indicate position of absorption maxima, of circular dichroism extrema or of crossover points.
c) pH-difference spectra obtained by subtracting the spectrum of apometallothionein, measured in 0.02 M perchloric acid, pH 1.7, from the spectrum of metal-containing metallothionein measured at pH 7.0.
d) Evaluated from equine renal metallothionein (Kägi, unpublished observation).
e) Evaluated from human hepatic metallothionein-2 (Bühler and Kägi, 1979).
f) Yeast Cu-metallothionein (Weser et al., 1977; Rupp and Weser, 1978; Weser and Rupp, 1979).
g) Absorption shoulder.
h) Bovine fetal Cu-metallothionein (Rupp and Weser, 1974; Hartmann and Weser, 1977; Weser and Rupp, 1979).

TABLE 5: Absorption and Circular Dichroism of Apometallothionein[a)]

	Wavelength (nm)		
	190	199	220
ε_{MRW} [b)]	7'400 m[d)]		770
$[\Theta]_{MRW}$ [c)]		-15'000 m	-4'100 s [d),e)]

a) Human hepatic metallothionein-2 dissolved in 0.02 M perchloric acid, pH 1.7 (Bühler and Kägi, 1979).
b) Mean residue absorptivity based on mean residue weight (= MRW) of 99.
c) Mean residue ellipticity based on mean residue weight (= MRW) of 99.
d) m = abbreviation for maximum; s = abbreviation for shoulder.
e) Identical value was obtained for chicken apometallothionein, pH 1.6 (Rupp and Weser, 1978).

metal thiolate (mercaptide) complexes of the constituent metals. These absorptions are manifested as shoulders at about 250 nm (Cd), 220 nm (Zn), and 270 nm (Cu) (Kägi and Vallee, 1961a; Kägi, 1970; Nordberg et al., 1972; Weser et al., 1973b; Nordberg et al., 1975b; Prinz and Weser, 1975; Hartmann and Weser, 1977; Bühler and Kägi, 1979; Weser and Rupp, 1979). The positions of these metal thiolate charge-transfer bands and their molar absorption coefficients are listed in Table 4. On removal of the metal at low pH, these bands are abolished. The remaining large absorption band at 190 nm (Table 5) can be attributed to the primary and secondary amide transitions and the cysteinyl side chains (Bühler and Kägi, 1979).

The binding of the metal also induces characteristic optical activity and circular dichroism features indicative of asymmetric coordination (Ulmer et al., 1962; Pulido et al., 1966; Weser et al., 1973b; Rupp and Weser, 1978; Bühler and Kägi, 1979) with the cysteinyl side chains and perhaps other as yet unidentified ligands (Table 4). Circular dichroic spectra of (cadmium, zinc)-metallothionein were simulated using the difference circular dichroic spectra of cadmium-metallothionein and zinc-metallothionein, and the circular dichroic spectrum of apometallothionein (Rupp and Weser, 1978). Removal of the metal at low pH abolishes these metal-dependent bands and results in a circular dichroism spectrum similar to that of random coil polyamino acids and of fully denatured proteins (Table 5) (Rupp and Weser, 1978; Bühler and Kägi, 1979).

Spectrophotometric titrations indicate that all cysteinyl side chains of the protein are involved in the formation of metal thiolate complexes and, hence, that in accordance with the ratio of the cysteinyl residues to the sum of bivalent metal ions bound (Section 4.2), each ion is coordinated to three thiolate ligands on the average (Kägi and Vallee, 1961a; Kägi, 1970). The exclusive presence of thiolate sulfur in native metallothionein was confirmed also by X-ray photoelectron spectrometry (XPS) (Weser et al., 1973b). Neither disulfide nor free sulfhydryl (-SH) groups were detectable. Higher oxidation states of sulfur were observed only upon oxidative destruction of copper-metallothionein (Hartmann and Weser, 1977; Weser et al., 1977). The metal sulfur coordination was also documented by ^{1}H NMR (nuclear magnetic resonance) (Rupp et al., 1974; Galdes et al., 1978, 1979; Rupp and Weser, 1979) and ^{113}Cd NMR spectroscopy (Otvos and Armitage, 1979). The chemical shifts of the ^{113}Cd resonances suggest that the metal ions are coordinated to at least three sulfur ligands.

5.3 Metal-binding site

The -Cys-x-Cys- sequences have been suggested by Kojima et al. (1976) to constitute the primary metal-binding sites (Fig. 2). After chelation of the metal ions by these bidentate sites, most metal ions are thought to interact in addition with a third cysteinyl residue located elsewhere in the molecule. The resulting negatively charged trithiolate (trimercaptide) complexes, $[Metal^{2+} (Cys^{-})_3]^{-}$, are the principal determinants to the overall negative charge of the metal-containing protein (Kägi, 1970). Besides the cysteinyl ligands, each metal-binding site may include one or more additional as yet unidentified ligand(s). However, a participation of inorganic sulfide in the metal complexes has been excluded (Kägi and Vallee, 1961a).

Although the existence of polynuclear metal clusters involving three-coordinated cysteine-sulfur bridges cannot be ruled out at present (Rupp and Weser, 1978; Weser and Rupp, 1979), spectrophotometric and complexometric titration data favor the above proposed model of separate metal-binding sites (Kägi and Vallee, 1961a; Kägi, 1970). This is now also supported by dark field electron microscopy which revealed well separated metal domains (Fiskin et al., 1977; Brady et al., 1979).

Assuming binding of each metal ion to three cysteinyl residues, dissociation constants of approximately 10^{-25} M and 10^{-21} M were initially calculated for the cadmium and the zinc complexes, respectively (Kägi and Vallee, 1961a). At pH 7, the corresponding apparent dissociation constants were estimated to be of the order of 10^{-15} M and 10^{-11} M (Kägi, 1970). Copper is bound much more firmly than either cadmium or zinc (Bremner and Marshall, 1974b; Rupp and Weser, 1974; Rupp et al., 1975; Rupp and Weser, 1978; Weser and Rupp, 1979). In copper-metallothionein, the copper became detectable by electron paramagnetic resonance (EPR) measurements only after oxidative treatments. Thus, in accordance with X-ray photoelectron spectrometric (XPS) measurements (Weser et al., 1977), the presence of copper(I)-thiolate units in copper-metallothionein is most probable (Rupp and Weser, 1979).

5.4 Spatial structure

The existence of a unique molecular form of metallothionein is documented by dark field electron microscopy (Fiskin et al., 1977; Brady et al., 1979). Its elongated shape is in fair agreement with its hydrodynamic properties measured by sedimentation analysis and gel filtration (Kägi et al., 1974). That metallothioneins have well-defined tertiary structures is also indicated by the marked differences in the amide region of ^{1}H NMR spectra of the metal-containing forms from the apometallothioneins (Galdes et al., 1978, 1979), and by the observation of numerous discrete peptide proton resonances (Galdes et al., 1978,1979) and of well-resolved carbon resonances in the ^{13}C NMR spectrum (Otvos and Armitage, 1979). The existence of a rigid conformational structure of metallothionein is also reflected by the slow exchange of some peptide hydrogens with the solvent (Ulmer and Vallee, 1971; Galdes et al., 1978,1979), and by its resistance toward tryptic digestion (Kägi, 1970; Feldman et al., 1978a,b) and towards heat denaturation (Webb, 1972b; Weser et al., 1973b). There is, however, as shown by circular dichroism studies, no α-helical and little or no pleated sheet secondary structure in the molecule (Rupp and Weser, 1978; Bühler and Kägi, 1979; Rupp and Weser, 1979).

6. METABOLISM

6.1 Biosynthesis and induction

The metallothionein content of liver, kidney and intestines has been shown to be increased by the parenteral or dietary administration of cadmium, copper or zinc (Piscator, 1964; Shaikh and Lucis, 1970; Nordberg et al., 1971a,1972; Webb, 1972b; Bremner et al., 1973; Nordberg et al., 1974; Piotrowski et al., 1974a,b; Squibb and Cousins, 1974; Trojanowska and Piotrowski, 1974; Mogilnicka et al., 1975; Nordberg et al., 1975b; Richards and Cousins, 1975a; Sabbioni and Marafante, 1975b; Winge et al., 1975a; Bremner and Davies, 1976; Chen et al., 1977; Cherian, 1977; Hidalgo and Bryan, 1977; Piotrowski et al., 1977; Probst et al., 1977a,b; Brady et al., 1979; Evans, 1979). Nutritional status, environmental stresses, route of administration together with the strain and sex can influence hepatic metallothionein accumulation (Bremner and Davies, 1975; Richards and Cousins, 1976; Hata et al., 1978; Oh et al., 1978; Sobocinski et al., 1978). Biosynthesis, which shows a maximum rate between 4 and 10 hours after administration of metal ions such as cadmium or zinc (Squibb and Cousins, 1974; Cempel and Webb, 1976; Probst et al., 1977a; Shaikh and Smith, 1977; Squibb et al., 1977; Andersen et al., 1978), has been monitored by measuring the increases in metallothionein levels, metal content and/or incorporation of labelled amino acids. Metallothionein concentration was found to increase in proportion to the injected dose of cadmium (Piotrowski et al., 1974a,b; Sabbioni and Marafante, 1975b; Probst et al., 1977b). In addition to these whole animal experiments, metallothionein synthesis has been studied in the isolated perfused liver, in primary and established cell cultures, and in homologous and heterologous cell-free systems (discussed below). In all of these studies, metallothionein was characterized by gel filtration. In some cases, other techniques such as ion exchange and covalent thiol chromatography and polyacrylamide gel electrophoresis were also used.

It has been suggested that copper may also induce synthesis of a copper-containing protein that is not metallothionein (Premakumar et al., 1975a,b; Winge et al., 1975b) (Section 8.1). They reported the ability of injected copper to stimulate the synthesis of a copper protein which they named copper-chelatin. This work was confirmed and extended by Day et al. (1978).

In rats exposed to $HgCl_2$ (Piotrowski et al., 1973b,1974a,b; Shaikh and Smith, 1975b) or to mercury vapor (Sapota et al., 1974; Cherian and Clarkson, 1976; Khayat, 1977), the biosynthesis of a low molecular weight mercury-binding protein has been observed in the kidney, but not in the liver. Winge et al. (1975a) reported the synthesis of zinc- and ^{203}Hg-containing metallothionein in the liver of rats exposed to mercury. However, under similar conditions, Martin and Brady (1977) observed the synthesis of hepatic zinc-metallothionein which did not contain mercury.

Mogilnicka et al. (1975,1976) and Piotrowski and Szymańska (1976) have listed a number of metals which did not increase significantly the levels of metallothionein in either liver or kidney of rats following repeated administration. These included: Pb, Mn, V, Rb, Cs, Tl, Sb, U, Se, As, Be, Mg, Cr, Ba, Ni, Sr, Sn. Similar results were reported by Sabbioni and Marafante (1975a) who in rat liver found no effects with Ni, Ir, V, Te, As, Mn, Se, Co, Mo and Bi.

On the other hand, injection of high doses of Pb, Mn, Ni, Cr and In ions which do not bind to metallothionein in vivo may increase the hepatic zinc pool and, thus, indirectly the synthesis of metallothionein (Suzuki and Yoshikawa, 1976). It is also possible that the metallothioneins containing mercury and other cations of high affinities to thiol binding sites derive from endogenous zinc-metallothionein, through in vivo replacement of zinc by these metals.

Actinomycin D, cordycepin and cycloheximide, administered to animals prior to an injection of either zinc or cadmium, blocked the induction of synthesis of hepatic metallothionein in rats. When these inhibitors were administered after injection of each metal, however, the induction process was not inhibited (Squibb and Cousins, 1974; Richards and Cousins, 1975a,b,1976,1977b; Squibb et al., 1977). These results have been confirmed in an isolated liver perfusion system (Frazier and Kingsley, 1976; Panemangalore and Brady, 1978) and in primary liver parenchymal cell cultures (Failla and Cousins, 1978b). Bremner and Davies (1976) have shown that cycloheximide inhibited hepatic copper-metallothionein accumulation. Premakumar et al. (1975a) found similar effects of actinomycin D and cycloheximide on hepatic copper-chelatin. In contrast, Shaikh and Smith (1975a,b,1977) have demonstrated that actinomycin D did not inhibit the biosynthesis of renal metallothionein induced by cadmium. Induction of a low molecular weight mercury-binding protein in the kidney, however, has been shown to be inhibited by actinomycin D.

Hidalgo and Bryan (1977) reported that cadmium (^{115m}Cd) was detectable in the liver cell nuclei shortly after in vivo administration and well before a substantial increase in hepatic cadmium-metallothionein biosynthesis occurred. As the site(s) at which metals affect the rates of biosynthesis of metallothioneins and the mechanism of this induction are not known, the possibility cannot be excluded that chemically dissimilar metals may stimulate biosynthesis by different mechanisms.

Recently, metallothionein synthesis has been demonstrated in vitro with isolated polysomes in the presence of mRNA (messenger RNA) from the liver of rats previously dosed with cadmium or zinc. Both homologous polysomes

(Squibb and Cousins, 1977) and a heterologous wheat germ protein synthesizing system (Andersen and Weser, 1978; Shapiro et al., 1978) have been used to translate the polyadenylated mRNA. In the liver of the treated animals, the amount of this mRNA, which has been reported to have a sedimentation coefficient of 9S and to be associated primarily with polysomes carrying 3 or 4 ribosomes (Andersen and Weser, 1978), was found to parallel the rate of biosynthesis of (cadmium, zinc)-metallothionein or zinc-metallothionein observed in vivo (Squibb and Cousins, 1977; Andersen et al., 1978). The transient 2- to 4-fold increase in the concentration of metallothionein mRNA after zinc or cadmium administration suggests that the messenger has a short half-life in the cytoplasm. Although the basis of the increase in metallothionein mRNA concentration during induction of metallothionein biosynthesis is not known, these data suggest that zinc and cadmium are able to induce the synthesis of metallothionein through changes in the intracellular concentration of metallothionein mRNA. Administration of actinomycin D blocked the appearance of metallothionein mRNA in liver polysomes (Shapiro et al., 1978). However, some regulation of metallothionein biosynthesis at the translational level cannot be excluded.

The mechanism by which the metals are getting bound to the translation product of metallothionein mRNA is still open to question. Richards and Cousins (1975a) reported that the accumulation of metals (Cd, Zn) in the cytoplasm and their binding to metallothionein was coincidental in time.

6.2 Degradation

The biological half-life of rat hepatic zinc-metallothionein has been estimated from measurements of the rates of degradation of ^{35}S- and ^{65}Zn-labelled metallothioneins, induced by a single injection of zinc, to be 18 to 20 hours

(Feldman and Cousins, 1976). A measurement of the half-time of decay of the zinc-metallothionein formed on induction by zinc sulfate gave a value of 10 hours (Brady et al., 1979), whereas Andersen et al. (1978) reported half-lives of 20 to 30 hours, based on [^{35}S]cysteine turnover, and 4 to 5 hours, based on the turnover of ^{203}Hg bound in vivo to zinc-metallothionein. The half-life for hepatic copper-metallothionein was estimated at 12 hours (Zn-deficient rats) and at 17 hours (Zn-supplemented rats), using [^{35}S]cysteine labelling (Bremner, 1979). The half-life for the copper in copper-chelatin was found to be 8.6 hours (Brady et al., 1979).

In contrast, the half-life of the hepatic cadmium-induced protein has been reported to be 2.8 days (Shaikh and Smith, 1975a,1976), 3.5 days (Feldman et al., 1978b), 4.2 days (Chen et al., 1975a,b), and 6.8 days (Brady et al., 1979). Whilst these differences may be due, at least in part, to disparities in experimental design, it is also possible that the half-life of this metalloprotein may vary with metal content (i.e., Cd/Zn ratio). Shaikh and Smith (1975a,1976) found that after cadmium administration, the half-life of the renal metallothionein (5 days) was almost double that of the hepatic protein (2.8 days). The half-life of renal mercury-binding protein was assessed to be 4.4 days (Shaikh and Smith, 1975b).

In vitro studies of the stability of metallothionein gave results comparable to those obtained in vivo. Extracts of purified liver lysosomes were capable of degrading metallothionein. The rate of degradation was markedly influenced by metal binding following the order: apometallothionein > zinc-metallothionein > (cadmium, zinc)-metallothionein. In contrast, trypsin and pronase were very effective in degrading apometallothionein but had little effect on the metal-containing proteins (Feldman et al., 1978a). Webb and Etienne (1977) observed significantly higher rates of degradation of cadmium-metallothionein when incubated for

4.5 hours with homogenates of kidney cortex than with tissue slices. It is of interest that the rates of disappearance of copper and zinc from metallothionein were exactly the same as those for the catabolism of the proteins (Bremner, 1979; Cousins, 1979; Whanger and Oh, 1979).

7. ROLE IN METAL METABOLISM AND TOXICITY

7.1 Physiological aspects

The biological function of metallothionein and of similar metal-binding proteins has not yet fully been clarified, although it was speculated in one of the earliest publications (Kägi and Vallee, 1960) that it could be involved either in catalysis, storage, immune phenomena or detoxication of metals. The binding of several metals, both essential and nonessential, to the protein has led many investigators to study the involvement of the protein in the metabolism and toxicity of metals. However, it is recognized that it may be misleading to search for a single biological role of a component of living organisms. Metallothionein may participate also in other cellular functions. In particular, it will be of interest to search for a role of this protein in the metabolism of sulfur compounds such as cysteine and glutathione (see Part I, "Metallothionein: Queries and conjectures"). The recent observation that GTP binds to metallothionein could reflect a role in energy or nucleic acid metabolism (Vallee, 1979).

7.1.1 Studies on zinc metabolism

Kägi (1970), Webb (1972b), Bühler and Kägi (1974), and Chen et al. (1974) observed zinc-metallothionein in liver and emphasized its potential rolc in zinc metabolism. Richards and Cousins (1975a) have obtained experimental evidence which suggests that metallothionein participates in the metabolism of zinc in both intestinal and hepatic cells. In rats pretreated with actinomycin D, metallothionein synthesis was inhibited and there was a concomitant decrease in the hepatic uptake of zinc. In addition, absorption of oral ^{65}Zn was greater in rats pretreated with actinomycin D than in control animals. These results suggest that intestinal metallothionein regulates the efflux of zinc from the intestinal cell to the body while hepatic metallothionein is involved in the uptake of zinc from the blood to the liver.

Recently, Whanger and coworkers demonstrated that, in rats, hepatic zinc was increased and the biosynthesis of metallothionein was strongly stimulated in response to a variety of stresses, such as cold and hot environments, heat burn, strenuous exercise and carbon tetrachloride intoxication (Oh et al., 1978; Whanger and Oh, 1979). Similar enhancements in metallothionein content in liver were obtained in response to food restriction (Bremner and Davies, 1975; Richards and Cousins, 1976) and bacterial infection (Sobocinski et al., 1978). The accumulation of zinc by liver parenchymal cells in primary culture was shown to be stimulated by glucocorticoids (Failla and Cousins, 1978a,b) by a process that involved de novo metallothionein biosynthesis. On the basis of these latter data, a hormonal loop involving metallothionein in hepatic zinc metabolism was proposed (Failla and Cousins, 1978b; Cousins, 1979).

Zinc-metallothionein is present in very high concentrations in the liver of the developing sheep fetus, with highest concentrations (about 150 µg Zn bound to metallothionein per g liver) occurring just after midterm (Bremner et al., 1977). Webb (1979) and

Whanger and Oh (1979) have also noted considerable accumulation of this protein in the developing rat fetus with highest concentrations of zinc in the newborn animal. The significance of this fetal accumulation of zinc is not clear at present.

7.1.2 Studies on copper metabolism

Evans et al. (1970) and Evans (1979) suggested that metallothionein is an important ligand in the metabolism of copper in the intestines. Davies and Campbell (1977) demonstrated that ^{64}Cu absorption is decreased in rats fed supplementary cadmium. The decreased passage of ^{64}Cu was associated with an increased incorporation of ^{64}Cu into a low molecular weight copper-binding protein. These observations strengthen the view that metallothionein functions in the regulation of copper absorption.

Metallothionein may also function in the handling of copper by the liver. Following intravenous injection of ^{67}Cu, Terao and Owen (1973) demonstrated that radioactivity first appeared in a low molecular weight fraction of liver cytosol. Subsequently, the tracer was observed in the 30,000 dalton fraction and later in material of still larger size that was thought to be ceruloplasmin. It was also shown that when copper was administered in relatively large amounts to rats, about half of the metal taken up by the liver was incorporated into newly synthesized metallothionein (Bremner and Davies, 1976).

In neonatal liver, Porter (1974) showed that the higher copper levels (relative to those of the adult) were reflected in large amounts of a polymeric form of copper-metallothionein. The protein was found to be associated with heavy lysosomes. More recently, Rydén and Deutsch (1978) purified and characterized soluble copper-metallothionein from human fetal liver.

Copper-metallothionein also accumulates in large amounts in the kidneys of animals under both physiological and pathological conditions (Bremner, 1979). Although the function of the renal protein has not been established, it is possible that, as with cadmium, it is involved in the tubular reabsorption of the metal.

7.2 Toxicological aspects

As discussed in Sections 3 and 4, cadmium has been shown to occur bound to metallothionein in human and animal tissues under normal conditions. The protein can be induced by cadmium, zinc, and copper exposure of animals (Section 6). In the following, evidence related to a possible role of metallothionein in metal toxicology will be presented. This topic has also been reviewed previously (Task Group on Metal Accumulation, 1973; Task Group on Metal Toxicity, 1976).

7.2.1 Involvement in short-term toxicity of cadmium

After a single injection of cadmium salts in rats, testicular damage occurs (Pařízek, 1957). This effect as well as lethality can be prevented by pretreatment with doses lower than those producing the above effect (Terhaar et al., 1965; Ito and Sawauchi, 1966; Gabbiani et al., 1967; Nordberg, 1971, Webb, 1972a). In animals protected by pretreatment, a larger proportion of testicular cadmium as well as of cadmium in some other tissues is bound to a cadmium-binding protein with the same molecular weight as metallothionein (Nordberg, 1971; Nordberg et al., 1971a). When cadmium-containing metallothionein was injected, no testicular damage occurred in mice, but such treatment gave rise to renal damage that was not caused by the same doses of an inorganic salt of cadmium (Nordberg, 1971).

A protective effect of metallothionein occurring intracellularly in the liver was discussed by Wiśniewska-Knypl and Jabłońska (1970) based on the observation that the activity of certain SH-containing enzymes in the liver was not influenced by cadmium in animals exposed repeatedly to this metal, whereas the same enzymes were inhibited when cadmium ions were added to liver homogenate in vitro. An increase in the LD_{50} values for cadmium was noticed in animals pretreated with cadmium (Leber and Miya, 1976). This observation supports a protective effect of metallothionein. A number of studies have demonstrated protective effects of metallothionein against acute effects of cadmium intoxication on various functions of the liver (see review by Friberg et al., 1974).

As discussed in Section 6, induction of metallothionein has also been demonstrated to occur on exposure of the animals to zinc. It has been shown that acute effects of cadmium in testes can be prevented by pretreatment with zinc. However, the dose of zinc required for complete protection is one to two orders of magnitude higher than that of cadmium (Pařízek, 1957; Mason and Young, 1967; Webb, 1979). The protection afforded by zinc pretreatment has been suggested by Webb (1972a) to be due to the induction of metallothionein synthesis. However, the synthesis of metallothionein cannot explain all of the protective effects of zinc against either testicular damage or lethality (Webb, 1979).

Wiśniewska-Knypl et al. (1971) discussed the involvement of metallothionein in a case of acute cadmium poisoning in man.

7.2.2 Distribution and effects of metallothionein administered to experimental animals

Several reports concerning distribution of injected metallothionein and its effects have been published. Only in a few of these studies (Nordberg, 1971; Suda et al., 1974; Fowler and Nordberg, 1975; Nordberg and Nordberg, 1975a,b; Nordberg et al., 1975a; Fowler and Nordberg, 1978; Nordberg, 1979), the purity of the metallothionein was documented by amino acid analysis.

After intravenous injection of relatively low doses (0.08 µg - 0.08 mg Cd/kg body weight) of cadmium bound to metallothionein in mice or rats, cadmium is taken up selectively by the kidney and only very small amounts are excreted in urine (Cherian and Shaikh, 1975; Nordberg and Nordberg, 1975b; Tanaka et al., 1975; Nordberg, 1978). When higher doses are used, a dose-related increase in urinary excretion of cadmium bound to metallothionein occurs. The renal retention and urinary excretion of injected cadmium-metallothionein (MT-1 or MT-2) was found to be dependent upon the amount of protein; whereas at 0.04 mg metallothionein/kg injection dose (containing 1.7 µg cadmium), only 5% of the cadmium were excreted, at 8.5 mg metallothionein/kg, 60% of the dose were excreted in the urine (Shaikh and Hirayama, 1979a). The renal handling of injected metallothionein in the rabbit was studied by Nomiyama and Foulkes (1977). At a filtered load of less than 1.2 µg Cd/min/kidney (corresponding to about 15 µg metallothionein/min/kidney), there was complete reabsorption. At higher loads, cadmium appeared in the urine.

Foulkes (1978) found that at an average cadmium concentration in the renal cortex of 225 µg/g wet weight, clearances of p-aminohippurate (PAH) and creatinine did not differ from control values, whereas metallothionein reabsorption was depressed. Furthermore, the data by Foulkes suggested that there was no peritubular transport of metallothionein. Dog liver metallothionein infused into the renal artery of dogs produced no response in blood pressure, renal plasma flow or glomerular filtration rate (Vostal and Cherian, 1974). The doses used, 0.034 to 0.09 mg Cd/kg body weight, had no effect on renal tubular transport of sodium. In contrast to experiments in which cadmium chloride was infused, a large portion of the cadmium bound to metallothionein was retained equally by both kidneys. More than 30% of infused metallothionein were excreted into the urine.

Preferential accumulation of cadmium in the kidney occurs also after oral administration of cadmium-metallothionein (Cherian et al., 1978; Cherian, 1979). Judged from the greater toxicity of cadmium-metallothionein compared to an equivalent amount of $CdCl_2$, it was suggested that cadmium-metallothionein is absorbed intact from the gastrointestinal tract (Cherian, 1979). This absorption may have some relevance to the consumption of cadmium bound to metallothionein in food products from liver and kidney, since the protein is relatively heat stable.

When metallothionein labelled in vivo with [^{35}S]cysteine was injected into rats, radioactivity appeared within two to three hours after injection, mainly in the high molecular weight fractions of the kidney 105,000 x g supernatant, indicating degradation of the labelled metallothionein (Cherian and Shaikh, 1975; Bremner et al., 1978; Shaikh and Hirayama, 1979a). The rate of disappearance of this material is much higher than that measured for metallothionein that was synthesized in renal tissue (see Section 6).

The renal damage induced by injected hepatic (cadmium, zinc)-metallothionein (Nordberg et al., 1975a) is of a tubular nature. Large doses of cadmium-metallothionein cause severe renal damage followed by death and are several times more toxic, as measured by the lethality after two weeks, than equivalent doses of cadmium salts (Nordberg et al., 1975a; Webb and Etienne, 1977). This is in marked contrast to the innocuousness of intravenously injected zinc-metallothionein that produces no toxic effect and results in no accumulation of zinc in the kidney (Cherian, 1977; Webb and Etienne, 1977).

It has been suggested that the acute renal toxicity of injected cadmium-metallothionein is due to the intracellular release of cadmium ions on degradation of the protein (Nordberg et al., 1975a). In the kidney of rabbits injected with cadmium-metallothionein (0.1 - 0.4 mg Cd/kg), electron dispersive X-ray microanalytical studies showed a higher cadmium concentration in lysosomes than in the surrounding cytoplasm as well as increased volume densities of lysosomes and enhanced lysosomal enzyme activity.

In the highest dose group, cadmium concentration in renal cortex was as high as 11 µg/g wet weight and serum creatinine concentration was increased (Fowler and Nordberg, 1978). Recent studies by Squibb et al. (1979) indicated uptake of cadmium in lysosomes of renal tubular cells already a few hours after injection of cadmium-metallothionein. The cadmium released on degradation can induce further synthesis of metallothionein in the kidney (Nordberg and Nordberg, 1975a; Cherian, 1978; Nordberg, 1978; Shaikh and Hirayama, 1979a).

An alternative explanation for the toxic effects of injected cadmium-metallothionein was put forward by

Cherian et al. (1976) who used 1.2 mg Cd/kg as metallothionein in rats with observation times up to 12 hours. These authors found that the injected metallothionein with high cadmium content was actively taken up by tubular kidney cells by pinocytosis and suggested that cell membranes were ruptured during this process. After entering the cell, cadmium-metallothionein caused irreversible damage to mitochondria with an effect on energy metabolism.

It is difficult to conclude from the available evidence which hypothesis is more plausible to account for the renal injury. Lysosomal uptake and degradation of the protein moiety of cadmium-metallothionein with subsequent release of cadmium ions at sensitive sites in the cell is supported by some of the data. However, a direct effect of the cadmium-metallothionein on cell membranes cannot be ruled out at the present time.

7.2.3 Involvement of metallothionein in chronic cadmium toxicity

A role for metallothionein in chronic cadmium poisoning, particularly in relation to the renal pathology encountered in this condition, was first suggested by Piscator (1964). In livers of rabbits that were repeatedly injected with cadmium salts, increased amounts of cadmium-metallothionein were found. It was proposed that the cadmium accumulating in renal tubular cells originated from hepatic cadmium-metallothionein which was released into the blood stream and, after glomerular filtration, was reabsorbed by the renal tubules. This hypothesis gained support from the finding of Nordberg et al. (1971b) and Nordberg (1972, 1978) that part of the cadmium in plasma and in erythrocytes is bound to a protein fraction of the same size as metallothionein. In mice in which chronic exposure to $CdCl_2$ (0.5 mg/kg, 5 days per week for 6 months) produced renal tubular dysfunction and a marked increase in urinary excretion of cadmium during the last months of treatment, Nordberg and Piscator (1972) found that the excreted cadmium was bound to a urinary protein fraction of the same molecular weight as metallothionein.

Increased binding of cadmium in blood of rats to a protein of the size of metallothionein was also observed by Cherian et al. (1977) during the last part of an 8 week experiment in which the animals were injected intraperitoneally with $CdCl_2$ (0.6 mg Cd/kg, 5 days per week). Renal damage was demonstrated by histological techniques in this phase of the experiment. Comparable results were also obtained (Shaikh and Hirayama, 1979b) when rats were injected subcutaneously with $CdCl_2$ (0.55 mg/kg, 5 days per week) for 14 weeks. At 4 weeks and later, cadmium in plasma was shown to be bound in part to a protein of the size of metallothionein. Excretion of cadmium in urine above normal values started at 10 weeks and increased thereafter. Urinary cadmium was bound to a protein fraction of the size of metallothionein. The animals exhibited pathological proteinuria and glucosuria. Even after a single subcutaneous injection of $CdCl_2$ in mice (1 mg Cd/kg body weight), cadmium in plasma was bound to a protein having the same size as metallothionein (Nordberg, 1978).

In spite of this suggestive evidence implicating metallothionein in the cadmium transport to the kidney, its importance for the pathogenesis of the renal manifestations of chronic cadmium exposure is still uncertain. Whereas renal tubular damage may be induced at a concentration of about 10 µg Cd/g wet weight in renal cortex after a single injection of cadmium-metallothionein, no renal damage is seen after repeated exposure to cadmium salts until the concentration in the kidney cortex reaches about 200 µg Cd/g wet weight. The significance of de novo synthesis of metallothionein in renal tissue in modulating toxic effects of cadmium in this organ also remains to be elucidated.

Metallothionein has also been discussed in connection with toxic effects of chronic cadmium exposure on bone. However, Kimura et al. (1974) showed that in bone of rats fed a cadmium-containing diet, only a small proportion of cadmium was bound to a soluble fraction of apparent molecular weight 10,000. Chronic exposure to cadmium is thought to be a causative factor in the skeletal manifestations of Itai-itai disease (see review by Friberg et al., 1974). Whether these effects on bone are related to the renal damage encountered in chronic cadmium poisoning or whether they are a result of a direct action of cadmium on bone is not entirely clear. Some observations indicate an effect of cadmium on vitamin D activation in kidney. Thus, Feldman and Cousins (1973) showed that metallothionein added in vitro prevented the inhibition of vitamin D activation by cadmium in renal mitochondria of chicks. Similar effects of metallothionein on in vitro as well as on in vivo inhibition of vitamin D activation by cadmium were reported by Suda et al. (1974).

7.2.4 Aspects related to other metals

Copper. The possible involvement of metallothionein at the onset of renal damage in copper-poisoned sheep has been considered by Bremner (1979). The sudden increase in kidney concentrations of copper-metallothionein when blood copper concentrations were increased, together with the observation that analogous renal tubular lesions occurred in copper-poisoned sheep (Gopinath et al., 1974) as in cadmium-intoxicated animals (see review by Friberg et al., 1974), suggests that similar mechanisms may be involved in the development of kidney damage in these two conditions (Bremner and Young, 1977). In support of this view, parenterally administered copper-metallothionein (0.25 mg Cu/kg) was found to be rapidly taken up by rat kidneys and also to cause a slight degree of renal tubular damage as assessed by histology (Bremner et al., 1978). Although no copper-metallothionein was found in the blood of the copper-poisoned sheep (Bremner et al., 1976), this may reflect the inadequacy of the present methods available for its detection (see Section 4.1).

The extreme susceptibility of sheep and young calves to chronic copper toxicity has been related to the difference in the hepatic distribution of copper in these animals in comparison with, for example, pigs and rats which have a much greater tolerance of copper. Only a limited amount of the hepatic copper in the susceptible species is present as metallothionein (Bremner, 1979).

Mercury. Renal toxicity of in vitro labelled and characterized mercury-metallothionein was demonstrated in mice to be similar to the renal toxicity already known for cadmium-metallothionein (Nordberg, 1979). Similar results were also obtained with an in vivo labelled mercury-induced low molecular weight protein from rat kidney (Hirayama and Shaikh, 1978). Its metabolism was very similar to that of cadmium-metallothionein administered intravenously to rats. A protective effect of the mercury-induced protein against subsequent doses of mercury was suggested by Piotrowski et al. (1973b).

It should be noted that methylmercury does not bind to any low molecular weight protein (Chen et al., 1973) and only in long term experiments could some interaction be observed due to methylmercury degradation (Chmielnicka and Brzeznicka, 1978).

Silver. The synthesis of metallothionein in the liver reported in animals exposed to silver salts (Winge et al., 1975a; Mogilnicka et al., 1976) is of minor toxicological interest since greater amounts of silver are deposited in the liver by other mechanisms (Mogilnicka et al., 1978).

8. OTHER LOW MOLECULAR WEIGHT METAL-BINDING PROTEINS

8.1 Copper-binding proteins

In addition to the copper-metallothioneins discussed in Section 3, other low molecular weight copper-binding proteins have been isolated and characterized from various sources. In 1961, Shapiro et al. isolated a protein which they termed L-6D, the amount of which was elevated in livers of patients with Wilson's disease. Porter and his coworkers purified a copper-rich protein from neonatal bovine liver (1962) and from neonatal human liver (1964). These proteins were first isolated in insoluble high molecular weight forms from the mitochondrial fraction and were designated "mitochondrocuprein" (Porter et al., 1962). They contained an extraordinarily large proportion of half-cystine (Porter, 1966). This material was later suggested to be a copper-rich polymerized form of metallothionein (Porter, 1974). It was also shown that it is localized in a distinct population of heavy lysosomes rather than in mitochondria (Porter, 1974).

The low molecular weight copper-binding protein which occurs naturally in the liver and kidney of adult, neonatal, and fetal animals has been shown to be copper-metallothionein, as assessed by amino acid analysis of the purified proteins (Bremner and Young, 1976a,1977; Hartmann and Weser, 1977; Rydén and Deutsch, 1978). Bremner and Young (1976b) and Evans and Johnson (1978) were able to isolate (copper, zinc)-metallothioneins from the liver and the intestines, respectively, of copper-injected rats. These proteins were identified on the basis of their chromatographic properties and their amino acid composition. They accounted for over 50% of the copper present in the low molecular weight copper-containing fraction eluted from Sephadex G-75.

In contrast to these findings, Premakumar et al. (1975a,b), Winge et al. (1975b), and Day et al. (1978) have stated that although copper injection induces the formation of zinc-metallothionein containing little copper, in rat liver, copper becomes associated with an induced protein which differs from metallothionein in amino acid composition and electrophoretic properties. This protein has been named "copper-chelatin" (Winge et al., 1975b). Copper-binding proteins with low molecular weight and amino acid compositions different from those of metallothionein and from copper-chelatin have been isolated by several other investigators (Riordan and Gower, 1975a,b; Evans and LeBlanc, 1976; Irons and Smith, 1977). Bremner and Young (1976b) have suggested that these proteins may be derivatives of copper-metallothionein which are formed during the isolation of the copper proteins, as they, like Hartmann and Weser (1977) and Evans and Johnson (1978), found that the maintenance of anaerobic conditions during the isolation of copper-metallothionein was essential. In those cases where copper proteins with other properties were isolated, this was not done. Antibodies raised against a copper-binding protein from rat liver, different from metallothionein, have been reported to crossreact strongly with cadmium-metallothionein from the same source (Madapallimattam and Riordan, 1977).

It is possible that the low cysteine content of copper-chelatin and the other purified copper proteins may result from the selective loss of cysteinyl residues during acid hydrolysis of the copper-containing protein; however, the differences in composition with respect to other amino acids between metallothionein and copper-chelatin and other copper proteins are not explainable. Further studies must be carried out exploring the possibilities of a conversion of copper-metallothionein into a protein(s) with the properties of copper-chelatin and other copper proteins before the question of whether copper induces the synthesis of more than one type of low molecular weight protein can be resolved.

Evidence for the existence of a low molecular weight copper-binding protein (molecular weight 7000) in cultured human fibroblasts and lymphoblasts was obtained by Riordan and Madapallimattam (1979). Its amino acid composition is remarkable by its low half-cystine (5 residue %) and its high glycine (30 - 35 residue %) and glutamic acid content (28 - 32 residue %). The protein contains 1 g atom of copper per mole.

A copper tolerant strain of the green alga Scenedesmus acutiformis was shown to contain a protein with an apparent molecular weight of approximately 10,000 and a copper content of 2.5 g atoms per mole (Stokes et al., 1977).

8.2 Zinc-binding ligands in milk and intestine

A zinc-binding fraction with an estimated molecular weight of 8700 was isolated from human and rat milk and from the intestinal mucosa of suckling rats (Eckhert et al., 1977; Hurley et al., 1977). No compositional data have been reported. This fraction which is absent from bovine milk is thought by the authors to be related to zinc nutrition or transport in the neonatal period before the development of intestinal mechanisms for zinc absorption.

8.3 Renal mercury-, gold-, and bismuth-binding proteins

In response to the administration of mercury salts (Wiśniewska et al., 1970; Nordberg et al., 1974; Piotrowski et al., 1974a,b; Shaikh and Smith, 1975b) or elemental mercury vapor (Sapota et al., 1974; Cherian and Clarkson, 1976), a low molecular weight mercury-binding protein is formed in the kidneys of the rat. This protein was tentatively classified as metallothionein.

Under combined exposure to selenium, less mercury is found in the kidney (Pařízek et al., 1971,1974) and less mercury is bound to this protein (Komsta-Szumska et al., 1976; Komsta-Szumska and Chmielnicka, 1977; Piotrowski et al., 1977).

In response to the administration of bismuth salts, low molecular weight bismuth-binding proteins are formed in the kidney of the rat (Piotrowski and Szymańska, 1976; Szymańska et al., 1977). These proteins are resolvable into at least two fractions and the main fraction could be distinguished from metallothionein by its different amino acid composition and spectral characteristics. The term "renal chromochelatin" was proposed for this protein (Piotrowski et al., 1979).

As with mercury, combined exposure to selenium and bismuth gives a decrease in the amount of bismuth bound to this protein (Szymańska et al., 1978). Mechanisms of interactions of selenium with cadmium and mercury were discussed elsewhere (The Task Group on Metal Interaction, 1978).

Administration of gold salts results in formation of low molecular weight gold-binding protein in the kidneys of the rat (Mogilnicka and Piotrowski, 1977). The protein appears in several fractions separable by ion exchange chromatography and is subject to aggregation during purification. The renal metal-binding proteins formed in response to mercury, bismuth and gold have apparent molecular weights of 7500 - 11,700 and contain copper in amounts of 1.8% - 4.5%. Their induction by the metals has been demonstrated by means of cysteine incorporation studies. No full amino acid composition has been published so far for any of these proteins (Piotrowski et al., 1979).

8.4 Lead-binding protein of low molecular weight in human erythrocytes

A protein with molecular weight of 10,000 was isolated from erythrocytes of lead-exposed workers (Raghavan and Gonick, 1977). No data are available on its relationship to metallothionein.

9. SUMMARY AND GENERAL CONCLUSIONS

Current knowledge concerning metallothionein and related proteins has been reviewed. A definition of metallothionein and guidelines for its nomenclature have been adopted. From its wide occurrence in animals and in microorganisms it is clear that metallothionein is a widespread and ancient protein. This conclusion is fully borne out by the determination of ten complete amino acid sequences. The remarkable conservation of primary structure from Neurospora to man is an unambiguous indication of very strong evolutional constraints acting on this molecule, suggesting its involvement in and its indispensability to some fundamental biological processes in the cell. Both biochemical and physicochemical data are in accordance with the view that metallothionein is a compact molecule containing characteristic metal thiolate groups.

There is good evidence that the turnover of metallothionein is fairly rapid and that the increased formation of metallothionein upon administration of cadmium and/or zinc is correlated with an increase in the concentration of mRNA for metallothionein in the cell.

It is recognized that our knowledge of the biochemical function of the protein is far from complete. However, present knowledge indicates that there may be an important involvement in the metabolism of zinc, cadmium and copper. At increased exposure of an organism to cadmium, it is thought that metallothionein modulates the toxic effects of the metal.

10. ACKNOWLEDGEMENTS

The financial sponsorship of the Federation of European Biochemical Societies (FEBS), the Schweizerische Naturforschende Gesellschaft, the Department of Education of the Kanton Zürich, Prof. Dr. F. Leuthardt, and the Permanent Commission and International Association on Occupational Health as well as of the Swedish companies Draco AB and NIFE Jungner AB for the "First International Meeting on Metallothionein and Other Low Molecular Weight Metal-binding Proteins" and for the publication of this report is acknowledged. The excellent technical and editorial assistance given throughout this project by Mss. Genia de Vallier and Helena Hasselgren is greatly appreciated.

11. REFERENCES

Numbers in italics indicate on what page in the report references are cited.

Amacher, D.E., and Ewing, K.L. (1975). A soluble cadmium-binding component in rat and dog spleen. Arch. Environ. Health 30, 510-513.
53,54.

Ammer, D., Budry, R., and Lerch, K. (1978). Structure and function of a copper-metallothionein from Neurospora crassa. Experientia 34, 905.
55,60,61.

Andersen, R.D., and Weser, U. (1978). Partial purification, characterization and translation in vitro of rat liver metallothionein messenger ribonucleic acid. Biochem. J. 175, 841-852.
53,74.

Andersen, R.D., and Weser, U. (1979). Partial purification and translation in vitro of metallothionein messenger RNA. This volume, pp. 303-310.
53.

Andersen, R.D., Winter, W.P., Maher, J.J., and Bernstein, I.A. (1978). Turnover of metallothioneins in rat liver. Biochem. J. 174, 327-338.
53,71,74,75.

Bethune, J.L., Budreau, A.J., Kägi, J.H.R., and Vallee, B.L. (1979). Determination of the charge of horse kidney metallothionein by free boundary electrophoresis. This volume, pp. 207-210.
65.

Bouquegneau, J.M., Gerday, C., and Disteche, A. (1975). Fish mercury-binding thionein related to adaptation mechanisms. FEBS Lett. 55, 173-177.
54.

Brady, F.O., Panemangalore, M., Day, F.A., Fiskin, A.M., and Peterson, G. (1979). In vivo and ex vivo induction of rat liver metallothionein. This volume, pp. 261-271.
57,69,70,71,75.

Bremner, I. (1979). Factors influencing the occurrence of copper-thioneins in tissues. This volume, pp. 273-280.
75,76,79,86,87.

Bremner, I., and Davies, N.T. (1975). The induction of metallothionein in rat liver by zinc injection and restriction of food intake. Biochem. J. 149, 733-738.
53,71,77.

Bremner, I., and Davies, N.T. (1976). Studies on the appearance of a hepatic copper-binding protein in normal and zinc-deficient rats. Br. J. Nutr. 36, 101-112.
71,73,78.

Bremner, I., and Marshall, R.B. (1974a). Hepatic copper- and zinc-binding proteins in ruminants. 1. Distribution of Cu and Zn among soluble proteins of livers of varying Cu and Zn content. Br. J. Nutr. 32, 283-291.
52.

Bremner, I., and Marshall, R.B. (1974b). Hepatic copper- and zinc-binding proteins in ruminants. 2. Relationship between Cu and Zn concentrations and the occurrence of a metallothionein-like fraction. Br. J. Nutr. 32, 293-300.
52,69.

Bremner, I., and Young, B.W. (1976a). Isolation of (copper,zinc)-thioneins from pig liver. Biochem. J. 155, 631-635.
52,57,58,89.

Bremner, I., and Young, B.W. (1976b). Isolation of (copper,zinc)-thioneins from the livers of copper-injected rats. Biochem. J. 157, 517-520.
53,57,58,89

Bremner, I., and Young, B.W. (1977). Copper thionein in the kidneys of copper-poisoned sheep. Chem. Biol. Interact. 19, 13-23.
52,56,87,89.

Bremner, I., Davies, N.T., and Mills, C.F. (1973). The effect of zinc deficiency and food restriction on hepatic zinc proteins in the rat. Biochem. Soc. Trans. 1, 982-985.
71.

Bremner, I., Young, B.W., and Mills, C.F. (1976). Protective effect of zinc supplementation against copper toxicosis in sheep. Br. J. Nutr. 36, 551-561.
87.

Bremner, I., Williams, R.B., and Young, B.W. (1977). Distribution of copper and zinc in the liver of the developing sheep foetus. Br. J. Nutr. 38, 87-92.
51,52,58,77.

Bremner, I., Hoekstra, W.G., Davies, N.T., and Williams, R.B. (1978). Renal accumulation of (copper, zinc)-thioneins in physiological and pathological states. In: Trace Element Metabolism in Man and Animals-3. (Kirchgessner, M., ed), pp. 44-51, Arbeitskreis für Tierernährungsforschung Weihenstephan Inst. Ernährungsphysiol., Freising-Weihenstephan, Federal Republic of Germany.
54,82,87.

Bryan, S.E., and Hidalgo, H.A. (1976). Nuclear ^{115}cadmium: uptake and disappearance correlated with cadmium-binding protein synthesis. Biochem. Biophys. Res. Commun. 68, 858-866.
53.

Bühler, R.H.O., and Kägi, J.H.R. (1974). Human hepatic metallothioneins. FEBS Lett. 39, 229-234.
52,57,65,76.

Bühler, R.H.O., and Kägi, J.H.R. (1979). Spectroscopic properties of zinc-metallothionein. This volume, pp. 211-220.
66,67,68,70.

Cempel, M., and Webb, M. (1976). The time-course of cadmium-thionein synthesis in the rat. Biochem. Pharmacol. 25, 2067-2071.
71.

Chen, R.W., and Ganther, H.E. (1975). Relative cadmium binding capacity of metallothionein and other cytosolic fractions in various tissues of the rat. Environ. Physiol. Biochem. 5, 378-388.
53,54,56.

Chen, R.W., Ganther, H.E., and Hoekstra, W.G. (1973). Studies on the binding of methylmercury by thionein. Biochem. Biophys. Res. Commun. 51, 383-390.
88.

Chen, R.W., Eakin, D.J., and Whanger, P.D. (1974). Biological function of metallothionein. II. Its role in zinc metabolism in the rat. Nutr. Rep. Int. 10, 195-200.
76.

Chen, R.W., Whanger, P.D., and Weswig, P.H. (1975a). Biological function of metallothionein. I. Synthesis and degradation of rat liver metallothionein. Biochem. Med. 12, 95-105.
75.

Chen, R.W., Whanger, P.D., and Weswig, P.H. (1975b). Rat liver metallothionein: relationship to metal metabolism. Fed. Proc. Fed. Am. Soc. Exp. Biol. 34, 927.
75.

Chen, R.W., Vasey, E.J., and Whanger, P.D. (1977). Accumulation and depletion of zinc in rat liver and kidney metallothioneins. J. Nutr. 107, 805-813.
71.

Cherian, M.G. (1974). Isolation and purification of cadmium binding proteins from rat liver. Biochem. Biophys. Res. Commun. 61, 920-926.
53.65.

Cherian, M.G. (1977). Studies on the synthesis and metabolism of zinc-thionein in rats. J. Nutr. 107, 965-972.
71,83.

Cherian, M.G. (1978). Induction of renal metallothionein synthesis by parenteral cadmium-thionein in rats. Biochem. Pharmacol. 27, 1163-1166.
53,83.

Cherian, M.G. (1979). Metabolism and potential toxic effects of metallothionein. This volume, pp. 337-345.
82.

Cherian, M.G., and Clarkson, T.W. (1976). Biochemical changes in rat kidney on exposure to elemental mercury vapor: effect on biosynthesis of metallothionein. Chem. Biol. Interact. 12, 109-120.
72,91.

Cherian, M.G., and Shaikh, Z.A. (1975). Metabolism of intravenously injected cadmium-binding protein. Biochem. Biophys. Res. Commun. 65, 863-869.
81,82.

Cherian, M.G., Goyer, R.A., and Delaquerriere-Richardson, L. (1976). Cadmium-metallothionein-induced nephropathy. Toxicol. Appl. Pharmacol. 38, 399-408.
84.

Cherian, M.G., Goyer, R.A., and Delaquerriere-Richardson, L. (1977). Relationship between plasma cadmium-thionein and cadmium-induced nephropathy. Toxicol. Appl. Pharmacol. 41, 145-146.
85.

Cherian, M.G., Goyer, R.A., and Valberg, L.S. (1978). Gastrointestinal absorption and organ distribution of oral cadmium chloride and cadmium-metallothionein in mice. J. Toxicol. Environ. Health 4, 861-868.
53,82.

Chmielnicka, J., and Brzeżnicka, E.A. (1978). The influence of selenium on the level of mercury and metallothionein in rat kidneys in prolonged exposure to different mercury compounds. Bull. Environ. Contam. Toxicol. 19, 183-190.
88.

Cousins, R.J. (1979). Synthesis and degradation of liver metallothionein. This volume, pp. 293-301.
76,77.

Cousins, R.J., Barber, A.K., and Trout, J.R. (1973). Cadmium toxicity in growing swine. J. Nutr. 103, 964-972.
52.

Davies, N.T., and Campbell, J.K. (1977). The effect of cadmium on intestinal copper absorption and binding in the rat. Life Sci. 20, 955-960.
78.

Day, F.A., Coles, B.J., and Brady, F.O. (1978). Postinductive actinomycin D effects on the concentrations of cadmium thionein, zinc thionein, and copper chelatin in rat liver. Bioinorg. Chem. 8, 93-105.
72,89.

Eckhert, C.D., Sloan, M.V., Duncan, J.R., and Hurley, L.S. (1977). Zinc binding: a difference between human and bovine milk. Science 195, 789-790.
90.

Evans, G.W. (1979). Metallothionein in intestinal copper metabolism. This volume, pp. 321-329.
54,71,78.

Evans, G.W., and Johnson, P.E. (1978). Copper- and zinc-binding ligands in the intestinal mucosa. In: Trace Element Metabolism in Man and Animals-3. (Kirchgessner, M., ed), pp. 98-105, Arbeitskreis für Tierernährungsforschung Weihenstephan Inst. Ernährungsphysiol., Freising-Weihenstephan, Federal Republic of Germany.
54,89.

Evans, G.W., and LeBlanc, F.N. (1976). Copper-binding protein in rat intestine: amino acid composition and function. Nutr. Rep. Int. 14, 281-288.
89.

Evans, G.W., Majors, P.F., and Cornatzer, W.E. (1970). Mechanism for cadmium and zinc antagonism of copper metabolism. Biochem. Biophys. Res. Commun. 40, 1142-1148.
52,78.

Failla, M.L., and Cousins, R.J. (1978a). Zinc uptake by isolated rat liver parenchymal cells. Biochim. Biophys. Acta 538, 435-444.
53,77.

Failla, M.L., and Cousins, R.J. (1978b). Zinc accumulation and metabolism in primary cultures of adult rat liver cells. Regulation by glucocorticoids. Biochim. Biophys. Acta 543, 293-304.
73,77.

Feldman, S.L., and Cousins, R.J. (1973). Influence of cadmium on the metabolism of 25-hydroxycholecalciferol in chicks. Nutr. Rep. Int. 8, 251-260.
86.

Feldman, S.L., and Cousins, R.J. (1976). Degradation of hepatic zinc-thionein after parenteral zinc administration. Biochem. J. 160, 583-588.
53,75.

Feldman, S.L., Failla, M.L., and Cousins, R.J. (1978a). Degradation of rat liver metallothioneins in vitro. Biochim. Biophys. Acta 544, 638-646.
70,75.

Feldman, S.L., Squibb, K.S., and Cousins, R.J. (1978b). Degradation of cadmium-thionein in rat liver and kidney. J. Toxicol. Environ. Health 4, 805-813.
70,75.

Fiskin, A.M., Peterson, G., and Brady, F.O. (1977). Cd-metallothionein. A test object for dark-field studies of protein structure. Ultramicroscopy 2, 389-395.
69,70.

Foulkes, E.C. (1978). Renal tubular transport of cadmium-metallothionein. Toxicol. Appl. Pharmacol. 45, 505-512.
81.

Fowler, B.A., and Nordberg, G.F. (1975). The renal toxicity of cadmium metallothionein. International Conference on Heavy Metals in the Environment, Abstracts, Toronto, Canada, October, 1975, B52-53.
81.

Fowler, B.A., and Nordberg, G.F. (1978). The renal toxicity of cadmium metallothionein: morphometric and X-ray microanalytical studies. Toxicol. Appl. Pharmacol. 46, 609-623.
81,83.

Frazier, J.M., and Kingsley, B.S. (1976). Kinetics of cadmium transport in the isolated perfused rat liver. Toxicol. Appl. Pharmacol. 38, 583-593.
73.

Friberg, L., Piscator, M., Nordberg, G.F., and Kjellström, T. (1974). Cadmium in the Environment, 2nd Ed, CRC Press, Cleveland, Ohio.
80,86,87.

Gabbiani, G., Baic, D., and Déziel, C. (1967). Studies on tolerance and ionic antagonism for cadmium or mercury. Can. J. Physiol. Pharmacol. 45, 443-450.
79.

Galdes, A., Vasak, M., Hill, H.A.O., and Kägi, J.H.R. (1978). ^{1}H NMR spectra of metallothioneins. FEBS Lett. 92, 17-21.
68,70.

Galdes, A., Hill, H.A.O., Kägi, J.H.R., Vašák, M., Bremner, I., and Young, B.W. (1979). The ^{1}H N.M.R. spectra of metallothioneins. This volume, pp. 241-248.
68,70.

Gopinath, C., Hall, G.A., and Howell, J. McC. (1974). The effect of chronic copper poisoning on the kidneys of sheep. Res. Vet. Sci. 16, 57-69.
87.

Hartmann, H.-J., and Weser, U. (1977). Copper-thionein from fetal bovine liver. Biochim. Biophys. Acta 491, 211-222.
51,52,56,58,66,67,68,89.

Hata, A., Tsunoo, H., Nakajima, H., and Kimura, M. (1978). Strain differences in susceptibility of mice to cadmium-induced metallothionein. Toxicol. Lett. 2, 45-49.
71.

Hidalgo, H.A., and Bryan, S.E. (1977). Cadmium-115 bound to nuclear and cytoplasmic proteins. Toxicol. Appl. Pharmacol. 42, 319-327.
71,73.

Hirayama, K., and Shaikh, Z.A. (1978). Metabolism of intravenously injected renal mercury-thionein in rats. Toxicol. Appl. Pharmacol. 45, 347.
87.

Howard, A.G., and Nickless, G. (1975). Protein binding of cadmium, zinc and copper in environmentally insulted limpets Patella vulgata. J. Chromatogr. 104, 457-459.
51.

Huang, I-Y., Yoshida, A., Tsunoo, H., and Nakajima, H. (1977). Mouse liver metallothioneins. Complete amino acid sequence of metallothionein-I. J. Biol. Chem. 252, 8217-8221.
54,60,61.

Huang, I-Y., Tsunoo, H., Kimura, M., Nakashima, H., and Yoshida, A. (1979). Primary structure of mouse liver metallothionein -I and -II. This volume, pp. 169-172.
54,56,60,61.

Hurley, L.S., Duncan, J.R., Sloan, M.V., and Eckhert, C.D. (1977). Zinc-binding ligands in milk and intestine: A role in neonatal nutrition? Proc. Natl. Acad. Sci. USA 74, 3547-3549.
90.

Irons, R.D., and Smith, J.C. (1977). Isolation of a non-thionein copper-binding protein from liver of copper-injected rats. Chem. Biol. Interact. 18, 83-89.
89.

Ito, T., and Sawauchi, K. (1966). Inhibitory effects on cadmium-induced testicular damage by pretreatment with smaller cadmium dose. Okajimas Folia Anat. Jpn. 42, 107-117.
79.

IUPAC-IUB Commission on Biochemical Nomenclature (CBN) (1977). Nomenclature of multiple forms of enzymes. Recommendations (1976). J. Biol. Chem. 252, 5939-5941.
50.

Jakubowski, M., Piotrowski, J., and Trojanowska, B. (1970). Binding of mercury in the rat: studies using $^{203}HgCl_2$ and gel filtration. Toxicol. Appl. Pharmacol. 16, 743-753.
53.

Kägi, J.H.R. (1960). Properties of a cadmium protein from equine kidney cortex. Fed. Proc. Fed. Am. Soc. Exp. Biol. 19, 340.
47.

Kägi, J.H.R. (1970). Hepatic metallothionein. Int. Congr. Biochem. Abstr., 8th, Switzerland, September, 1970, pp. 130-131.
52,57,67,68,69,70,76.

Kägi, J.H.R., and Kojima, Y. (1979). Nomenclature of metallothionein: a proposal. This volume, pp. 141-142.
48.

Kägi, J.H.R., and Vallee, B.L. (1960). Metallothionein: a cadmium- and zinc-containing protein from equine renal cortex. J. Biol. Chem. 235, 3460-3465.
47,48,52,57,58,65,76.

Kägi, J.H.R., and Vallee, B.L. (1961a). Metallothionein: a cadmium and zinc-containing protein from equine renal cortex. II. Physicochemical properties. J. Biol. Chem. 236, 2435-2442.
47,52,56,57,65,67,68,69.

Kägi, J.H.R., and Vallee, B.L. (1961b). Cadmium- and zinc-bonds in metallothionein. Int. Congr. Biochem. Abstr., 5th, Moscow, August, 1961, p. 35.
47.

Kägi, J.H.R., Himmelhoch, S.R., Whanger, P.D., Bethune, J.L., and Vallee, B.L. (1974). Equine hepatic and renal metallothioneins. Purification, molecular weight, amino acid composition, and metal content. J. Biol. Chem. 249, 3537-3542.
48,52,57,58,64,65,70.

Kaszper, B.W., Piotrowski, J.K., Marciniak, W., Sielczyńska, M. (1976). Comparative study of the level of metallothionein-like proteins, cadmium, and zinc in the liver and kidneys of rats. Bromatol. Chem. Toksykol. 9, 315-326 (In Polish).
53.

Khayat, A.I. (1977). Alteration of mercury vapor metabolism by ethyl alcohol in rat tissues in vivo and in vitro. M.S. Thesis, University of Rochester.
72.

Kimura, M., and Otaki, N. (1975). Cadmium accumulation in liver and kidney of rabbit with single injection. Ind. Health 13, 203-210.
53.

Kimura, M., Otaki, N., Yoshiki, S., Suzuki, M., Horiuchi, N., and Suda, T. (1974). The isolation of metallothionein and its protective role in cadmium poisoning. Arch. Biochem. Biophys. 165, 340-348.
54,86.

Kimura, M., Otaki, N., and Imano, M. (1979a). Rabbit liver metallothionein. Tentative amino acid sequence of metallothionein-B. This volume, pp. 163-168.
53,56,60,61.

Kimura, M., Otaki, N., and Kakefuda, T. (1979b). Characterization of cadmium-induced metallothionein in African green monkey kidney cells in vitro. This volume, pp. 187-196.
52.

Kissling, M.M., and Kägi, J.H.R. (1977). Primary structure of human hepatic metallothionein. FEBS Lett. 82, 247-250.
52,60,61.

Kissling, M.M., and Kägi, J.H.R. (1979). Amino acid sequence of human hepatic metallothioneins. This volume, pp. 145-151.
52,56,60,61.

Kissling, M.M., Berger, C., Kägi, J.H.R., Andersen, R.D., and Weser, U. (1979). The amino-terminal sequence of a rat liver metallothionein (MT-2). This volume, pp. 181-185.
53,60,61.

Kojima, Y., and Kägi, J.H.R. (1978). Metallothionein. Trends Biochem. Sci. 3, 90-93.
56.

Kojima, Y., Berger, C., Vallee, B.L., and Kägi, J.H.R. (1976). Amino-acid sequence of equine renal metallothionein-1B. Proc. Natl. Acad. Sci. USA 73, 3413-3417.
48,52,58,60,61,64,69.

Kojima, Y., Berger, C., and Kägi, J.H.R. (1979). The amino acid sequence of equine metallothioneins. This volume, pp. 153-161.
52,56,60,61.

Komsta-Szumska, E., and Chmielnicka, J. (1977). Binding of mercury and selenium in subcellular fractions of rat liver and kidneys following separate and joint administration. Arch. Toxicol. 38, 217-228.
91.

Komsta-Szumska, E., Chmielnicka, J., and Piotrowski, J.K. (1976). The influence of selenium on binding of inorganic mercury by metallothionein in the kidney and liver of the rat. Biochem. Pharmacol. 25, 2539-2540.
91.

Kotsonis, F.N., and Klaassen, C.D. (1977). Comparison of methods for estimating hepatic metallothionein in rats. Toxicol. Appl. Pharmacol. 42, 583-588.
56.

Lafont, J., Rouanet, J.-M., Besançon, P., and Moretti, J. (1976). Existence d'une métallothionéine dans le placenta. C.R. Acad. Sci. Ser. C. 283, 417-420.
53,54.

Leber, A.P., and Miya, T.S. (1976). A mechanism for cadmium- and zinc-induced tolerance to cadmium toxicity: involvement of metallothionein. Toxicol. Appl. Pharmacol. 37, 403-414.
80.

Lerch, K. (1979). Amino-acid sequence of copper-metallothionein from Neurospora crassa. This volume, pp. 173-179.
55,58,60,61.

Lucis, O.J., Shaikh, Z.A., and Embil, J.A., Jr. (1970). Cadmium as a trace element and cadmium binding components in human cells. Experientia 26, 1109-1110.
52.

Lucis, O.J., Lucis, R., and Shaikh, Z.A. (1972). Cadmium and zinc in pregnancy and lactation. Arch. Environ. Health 25, 14-22.
53,54.

Maclean, F.I., Lucis, O.J., Shaikh, Z.A., and Jansz, E.R. (1972). The uptake and subcellular distribution of Cd and Zn in microorganisms. Fed. Proc. Fed. Am. Soc. Exp. Biol. 31, 699.
55.

Madapallimattam, G., and Riordan, J.R. (1977). Antibodies to the low molecular weight copper binding protein from liver. Biochem. Biophys. Res. Commun. 77, 1286-1293.
57,90.

Marafante, E. (1976). Binding of mercury and zinc to cadmium-binding protein in liver and kidney of goldfish (Carassius auratus L.). Experientia 32, 149-150.
54.

Margoshes, M., and Vallee, B.L. (1957). A cadmium protein from equine kidney cortex. J. Am. Chem. Soc. 79, 4813-4814.
47,52,57.

Martin, M., and Brady, F.O. (1977). Induction of rat liver zinc metallothionein by mercuric chloride. Proc. S. D. Acad. Sci. 56, 72-76.
72.

Mason, K.E., and Young, J.O. (1967). Effectiveness of selenium and zinc in protecting against cadmium-induced injury of the rat testis. In: Selenium in Biomedicine, 1st International Symposium, Oregon State University, 1966 (Muth, O.H., ed), pp. 383-394, The AVI Publishing Co., Westport, Conn.
80.

Mogilnicka, E.M., and Piotrowski, J.K. (1977). Binding of gold in the kidney of the rat. Biochem. Pharmacol. 26, 1819-1820.
92.

Mogilnicka, E.M., Piotrowski, J.K., and Trojanowska, B. (1975). Effect of certain metals (cadmium, mercury, zinc, copper, lead, manganese, vanadium) on the metallothionein level in rat liver and kidneys. Med. Pr. 26, 147-155 (In Polish).
71,72.

Mogilnicka, E.M., Piotrowski, J.K., and Tomaszewski, A. (1976). Effect of some metals (rubidium, cesium, silver, thallium, antimony, uranium) on the metallothionein level in the liver and kidneys of the rat. Bromatol. Chem. Toksykol. 9, 357-359 (In Polish).
72,88.

Mogilnicka, E.M., Milaszewicz, M., and Piotrowski, J.K. (1978). Binding of silver in the rat liver. Bromatol. Chem. Toksykol. 11, 59-65 (In Polish).
88.

Noël-Lambot, F. (1976). Distribution of cadmium, zinc and copper in the mussel Mytilus edulis. Existence of cadmium-binding proteins similar to metallothioneins. Experientia 32, 324-326.
55.

Nomiyama, K., and Foulkes, E.C. (1977). Reabsorption of filtered cadmium-metallothionein in the rabbit kidney. Proc. Soc. Exp. Biol. Med. 156, 97-99.
81.

Nordberg, G.F. (1971). Effects of acute and chronic cadmium exposure on the testicles of mice. With special reference to protective effects of metallothionein. Environ. Physiol. (Environ. Physiol. Biochem.) 1, 171-187.
54,79,80,81.

Nordberg, G.F. (1972). Cadmium metabolism and toxicity. Experimental studies on mice with special reference to the use of biological materials as indices of retention and the possible role of metallothionein in transport and detoxification of cadmium. Environ. Physiol. Biochem. 2, 7-36.
84.

Nordberg, see also Nordberg, M.

Nordberg, G.F. (1979). Renal effects of mercury- and cadmium-metallothionein. This volume, pp. 347-350.
81,87.

Nordberg, G.F., and Nordberg, M. (1975a). Metabolism and toxicity of metallothionein-bound cadmium. International Conference on Heavy Metals in the Environment, Abstracts, Toronto, Canada, October, 1975, B50-51.
81,83.

Nordberg, G.F., and Piscator, M. (1972). Influence of long-term cadmium exposure on urinary excretion of protein and cadmium in mice. Environ. Physiol. Biochem. 2, 37-49.
85.

Nordberg, G.F., Piscator, M., and Lind, B. (1971a). Distribution of cadmium among protein fractions of mouse liver. Acta Pharmacol. Toxicol. 29, 456-470.
54,71,79.

Nordberg, G.F., Piscator, M., and Nordberg, M. (1971b). On the distribution of cadmium in blood. Acta Pharmacol. Toxicol. 30, 289-295.
54,84.

Nordberg, G.F., Nordberg, M., Piscator, M., and Vesterberg, O. (1972). Separation of two forms of rabbit metallothionein by isoelectric focusing. Biochem. J. 126, 491-498.
53,56,57,65,67,71.

Nordberg, G.F., Goyer, R.A., and Nordberg, M. (1975a). Comparative toxicity of cadmium-metallothionein and cadmium chloride on mouse kidney. Arch. Pathol. 99, 192-197.
81,83.

Nordberg, M. (1978). Studies on metallothionein and cadmium. Environ. Res. 15, 381-404.
56,81,83,84,85.

Nordberg, M., and Nordberg, G.F. (1973). Isolation and identification of metallothionein from mice. Int. Congr. Biochem. Abstr., 9th, Stockholm, July, 1973, p. 82.
54.

Nordberg, M., and Nordberg, G.F. (1975b). Distribution of metallothionein-bound cadmium and cadmium chloride in mice: preliminary studies. Environ. Health Perspect. 12, 103-108.
81.

Nordberg, M., Trojanowska, B., and Nordberg, G.F. (1974). Studies on metal-binding proteins of low molecular weight from renal tissue of rabbits exposed to cadmium and mercury. Environ. Physiol. Biochem. 4, 149-158.
53,71,91.

Nordberg, see also Nordberg, G.F.

Nordberg, M., Nordberg, G.F., and Piscator, M. (1975b). Isolation and characterization of a hepatic metallothionein from mice. Environ. Physiol. Biochem. 5, 396-403.
54,57,65,67,71.

Norheim, G., and Steinnes, E. (1976). Distribution of some protein-bound trace elements among soluble protein fractions from human liver. Acta Pharmacol. Toxicol. 38, 137-144.
52.

Oh, S.H., Deagen, J.T., Whanger, P.D., and Weswig, P.H. (1978). Biological function of metallothionein. V. Its induction in rats by various stresses. Am. J. Physiol. 234, E282-E285.
71,77.

Olafson, R.W., and Thompson, J.A.J. (1974). Isolation of heavy metal binding proteins from marine vertebrates. Mar. Biol. 28, 83-86.
54.

Olafson, R.W., Sim, R.G., and Kearns, A. (1979). Physiological and chemical characterization of invertebrate metallothionein-like proteins. This volume, pp. 197-204.
51,55,56,64.

Otvos, J.D., and Armitage, I.M. (1979). Structural characterization of metallothionein by multinuclear NMR. This volume, pp. 249-257.
68,70.

Overnell, J., Davidson, I.A., and Coombs, T.L. (1977). A cadmium-binding glycoprotein from the liver of the plaice (Pleuronectes platessa). Biochem. Soc. Trans. 5, 267-269.
54.

Panemangalore, M., and Brady, F.O. (1978). Induction and synthesis of metallothionein in isolated perfused rat liver. J. Biol. Chem. 253, 7898-7904.
73.

Pařízek, J. (1957). The destructive effect of cadmium ion on testicular tissue and its prevention by zinc. J. Endocrinol. 15, 56-63.
79,80.

Pařízek, J., Ošťádalová, I., Kalousková, J., Babický, A., and Beneš, J. (1971). The detoxifying effects of selenium. Interrelations between compounds of selenium and certain metals. In: Newer Trace Elements in Nutrition (Mertz, W., and Cornatzer, W.E., eds), pp. 85-122, Marcel Dekker, New York, N.Y.
91.

Pařízek, J., Kalousková, J., Babický, A., Beneš, J., and Pavlík, L. (1974). Interaction of selenium with mercury, cadmium, and other toxic metals. In: Trace Element Metabolism in Animals-2 (Hoekstra, W.G., Suttie, J.W., Ganther, H.E., and Mertz, W., eds), pp. 119-131, University Park Press, Baltimore, Md.
91.

Piotrowski, J.K., and Bolanowska, W. (1970). Binding of phenyl-mercury acetate (^{203}Hg) in the body of rat, studied by molecular filtration technique. Med. Pr. 21, 338-348 (In Polish).
53.

Piotrowski, J.K., and Szymańska, J.A. (1976). Influence of certain metals on the level of metallothionein-like proteins in the liver and kidneys of rats. J. Toxicol. Environ. Health 1, 991-1002.
72,91.

Piotrowski, J.K., Bolanowska, W., and Sapota, A. (1973a). Evaluation of metallothionein content in animal tissues. Acta Biochim. Pol. 20, 207-215.
56.

Piotrowski, J.K., Trojanowska, B., Wisniewska-Knypl, J.M., and Bolanowska, W. (1973b). Further investigations on binding and release of mercury in the rat. In: Mercury, Mercurials and Mercaptans (Miller, M.W., and Clarkson, T.W., eds), pp. 247-263, Charles C. Thomas Publisher, Springfield, Ill.
72,87.

Piotrowski, J.K., Trojanowska, B., and Sapota, A. (1974a). Binding of cadmium and mercury by metallothionein in the kidneys and liver of rats following repeated administration. Arch. Toxicol. 32, 351-360.
51,53,54,71,72,91.

Piotrowski, J.K., Trojanowska, B., Wiśniewska-Knypl, J.M., and Bolanowska, W. (1974b). Mercury binding in the kidney and liver of rats repeatedly exposed to mercuric chloride: induction of metallothionein by mercury and cadmium. Toxicol. Appl. Pharmacol. 27, 11-19.
53,54,71,72,91.

Piotrowski, J.K., Bem, E.M., and Werner, A. (1977). Cadmium and mercury binding to metallothionein as influenced by selenium. Biochem. Pharmacol. 26, 2191-2192.
71,91.

Piotrowski, J.K., Szymańska, J.A., Mogilnicka, E.M., and Żelazowski, A.J. (1979). Renal metal binding proteins. This volume, pp. 363-371.
91,92.

Piscator, M. (1964). Om kadmium i normala människonjurar samt redogörelse för isolering av metallothionein ur lever från kadmiumexponerade kaniner. (On cadmium in normal human kidneys together with a report on the isolation of metallothionein from livers of cadmium-exposed rabbits.) Nord. Hyg. Tidskr. 45, 76-82.
47,53,71,84.

Porter, H. (1966). The tissue copper proteins: cerebrocuprein, erythrocuprein, hepatocuprein, and neonatal hepatic mitochondrocuprein. In: The Biochemistry of Copper (Peisach, J., Aisen, P., and Blumberg, W.E., eds), pp. 159-174, Academic Press, New York and London.
88.

Porter, H. (1974). The particulate half-cystine-rich copper protein of newborn liver. Relationship to metallothionein and subcellular localization in non-mitochondrial particles possibly representing heavy lysosomes. Biochem. Biophys. Res. Commun. 56, 661-668.
78,88.

Porter, H., Johnston, J., and Porter, E.M. (1962). Neonatal hepatic mitochondrocuprein. I. Isolation of a protein fraction containing more than 4% copper from mitochondria of immature bovine liver. Biochim. Biophys. Acta 65, 66-73.
88.

Porter, H., Sweeney, M., and Porter, E.M. (1964). Neonatal hepatic mitochondrocuprein. II. Isolation of the copper-containing subfraction from mitochondria of newborn human liver. Arch. Biochem. Biophys. 104, 97-101.
88.

Premakumar, R., Winge, D.R., Wiley, R.D., and Rajagopalan, K.V. (1975a). Copper-induced synthesis of copper-chelatin in rat liver. Arch. Biochem. Biophys. 170, 267-277.
72,73,89.

Premakumar, R., Winge, D.R., Wiley, R.D., and Rajagopalan, K.V. (1975b). Copper-chelatin: isolation from various eucaryotic sources. Arch. Biochem. Biophys. 170, 278-288.
72,89.

Prinz, R., and Weser, U. (1975). A naturally occurring Cu-thionein in Saccharomyces cerevisiae. Hoppe-Seyler's Z. Physiol. Chem. 356, 767-776.
55,67.

Probst, G.S., Bousquet, W.F., and Miya, T.S. (1977a). Kinetics of cadmium-induced hepatic and renal metallothionein synthesis in the mouse. Toxicol. Appl. Pharmacol. 39, 51-60.
71.

Probst, G.S., Bousquet, W.F., and Miya, T.S. (1977b). Correlation of hepatic metallothionein concentrations with acute cadmium toxicity in the mouse. Toxicol. Appl. Pharmacol. 39, 61-69.
71.

Pulido, P., Kägi, J.H.R., and Vallee, B.L. (1966). Isolation and some properties of human metallothionein. Biochemistry 5, 1768-1777.
47,52,56,68.

Raghavan, S.R.V., and Gonick, H.C. (1977). Isolation of low-molecular-weight lead-binding protein from human erythrocytes. Proc. Soc. Exp. Biol. Med. 155, 164-167.
92.

Richards, M.P., and Cousins, R.J. (1975a). Mammalian zinc homeostasis: requirement for RNA and metallothionein synthesis. Biochem. Biophys. Res. Commun. 64, 1215-1223.
53,71,73,74,77.

Richards, M.P., and Cousins, R.J. (1975b). Influence of parenteral zinc and actinomycin D on tissue zinc uptake and the synthesis of a zinc-binding protein. Bioinorg. Chem. 4, 215-224.
53,73.

Richards, M.P., and Cousins, R.J. (1976). Metallothionein and its relationship to the metabolism of dietary zinc in rats. J. Nutr. 106, 1591-1599.
71,73,77.

Richards, M.P., and Cousins, R.J. (1977a). Isolation of an intestinal metallothionein induced by parenteral zinc. Biochem. Biophys. Res. Commun. 75, 286-294.
54.

Richards, M.P., and Cousins, R.J. (1977b). Influence of inhibitors of protein synthesis on zinc metabolism. Proc. Soc. Exp. Biol. Med. 156, 505-508.
73.

Riordan, J.R., and Gower, I. (1975a). Small copper-binding proteins from normal and copper-loaded liver. Biochim. Biophys. Acta 411, 393-398.
53,89.

Riordan, J.R., and Gower, I. (1975b). Purification of low molecular weight copper proteins from copper loaded liver. Biochem. Biophys. Res. Commun. 66, 678-686.
53,89.

Riordan, J.R., and Madapallimattam, G. (1979). Low molecular weight copper-binding proteins in cultured human cells. This volume, pp. 353-361.
52,90.

Roosemont, J.L. (1972). Properties of metallothionein-like fractions from equine renal cortex. Arch. Int. Physiol. Biochim. 80, 407-409.
52.

Rudd, C.J., and Herschman, H.R. (1978). Metallothionein accumulation in response to cadmium in a clonal rat liver cell line. Toxicol. Appl. Pharmacol. 44, 511-521.
53.

Rugstad, H.E., and Norseth, T. (1975). Cadmium resistance and content of cadmium-binding protein in cultured human cells. Nature (London) 257, 136-137.
52.

Rugstad, H.E., and Norseth, T. (1978). Cadmium resistance and content of cadmium-binding protein in two enzyme-deficient mutants of mouse fibroblasts (L-cells). Biochem. Pharmacol. 27, 647-650.
54.

Rupp, H., and Weser, U. (1974). Conversion of metallothionein into Cu-thionein, the possible low molecular weight form of neonatal hepatic mitochondrocuprein. FEBS Lett. 44, 293-297.
58,66,69.

Rupp, H., and Weser, U. (1978). Circular dichroism of metallothioneins. A structural approach. Biochim. Biophys. Acta 533, 209-226.
57,65,66,67,68,69,70.

Rupp, H., and Weser, U. (1979). Structural aspects and reduction oxidation reactions of metallothionein. This volume, pp. 231-240.
68,70.

Rupp, H., Voelter, W., and Weser, U. (1974). 270 MHz proton magnetic resonance spectra of metallothionein. FEBS Lett. 40, 176-179.
68.

Rupp, H., Voelter, W., and Weser, U. (1975). Molecular biology of copper. A circular dichroism study on copper complexes of thionein and penicillamine. Hoppe-Seyler's Z. Physiol. Chem. 356, 755-765.
69.

Rydén, L., and Deutsch, H.F. (1978). Preparation and properties of the major copper-binding component in human fetal liver. Its identification as metallothionein. J. Biol. Chem. 253, 519-524.
51,52,56,58,60,61,78,89.

Sabbioni, E., and Marafante, E. (1975a). Heavy metals in rat liver cadmium binding protein. Environ. Physiol. Biochem. 5, 132-141.
53,72.

Sabbioni, E., and Marafante, E. (1975b). Accumulation of cadmium in rat liver cadmium binding protein following single and repeated cadmium administration. Environ. Physiol. Biochem. 5, 465-473.
53,71.

Sabbioni, E., Marafante, E., Amantini, L., Ubertalli, L., and Pietra, R. (1978). Cadmium toxicity studies under long term-low level exposure (LLE) conditions. I. Metabolic patterns in rats exposed to present environmental dietary levels of cadmium for two years. Sci. Total Environ. 10, 135-161.
54.

Sapota, A., Piotrowski, J.K., and Barański, B. (1974). Level of metallothionein in tissues of pregnant rats and foetuses exposed to mercury vapours. Med. Pr. 25, 129-135.
72,91.

Shaikh, Z.A. (1979). The low molecular weight cadmium-, mercury- and zinc-binding proteins (metallothioneins): biosynthesis, metabolism, and possible role in metal toxicity. This volume, pp. 331-336.
53.

Shaikh, Z.A., and Hirayama, K. (1979a). Metallothionein and metabolism of cadmium. In: Kadmium-Symposium (Anke, M., and Schneider, H.-J., eds), pp. 95-101, Wissenschaftliche Beiträge der Friedrich-Schiller-Universität, Jena, German Democratic Republic.
81,82,83.

Shaikh, Z.A., and Hirayama, K. (1979b). Metallothionein in the extra cellular fluids as an index of cadmium toxicity. Environ. Health Perspect. 28, 267-271.
85.

Shaikh, Z.A., and Lucis, O.J. (1969). Distribution and binding of ^{65}Zn and ^{109}Cd in experimental animals. Proc. Canad. Fed. Biol. Soc. 12, 101.
53,54.

Shaikh, Z.A., and Lucis, O.J. (1970). Induction of cadmium binding protein. Fed. Proc. Fed. Am. Soc. Exp. Biol. 29, 298.
71.

Shaikh, Z.A., and Lucis, O.J. (1971a). The nature and biosynthesis of cadmium binding proteins. Fed. Proc. Fed. Am. Soc. Exp. Biol. 30, 238.
54,56.

Shaikh, Z.A., and Lucis, O.J. (1971b). Isolation of cadmium-binding proteins. Experientia 27, 1024-1025.
56.

Shaikh, Z.A., and Lucis, O.J. (1972). Cadmium and zinc binding in mammalian liver and kidneys. Arch. Environ. Health 24, 419-425.
52,54.

Shaikh, Z.A., and Smith, J.C. (1975a). Cadmium induced synthesis of hepatic and renal metallothionein. Fed. Proc. Fed. Am. Soc. Exp. Biol. 34, 266.
73,75.

Shaikh, Z.A., and Smith, J.C. (1975b). Mercury-induced synthesis of renal metallothionein. International Conference on Heavy Metals in the Environment, Abstracts, Toronto, Canada, October, 1975, pp. B108-110.
72,73,75,91.

Shaikh, Z.A., and Smith, J.C. (1976). The biosynthesis of metallothionein in rat liver and kidney after administration of cadmium. Chem. Biol. Interact. 15, 327-336.
75.

Shaikh, Z.A., and Smith, J.C. (1977). The mechanisms of hepatic and renal metallothionein biosynthesis in cadmium-exposed rats. Chem. Biol. Interact. 19, 161-171.
71,73.

Shapiro, J., Morell, A.G., and Scheinberg, I.H. (1961). A copper-protein of human liver. J. Clin. Invest. 40, 1081.
88.

Shapiro, S.G., Squibb, K.S., Markowitz, L.A., and Cousins, R.J. (1978). Cell-free synthesis of metallothionein directed by rat liver polyadenylated messenger ribonucleic acid. Biochem. J. 175, 833-840.
74.

Sobocinski, P.Z., Canterbury, W.J., Jr., Mapes, C.A., and Dinterman, R.E. (1978). Involvement of hepatic metallothioneins in hypozincemia associated with bacterial infection. Am. J. Physiol. 234, E399-E406.
71,77.

Sokolowski, G., and Weser, U. (1975). Formation, circular dichroism and X-ray photoelectron spectroscopy of hepatic Zn-thionein. Hoppe-Seyler's Z. Physiol. Chem. 356, 1715-1726.
53.

Sonawane, B.R., Nordberg, M., Nordberg, G.F., and Lucier, G.W. (1975). Placental transfer of cadmium in rats: influence of dose and gestational age. Environ. Health Perspect. 12, 97-102.
54.

Squibb, K.S., and Cousins, R.J. (1974). Control of cadmium binding protein synthesis in rat liver. Environ. Physiol. Biochem. 4, 24-30.
71,73.

Squibb, K.S., and Cousins, R.J. (1977). Synthesis of metallothionein in a polysomal cell-free system. Biochem. Biophys. Res. Commun. 75, 806-812.
74.

Squibb, K.S., Cousins, R.J., and Feldman, S.L. (1977). Control of zinc-thionein synthesis in rat liver. Biochem. J. 164, 223-228.
53,71,73.

Squibb, K.S., Ridlengton, J.W., Carmichael, N.G., and Fowler, B.A. (1979). Early cellular effects of circulating cadmium-thionein on kidney proximal tubules. Environ. Health Perspect. 28, 287-296.
83.

Stokes, P.M., Maler, T., and Riordan, J.R. (1977). A low molecular weight copper binding protein in a copper tolerant strain of Scenedesmus acutiformis. In: Trace Substances in Environmental Health-XI (Hemphill, D.D., ed), pp. 146-155.
90.

Suda, T., Horiuchi, N., Ogata, E., Ezawa, I., Otaki, N., and Kimura, M. (1974). Prevention by metallothionein of cadmium-induced inhibition of vitamin D activation reaction in kidney. FEBS Lett. 42, 23-26.
53, 81, 86.

Suzuki, Y., and Yoshikawa, H. (1976). Induction of hepatic zinc-binding proteins of rats by various metals. Ind. Health 14, 25-31.
72.

Szymańska, J.A., Mogilnicka, E.M., and Kaszper, B.W. (1977). Binding of bismuth in the kidneys of the rat: the role of metallothionein-like proteins. Biochem. Pharmacol. 26, 257-258.
91.

Szymańska, J.A., Zychowicz, M., Żelazowski, A.J., and Piotrowski, J.K. (1978). Effect of selenium on the organ distribution and binding of bismuth in rat tissues. Arch. Toxicol. 40, 131-141.
91.

Tanaka, K., Sueda, K., Onosaka, S., and Okahara, K. (1975). Fate of ^{109}Cd-labeled metallothionein in rats. Toxicol. Appl. Pharmacol. 33, 258-266.
81.

Task Group on Metal Accumulation (1973). Accumulation of toxic metals with special reference to their absorption, excretion and biological half-times. Report from a meeting in Buenos Aires, September, 1972. Environ. Physiol. Biochem. 3, 65-107.
79.

Task Group on Metal Toxicity (Nordberg, G.F., ed) (1976). Effects and dose-response relationships of toxic metals. A consensus report from a meeting in Tokyo, November, 1974. In: Effects and Dose-Response Relationships of Toxic Metals (Nordberg, G.F., ed), pp. 1-111, Elsevier Scientific Publishing Co., Amsterdam.
79.

Terao, T., and Owen, C.A., Jr. (1973). Nature of copper compounds in liver supernate and bile of rats: studies with ^{67}Cu. Am. J. Physiol. 224, 682-686.
78.

Terhaar, C.J., Vis, E., Roudabush, R.L., and Fassett, D.W. (1965). Protective effects of low doses of cadmium chloride against subsequent high oral doses in the rat. Toxicol. Appl. Pharmacol. 7, 500.
79.

The Task Group on Metal Interaction (Nordberg, G.F., ed) (1978). Factors influencing metabolism and toxicity of metals: a consensus report. Report from a meeting in Stockholm, July, 1977. Environ. Health Perspect. 25, 3-41.
91.

Tohyama, C., and Shaikh, Z.A. (1978). Cross-reactivity of metallothioneins from different origins with rabbit anti-rat hepatic metallothionein antibody. Biochem. Biophys. Res. Commun. 84, 907-913.
57.

Trojanowska, B., and Piotrowski, J.K. (1974). Binding of zinc by metallothionein in the liver and kidneys of rat. Bromatol. Chem. Toksykol. 7, 137-142 (In Polish).
51,53,54,71.

Tsunoo, H., Kino, K., Nakajima, H., Hata, A., Huang, I-Y., and Yoshida, A. (1978). Mouse liver metallothioneins. Purification, molecular weight, amino acid composition, and metal content. J. Biol. Chem. 253, 4172-4174.
54,65.

Ulmer, D.D., and Vallee, B.L. (1971). Structure and function of metalloenzymes. Adv. Chem. Ser. 100, 187-218.
70.

Ulmer, D.D., Kägi, J.H.R., and Vallee, B.L. (1962). Optically active metalloprotein chromophores. I. Metallothionein. Biochem. Biophys. Res. Commun. 8, 327-330.
47,68.

Vallee, B.L. (1960). Metalloenzymes: correlation of composition, structure and function. In: Fourth International Congress of Biochemistry, Vienna, 1958, Vol. VIII--Proteins, pp. 138-165, Pergamon Press, London.
47.

Vallee, B.L. (1979). Metallothionein: historical review and perspectives. This volume, pp. 19-40.
47,76.

Vander Mallie, R.J., and Garvey, J.S. (1978). Production and study of antibody produced against rat cadmium thionein. Immunochemistry 15, 857-868.
57.

Vostal, J.J., and Cherian, M.G. (1974). Effects of cadmium metallothionein on the renal tubular transport of sodium. Fed. Proc. Fed. Am. Soc. Exp. Biol. 33, 519.
53,82.

Waters, M.D., Nordberg, G.F., Cox, C.C., Vaughan, T.O., and Coffin, D.L. (1975). Cellular toxic effects of cadmium and other metals. International Conference on Heavy Metals in the Environment, Abstracts, Toronto, Canada, October, 1975, B54-56.
53.

Webb, M. (1972a). Biochemical effects of Cd^{2+}-injury in the rat and mouse testis. J. Reprod. Fertil. 30, 83-98.
79.

Webb, M. (1972b). Binding of cadmium ions by rat liver and kidney. Biochem. Pharmacol. 21, 2751-2765.
70,71,76.

Webb, M. (1979). Functions of hepatic and renal metallothioneins in the control of the metabolism of cadmium and certain other bivalent cations. This volume, pp. 313-320.
51,52,53,77,80.

Webb, M., and Daniel, M. (1975). Induced synthesis of metallothionein by pig kidney cells in vitro in response to cadmium. Chem. Biol. Interact. 10, 269-276.
52.

Webb, M., and Etienne, A.T. (1977). Studies on the toxicity and metabolism of cadmium-thionein. Biochem. Pharmacol. 26, 25-30.
75,83.

Weser, U., and Rupp, H. (1979). Copper-thionein and other metal-sulphur-proteins. This volume, pp. 221-230.
65,66,67,69.

Weser, U., Donay, F., and Rupp, H. (1973a). Cadmium-induced synthesis of hepatic metallothionein in chicken and rats. FEBS Lett. 32, 171-174.
53,54.

Weser, U., Rupp, H., Donay, F., Linnemann, F., Voelter, W., Voetsch, W., and Jung, G. (1973b). Characterization of Cd, Zn-thionein (metallothionein) isolated from rat and chicken liver. Eur. J. Biochem. 39, 127-140.
53,54,65,67,68,70.

Weser, U., Hartmann, H.-J., Fretzdorff, A., and Strobel, G.-J. (1977). Homologous copper(I)-(thiolate)$_2$-chromophores in yeast copper thionein. Biochim. Biophys. Acta 493, 465-477.
57,58,66,68,70.

Whanger, P.D., and Oh, S.H. (1979). Nutritional and environmental factors affecting metallothionein levels. This volume, pp. 281-291.
76,77,78.

Winge, D.R., and Rajagopalan, K.V. (1972). Purification and some properties of Cd-binding protein from rat liver. Arch. Biochem. Biophys. 153, 755-762.
53.

Winge, D.R., Premakumar, R., and Rajagopalan, K.V. (1975a). Metal-induced formation of metallothionein in rat liver. Arch. Biochem. Biophys. 170, 242-252.
53,71,72,88.

Winge, D.R., Premakumar, R., Wiley, R.D., and Rajagopalan, K.V. (1975b). Copper-chelatin: purification and properties of a copper-binding protein from rat liver. Arch. Biochem. Biophys. 170, 253-266.
72,89.

Winge, D.R., Premakumar, R., and Rajagopalan, K.V. (1978). Studies on the zinc content of Cd-induced thionein. Arch. Biochem. Biophys. 188, 466-475.
53.

Wiśniewska-Knypl, J.M., and Jabłońska, J. (1970). Selective binding of cadmium in vivo on metallothionein in rat's liver. Bull. Acad. Pol. Sci. Ser. Sci. Biol. 18, 321-327.
80.

Wiśniewska, J.M., Trojanowska, B., Piotrowski, J., and Jakubowski, M. (1970). Binding of mercury in the rat kidney by metallothionein. Toxicol. Appl. Pharmacol. 16, 754-763.
54,91.

Wiśniewska-Knypl, J.M., Jabłońska, J., and Myślak, Z. (1971). Binding of cadmium on metallothionein in man: an analysis of a fatal poisoning by cadmium iodide. Arch. Toxicol. 28, 46-55.
52,80.

Żelazowski, A.J., and Piotrowski, J.K. (1977a). A modified procedure for determination of metallothionein-like proteins in animal tissues. Acta Biochim. Pol. 24, 97-103.
56.

Żelazowski, A.J., and Piotrowski, J.K. (1977b). The levels of metallothionein-like proteins in animal tissues. Experientia 33, 1624-1625.
56.

12. AUTHOR INDEX

Numbers in italics indicate on what page in the report the author's name is cited; numbers in regular type denote the page where the reference to the cited article(s) is given. A citation in parentheses indicates the multiple-author group in which the author is given reference.

L

M

N

O

P

13. SUBJECT INDEX

Abbreviation: MT, metallothionein

Abbreviation: MT, metallothionein

Abbreviation: MT, metallothionein

Metallothionein
Jeremias H. R. Kägi and Monica Nordberg, eds.
Birkhäuser Verlag, Basel/Boston/Stuttgart, 1979

General discussion held in the plenary session of July 19, 1978, of the "First International Meeting on Metallothionein and Other Low Molecular Weight Metal-binding Proteins", held in Zürich, Switzerland, July 17 - 22, 1978

METALLOTHIONEIN: QUERIES AND CONJECTURES

Chairman: Bert L. Vallee

The chairman opens the session by calling on the participants to submit questions on metallothionein for discussion. The following specific questions were listed:

1. Does thionein exist as such in vivo?
2. Is thionein rather than metallothionein the active, functional agent?
3. What is the role of thionein? Has anybody studied it?
4. Does sulfur deficiency affect the production of metallothionein?
5. Does zinc deficiency cause changes in the tissue concentration of metallothionein and/or thionein?
6. Does metallothionein (or thionein) form disulfide bonds in vivo with cysteine or glutathione? Can the protein serve as a reservoir or in the transport of these and related metabolic agents?
7. Are there organic or inorganic derivatives of metallothionein which might be metabolic intermediates?
8. Does metallothionein serve as an intermediate in selenium metabolism?
9. What is the conclusive evidence that metallothionein serves as a protective agent against metal intoxication?
10. Is there endogenous production of metallothionein? What metal(s) induce(s) metallothionein preferentially?
11. What is the basis for the large differences in the natural abundance of metallothionein in different species?
12. Is metallothionein (or thionein) present in the extracellular space?
13. Does metallothionein (or thionein) occur in plasma?
14. Do prokaryotic organisms contain metallothionein (or thionein)?

Questions submitted in writing by M.D. Kamen:

15. Is there endogenous production of the apo-protein(s)?

16. What basis is there for the current continued preoccupation with the toxicology of metallothioneins in terms of definitive status or function? Obviously, the answer to (15) bears on this point too. The present status of knowledge, at least relative to biochemically meaningful data, indicates that one must keep an open mind on biological function(s) of metallothioneins.

17. Has the role of thionein been studied?

Chairman: In nearly all studies on record the question of the biological role of metallothionein has centered on its metal complement, first on cadmium and zinc, and more recently also on mercury and copper. The protein moiety, thionein, has been of surprisingly little interest to most workers in the field and the possibility that it may have functions other than metal binding has, to my knowledge, not been pursued. The recent finding that the primary structure has been preserved in evolution with relatively few alterations indicates that it is subject to strong evolutionary pressure and by this criterion important to fundamental metabolic processes. Initially, we looked for a cadmium protein. It turned out that it was only "half" a cadmium protein, also containing zinc and copper, and 20 years later we still don't know its biological function with respect to any of these metals. I think it is important, therefore, to enlarge the scope and to explore in depth also the possibility implied by the first three questions, namely that thionein rather than metallothionein might be the functional species.

Bannister: Has anyone attempted to isolate thionein directly?

Kägi: I am not aware of any systematic effort in this regard. The isolation of thionein from tissues would probably not be very easy on account of both its known instability towards oxidation on exposure to air and its marked susceptibility to proteases.

Chairman: If thionein had a role, what might it be? (Question No. 3.) An obvious possibility is an involvement in sulfur metabolism.

Ideas along this line and suggestions for experiments are contained in the following comment submitted by Dr. Meister for this discussion.

Meister: (written comment) The study of the potential role of thionein and/or metallothionein is in need of stressing research that might elucidate relationships between glutathione, cysteine, cystathionine metabolism and metallothioneins. As an example, isotope studies could be performed to see if there is a flow of sulfur from metallothionein to glutathione - or the reverse. In this regard, it should be noted that tissue and blood levels of cysteine and cystine are very low - 10-70 μM (CysH + $(Cys)_2$). Is it possible that metallothionein serves as a storage form of cysteine similar to glutathione?

Metallothionein might function to bind glutathione by mixed disulfide bonds. There is evidence that there are at least two pools of glutathione. One pool turns over rapidly ($t_{1/2} < 20$ min) while the other pool has a much larger $t_{1/2}$. Perhaps, the latter could be protein-bound, for example to metallothionein.

Some possible experiments that could serve as examples:

1) Explore the effect of sulfur starvation on metallothionein production by placing animals on a completely sulfur-free diet.

2) Administer ^{35}S-methionine and perform kinetic studies to identify its uptake into cysteine, cystathionine, glutathione, taurine, and metallothionein.

3) Search for possible mixed disulfide binding of glutathione to metallothionein. Theoretically, metallothionein might bind 20 glutathione molecules per mole of metallothionein.

Chairman: Does sulfur deficiency affect the production of metallothionein? (Question No. 4.)

Bremner: The working paper by Whanger describes an experiment in which this was studied. The accumulation of zinc-metallothionein in the livers of rats maintained on a low-sulfur diet was less than that in rats receiving a diet containing adequate amounts of sulfur.

Andersen: We find that the degradation of metallothioneins in the cell is rapid and that the biosynthesis induced by cadmium or zinc is transient, suggesting a rapidly degradable mRNA as well. This suggests that metallothionein may represent a rapidly expandable and contractable pool for the maintenance of zinc homeostasis.

Pařízek: The suggestion of Professor Meister on the possible significance of metallothionein as a "reserve pool" of cysteine could open up a new avenue for understanding the interrelation between zinc and sulfur-containing amino acids. Exposure to zinc - through changing the rate of metallothionein synthesis - would modify the size of this pool, and the effect of zinc (and other metals) on the stability of metallothionein would influence the availability of this highly labile amino acid. It would be of interest in this respect to discuss not only the effect of zinc administration but also the question of metallothionein synthesis and metallothionein levels in zinc-deficient animals.

Chairman: Does zinc deficiency cause changes in the tissue concentrations of metallothionein and/or thionein? (Question No.5.)

Bremner: We have been unable to detect zinc-metallothionein in the livers of zinc-deficient rats, pigs or sheep. Furthermore, even though the concentration of copper-metallothionein in the liver of monogastric animals has been found to be proportional to the liver copper content, this is not so in zinc-deficient animals, and little or no copper-metallothionein can be detected in the livers of such animals.

Hill: Could this be a reflection of more general effects of zinc deficiency on protein synthesis?

Bremner: I don't think so, as the zinc-deficient rats can still synthesize copper-metallothionein in response to copper injection. However, the biological half-life of copper-metallothionein is less in zinc-deficient than in zinc-supplemented animals, and we think this may explain the absence of copper-metallothionein in zinc-deficient animals receiving copper in the diet in long-term experiments. It is possible that zinc can stabilize copper-metallothionein in some way.

Chairman: Does metallothionein combine with cysteine or glutathione by disulfide interchange? Does it serve as a reservoir or to transport these or other metabolic agents as intermediates? (Question No. 6.)

Andersen: We find that in newly synthesized metallothioneins all of the incorporated ^{3}H-cysteine is part of the polypeptide chain. No ^{3}H-cysteine is bound to the protein through disulfide bridges.

Kägi: The cysteine content of pure metallothionein is remarkably constant and there is no spectroscopic evidence that native metallothioneins contain disulfide bonds. However, it is possible that we bias ourselves by the isolation methods we use. Disulfide derivatives of metallothionein are likely to differ from the fully reduced forms in their chromatographic behaviour on ion-exchangers.

Chairman: There is, of course, good evidence for the occurrence of mixed disulfides in many intracellular and extracellular proteins. King, T.P., Yphantis, D.A. and Craig, L.C. (J. Am. Chem. Soc. 82, 3350 (1960))reported long ago the occurrence of oxidized forms of human serum albumin containing both cysteine or glutathione.

Weser: Can oxidized thionein - "cystinethionein" - and the metal-loaded metallothioneins be differentiated by immunochemical means?

Bethune: Antibodies have only been prepared against metallothionein but not against thionein. However, thionein crossreacts with metallothionein.

Riordan: In fact, the only published work on antibodies (Madapallimatam, G. and Riordan, J.R., Biochem. Biophys. Res. Commun. 77, 1286 (1977)) has shown immunochemical identity between the copper binding protein of liver and cadmium thionein. The same antisera have more recently been shown to cross-react with the apoprotein as well.

Piscator: A way to study the role of metallothionein in sulfur metabolism might be to induce synthesis of metallothionein by zinc exposure, followed by exposure to compounds that are known to combine with glutathione, cysteine, etc. in detoxification processes. Examples of such compounds may be methyl mercury and vinylchloride. Comparisons between exposed groups with low and high metallothionein content may give information about the utilization of the S-compounds.

Chairman: Are there multiple species of both inorganically and organically derivatized forms of metallothionein and thionein? (Question No.7.)

Weser: We have looked for metal derivatives but not under in vivo conditions. It would be difficult to know which metal derivative is present initially owing to metal exchange, external and internal contamination.

Chairman: The question of the presence of selenium in metallothionein (Question No. 8) requires reexamination, especially in view of its accumulation in the organism as a function of the Hg content (J. Pařízek, I. Ošťádalová, J. Kalousková, A. Babický and J. Beneš, in "Newer Trace Elements in Nutrition" (W. Mertz and W.A. Cornatzer, eds), pp. 85, Marcel Dekker, New York, 1971; L. Kosta, A.R. Byrne and V. Zelenko, Nature 254, 238-239 (1975)). It is technically difficult to establish the selenium content.

Pařízek: The question of metallothionein involvement in selenium metabolism is certainly interesting. Two possibilities could be considered: (a) replacement of cysteine moieties in metallothionein by selenocysteine. This might be possible provided selenocysteine is available (its formation from inorganic selenium compounds in monogastric animals is still questionable). I do not recall any report on the detection of selenide groups in metallothionein. However, such metallothionein homologs, when prepared, could be of interest with respect to metal binding, stability, etc. ^{75}Se-Selenium would also have some advantage with respect to radioisotopic labelling of metallothionein. (b) The second possibility would be a reaction between -SH groups and selenium compounds (e.g., analogous to Painter's reaction). Even here, there does not seem to be any evidence available. Present knowledge does not allow an assignment of a direct role to metallothionein in selenium-metal interactions.

Weser: We have studied the relationships of selenium and metallothionein. Intraperitoneal injection of labeled selenium (given as SeO_2) does not show up in isolated metallothionein. On the other hand, in vitro SeO_2 does form a thionein-selenotrisulphide of the type RS-Se-SR. The oxidation state of selenium has been measured by X-ray photoelectron spectrometry to be approximately $+\frac{2}{3}$.

Chairman: Much emphasis has been given in recent years to the purported role of metallothionein as a protective agent against intoxication by metals, especially cadmium. In published data, most of the evidence has remained ambiguous, however, and hence one is led to wonder about the justification for the continued, manifold claims of a definitive, toxicological function of metallothionein. What is the conclusive evidence that metallothionein serves as a protective agent against metal poisoning? (Questions No. 9 and 16.)

Piscator: Most of the cadmium in the human body is stored and bound to metallothionein. If synthesis of metallothionein suddenly stopped, the amount of cadmium released in the kidneys would cause severe acute renal damage, even death.

Hill: The most effective method of avoiding the problem would simply be to excrete the cadmium. Binding so that it is retained does not seem the most effective method of detoxification.

Chairman: Let us just speculate that metallothionein happens to be a system involved in S-metabolism, and its purpose is primarily that. In that case, the metal binding properties would be incidental and quite misleading. I make this comment solely to play "devil's advocate" to induce different perspectives.

Piscator: If that were the case, why would the change in cadmium concentration from less than 1 μg cadmium per gram wet weight in the fetus and the newborn to more than 100 μg per gram bound in the form of cadmium metallothionein in the liver of exposed workers not be reflected in changes in liver function?

G. Nordberg: If cadmium-metallothionein would serve a physiological role in the body, it would be difficult to explain why there is such a striking age-dependent increase in cadmium metallothionein from very small amounts in the young to the large quantities in adult man. It is thus likely that, in relation to cadmium, metallothionein serves a role in modifying its toxicity and that cadmium-metallothionein does not serve a physiological role. This does not exclude the possibility that the primary role of metallothionein might be in S-metabolism.

Smith: Do we know if metallothionein is present at birth or is it thionein which is then present and only binds cadmium with age? Also, if cadmium-free forms of metallothionein are present at birth, what is their function?

Chairman: It was my impression that metallothionein is present at birth.

Bremner: Both zinc- and copper-metallothioneins but not cadmium-metallothionein are present in large amounts in the liver from fetal lambs and calves, and probably from other species also. It is important though to appreciate that considerable differences may occur between species in the amounts of metallothionein which are present in tissues. In our experience, this is generally a simple function of the concentration of metal in the tissue. The rat and other laboratory animals are perhaps somewhat atypical in this regard as they tend to have only very low concentrations of metallothionein in their livers under normal circumstances. This may simply be a reflection of homeostatic control mechanisms which limit the accumulation of zinc and copper in the livers of these animals.

Smith: If cadmium-free forms of metallothionein are present at birth, what is their function?

Piscator: One function is to bind cadmium when this toxic element is absorbed via lungs and gut. That's a fact. However, this does not exclude many other functions, nor does it mean necessarily that metallothionein was designed primarily as a detoxifying protein.

Brändén: I am given to understand that in some species the concentrations of metallothionein are much higher than in others. What is the basis for these differences? (Question No. 11.)

Chairman: I am not aware that anybody has studied this question which has important practical implications. In our initial survey for a cadmium protein we found the highest amounts of cadmium in horse kidney and therefore adopted this tissue for our efforts of isolation and characterization. In rats, the concentrations of the protein were prohibitively small. This posed major problems for the chemistry, since insufficient material would be available. It may well be that because of the low natural abundance of metallothionein in this species metabolic studies are also handicapped and may not be truly representative.

Kägi: The natural abundance of metallothionein varies also quite markedly within a particular species. Thus, in human liver the tissue concentration can vary by a factor of five (unpublished data) and similar variations are found in the horse. Obviously, the intracellular steady-state concentrations are determined by the rates of synthesis and degradation which are set by a variety of physiological factors. Recent studies from several laboratories indicate that changes in metallothionein concentration are affected not only by nutritional variables (metals) but also by endogenous changes in metabolic activity (Oh, S.H., Deagen, J.T., Whanger, P. D., and Weswig, P.H., Am. J. Physiol. 234, E282 (1978); Failla, M.L. and Cousins, R.J., Biochim. Biophys. Acta 543, 293 (1978)).

Hill: What is the distribution of metallothionein in the liver cell?

Lerch: Published reports state that the bulk of the protein is located in the cytoplasm. However, mitochondrocuprein which is now thought to be a polymerized form of metallothionein has been found to be associated mainly with heavy lysosomes (Porter, H., Biochem. Biophys. Res. Commun. 56, 661 (1974)).

Bannister: Does metallothionein occur in the extracellular space?
Does it occur in plasma? (Questions No. 12 and 13.)

Chairman: We have attempted to detect the protein in plasma but have been universally unsuccessful.

Piscator: My theoretical calculations on the concentration of metallothionein in plasma have shown that it would not contain more than 2 μg/l, even in a heavily exposed worker. In a normal human being, a metallothionein concentration of about 30 ng per liter would explain the normal accumulation of cadmium in the kidneys, assuming that metallothionein contains 5% cadmium.

G. Nordberg: At present, we cannot answer the question whether or not thionein is present in blood plasma. Cadmium-metallothionein as well as some other metallothioneins will quickly be filtered through the renal glomeruli and reabsorbed in tubuli if they are released into blood plasma. If we assume that thionein would behave similarly, it would also be expected to end up in the kidney.

Chairman: I think the question was whether or not metallothionein is a carrier either of metals or sulfur. It would be strange that none is present in plasma if this were the case. If it serves such purposes, it would have to be largely extracellular.

Brändén: Can we be sure that what is present in the cell is actually isolated in its native state?

Chairman: This is extremely difficult to answer for tissue constituents that do not exhibit a readily measurable function. For enzymes, activity measurements are a reliable criterion for the maintenance of the native state. In many metalloenzymes it is even possible to exchange the metal for another one and yet to retain activity.

Kägi: I think that in the case of metallothionein we can be reasonably sure that we are isolating the native protein. Owing to the numerous metal crosslinks cadmium- and zinc-metallothioneins are very sturdy molecules that can be denatured only under extreme conditions.

Chairman: The question has been asked whether or not metallothionein occurs only in eukaryotic organisms or also in prokaryotes (questions No. 14 and 17). Ferredoxins are the only proteins of similar size which serve in an enzymatic function. Other small proteins are often hormones and one would rather look at multicellular systems to detect hormonal effects. Metallothionein has now been found in unicellular eukaryotes but not yet in bacteria. We have looked at several bacterial strains, but without success. It would be well to choose a series of bacterial strains, to study their metal composition, and their dependence on those metals of interest here and possibly attempt to isolate metallothionein.
More detailed suggestions on how to proceed in such studies are given in the following comment which was prepared for this discussion by Dr. Kamen:

Kamen: (written comment) It is of great biological importance to establish if metallothionein is a uniquely eukaryotic protein or if it is found also in bacteria. Apparently, no systematic search of its occurrence in prokaryotes has been performed as yet. Such investigations should include:

(1) Careful assays of mixed soil and marine bacterial populations for heavy metal content. (Perhaps such data are already available from the relatively recent experiments on fall-out during atom bomb tests.) Obviously, zinc, cadmium, mercury, and copper contents should be mentioned in particular.

(2) Enrichment cultures on mixed soil and marine sediments using media relatively high in heavy metal content. As most microorganisms are likely to be sensitive to high levels of cadmium, mercury, etc., a negative result might be expected in short-term experiments. Therefore, studies should be designed over a wide range of concentrations and using long incubation periods or repeated transfers for microorganisms which appear to survive. It should be borne in mind that thioneins may not be wholly involved in detoxification.

(3) Sufficient quantities of microorganisms should be used to allow an estimate of upper limits where no heavy metals are detected. The choice of microorganisms should include all representative species, as there is no basis for prediction as to which bacteria will be good sources. If positive, the experiments in (1) and (2) should provide useful clues as to choice.

Olafson: We have done a preliminary experiment with the marine phytoplankter Dunaliella tertiolecta. The organism has been cultured in low concentrations of cadmium chloride for 10 days and 20g of broken, bacteria-free cells were mixed with a buffer solution containing cadmium-109 and saturating cold cadmium chloride. The broken cell homogenate was subjected to gel permeation chromatography and we monitored the degree of cadmium-109 binding and the sulfhydryl content in the 10,000 MW region. No evidence was obtained for a cadmium-binding protein near this molecular weight. Further experiments are obviously necessary for a conclusive answer to the question of the presence of metallothionein in phytoplankton.

Chairman: This is of course a tricky field. We looked at E. coli which happens to be one of the heartiest organisms when it comes to zinc deficiency. Different strains vary by 3 orders of magnitude relative to the ease with which zinc deficiency can be induced: the wild type strain becomes deficient only at 10^{-9} M zinc! I do not think one can say much from the few negative experiments which we have performed. Dr. Kamen's suggestions are very timely and pertinent.

* * *

Thank you all very much for your most interesting contributions.

PART II

DEFINITION, NOMENCLATURE AND OCCURRENCE OF METALLOTHIONEIN

Metallothionein
Jeremias H. R. Kägi and Monica Nordberg, eds.
Birkhäuser Verlag, Basel/Boston/Stuttgart, 1979

NOMENCLATURE OF METALLOTHIONEIN: A PROPOSAL

Jeremias H.R. Kägi and Yutaka Kojima

Biochemisches Institut
Universität Zürich
8028 Zürich, Switzerland

The term "metallothionein" introduced by Kägi and Vallee in 1960 (1) has been widely accepted. As a consequence of new knowledge accruing on this protein and of the large increase in the number of investigators in the field, there have been, however, numerous departures from this terminology resultin in considerable confusion as pointed out by some authors (2).

The "First International Meeting on Metallothionein and Other Low Molecular Weight Metal-binding Proteins" offers an opportunity to discuss and remedy such nomenclatural problems. As a basis for discussion, we would like to make the following propositions:

<u>Metallothionein</u>

1. The term "metallothionein" should be used as a broad term for low molecular weight proteins with an extraordinary high content of metals and cysteine. This designation should be employed for bibliographic classifications of such proteins.

2. The term "thionein" should be used only to specify the apoprotein of metallothionein. The designation "apometallothionein" may also be used in analogy to apoferritin, apotransferrin, <u>etc.</u> Terms such as "thionein", "zinc-thionein", "cadmium-thionein" should be avoided in publication titles and running titles and should also not be used as a key word for computer filing.

3. More specific terms such as "cadmium-metallothionein" could be employed when metallothionein contains only cadmium. For metallothionein containing another metal an analogous nomenclature should be adopted. The metal prefixed to the word metallothionein should not be used to designate the metal employed for metallothionein induction.

Submitted to the "First International Meeting on Metallothionein and Other Low Molecular Weight Metal-binding Proteins", Zürich, July 17 - 22, 1978

4. When metallothionein contains more than one metal, for example, 5.3 gram atoms of cadmium and 1.7 gram atoms of zinc, terms such as "(cadmium, zinc)-metallothionein" or "($Cd_{5.3}$, $Zn_{1.7}$)-metallothionein" could be used. The sequence in which the metals are named should indicate the order of their relative abundance.

5. In contexts where specification of the metal composition is not available or is of no special interest (for example, in presentations of the amino acid sequence), the term "metallothionein" should be used.

Multiple forms of metallothionein

In adaptation of recommendations (1976) of the "Nomenclature of Multiple Forms of Enzymes, IUPAC-IUB Commission on Biochemical Nomenclature (CBN)" (3), the following rules are proposed:

1. The term "multiple forms of metallothionein" or "isometallothioneins" should be used as a broad term covering all metallothioneins occurring naturally in a single species. These terms should apply only to forms of metallothioneins arising from genetically determined differences in primary structure and not to forms derived by modification of the same primary sequence nor to forms differing only in metal composition.

2. In naming isometallothioneins, the term "metallothionein" should be used, followed by a number. The numbers should be allotted consecutively, preferably on the basis of charge properties as manifested in specified chromatographic separations.

3. Where complex isometallothionein patterns occur, the numbers may be used to designate the major groups, with subscript lower case letters applied consecutively for the individual subforms separated by additional procedures (1_a, 1_b, 1_c; 2_a, 2_b, *etc*.).

REFERENCES

1. Kägi, J.H.R., and Vallee, B.L. (1960) J. Biol. Chem. 235, 3460-3465.

2. Webb, M., and Verschoyle, R.D. (1976) Biochem. Pharmacol. 25, 673-679.

3. IUPAC-IUB Commission on Biochemical Nomenclature (CBN) (1977) J. Biol. Chem. 252, 5939-5941.

BIOCHEMICAL PROPERTIES OF METALLOTHIONEIN

Metallothionein
Jeremias H. R. Kägi and Monica Nordberg, eds.
Birkhäuser Verlag, Basel/Boston/Stuttgart, 1979

AMINO ACID SEQUENCE OF HUMAN HEPATIC METALLOTHIONEINS

Margrit M. Kissling and Jeremias H.R. Kägi

Biochemisches Institut
Universität Zürich
CH-8028 Zürich, Switzerland

Metallothioneins are naturally occurring, primarily and often exclusively zinc-containing metalloproteins of mammalian liver. Their presence in human liver was first demonstrated by Kägi and co-workers (1,2), who found that these proteins are an abundant component of this tissue (0.002-0.01% of wet weight) and that they may contain as much as 10% of the total zinc of adult liver. This finding suggested that their primary physiological role in this organ should be sought in relation to the metabolism of zinc.

Except for their metal composition the hepatic metallothioneins are very similar to both cadmium- and zinc-containing proteins isolated previously from equine and human kidney (3,4). There are also two chromographically distinct forms of metallothionein in human liver. In this report, amino acid sequence data of these proteins are presented. The primary structures are highly homologous to those of other metallothioneins but differ fundamentally from those of all other known polypeptide sequences, suggesting that they belong to a new superfamily of proteins (5). Based on structural comparison, some evolutionary aspects of these proteins are discussed.

Amino acid sequences

The chromatographic separation of human hepatic metallothionein into two principal fractions, MT-1 and MT-2, is presented in Fig. 1. After a further ion exchange chromatographic step, the two fractions were obtained homogeneous on starch gel electrophoresis. Their amino acid compositions listed in Table 1 document the high cysteine content and the complete lack of aromatic amino acids typical for metallothioneins. The data

Submitted to the "First International Meeting on Metallothionein and Other Low Molecular Weight Metal-binding Proteins", Zürich, July 17 - 22, 1978

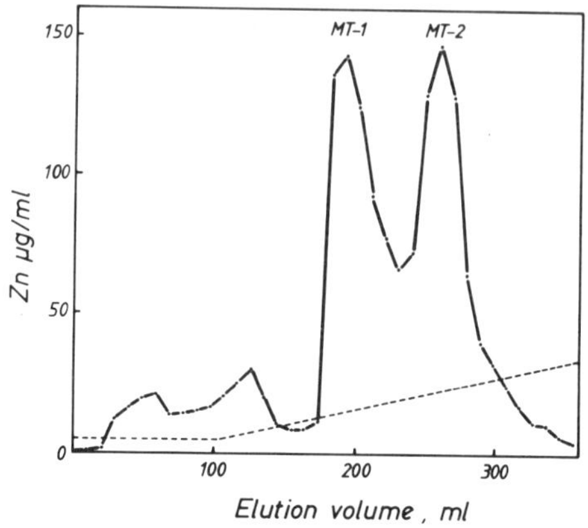

Fig. 1 Separation of principal forms of human hepatic metallothionein on DEAE-Sephadex A-25, pH 8.6, using a salt gradient as described by Bühler and Kägi (2). Starting buffer: 0.02 M Tris-0.005 M HCl; limit buffer: 0.2 M Tris-0.05 M HCl. Zinc (——), specific conductance (---).

indicate, however, differences in the number of Thr, Pro, Asx, Glx, Ala, Cys, Val and Leu. In addition, there is in MT-1 a large deviation from integral stoichiometry which implies the existence of some residual microheterogeneity in this fraction.

The amino acid sequence data of MT-2 (6) and of MT-1 are shown in Fig. 2. Both forms have their N-terminus blocked by an acetyl group. However, by selective cleavage of the penultimate acid-labile Asp-Pro bond, polypeptide chains susceptible to Edman degradation were obtained. The sequence analysis revealed clear differences between MT-2 and MT-1 in seven positions and implies thus strongly that in human tissues like in the horse (7) there are also two (or perhaps even more) cistrons coding for metallothionein. The structure of MT-2 was determined unambiguously by automated Edman degradation and carboxypeptidase digestion and was also confirmed in part by the analysis of tryptic peptides (8). In contrast to MT-2, some of the assignments in MT-1 remained equivocal. In five positions two and in one position even three different amino acid residues were found to occur in comparable amounts. These data substantiate thus the existence of microheterogeneity inferred from the amino acid composition and suggest that the MT-1 fraction may be composed of as much as three subforms. Since they do not differ in charge, they could not be separated by the chromatographic and electrophoretic procedures employed.

TABLE I

Amino acid compositions of human hepatic metallothioneins[a]

Residue	MT-1 Analysis[b](2)	MT-2 Analysis[b](2)	MT-2 Sequence (6)
Cys	18.50	20.55	20
Asx	2.89	4.06	
Asp			3
Asn			1
Thr	2.45	2.08	2
Ser	8.40	8.30	8
Glx	3.42	2.39	
Glu			1
Gln			1
Pro	2.40	2.07	2
Gly	4.87	5.14	5
Ala	5.76	6.81	7
Val	2.14	1.12	1
Met	0.83	0.87	1
Ile	1.18	0.95	1
Leu	0.68	-	-
Lys	7.80	7.78	8
Total	61.32	62.12	61

[a]Number of residues per molecule

[b]Based on the calculated minimum chain weights of 6113 and 6136 for MT-1 and MT-2, respectively (2).

Comparison of human isometallothioneins

In spite of the remaining uncertainties, the data indicate a remarkable similarity between the different isometallothioneins. The most striking feature is the identity in the distribution of the cysteinyl, lysyl and all but one seryl residues. In both MT-2 and MT-1 there are seven -Cys-X-Cys- oligopeptide sequences which are thought to constitute the primary chelation sites for the metal and which by the subsequent interaction with a third cysteinyl residue in all except one case are completed to the trithiolate (trimercaptide) metal-binding sites typical for metallothioneins (9). Whether or not the juxtaposed seryl and lysyl residues are also involved in the protein-metal interaction remains to be established.

The differences between the sequences of MT-2 and MT-1 are restricted largely to residues supposedly not involved in metal-binding and are, with one exception, accountable by single-base changes in the genetic code. The Gly-Asp substitu-

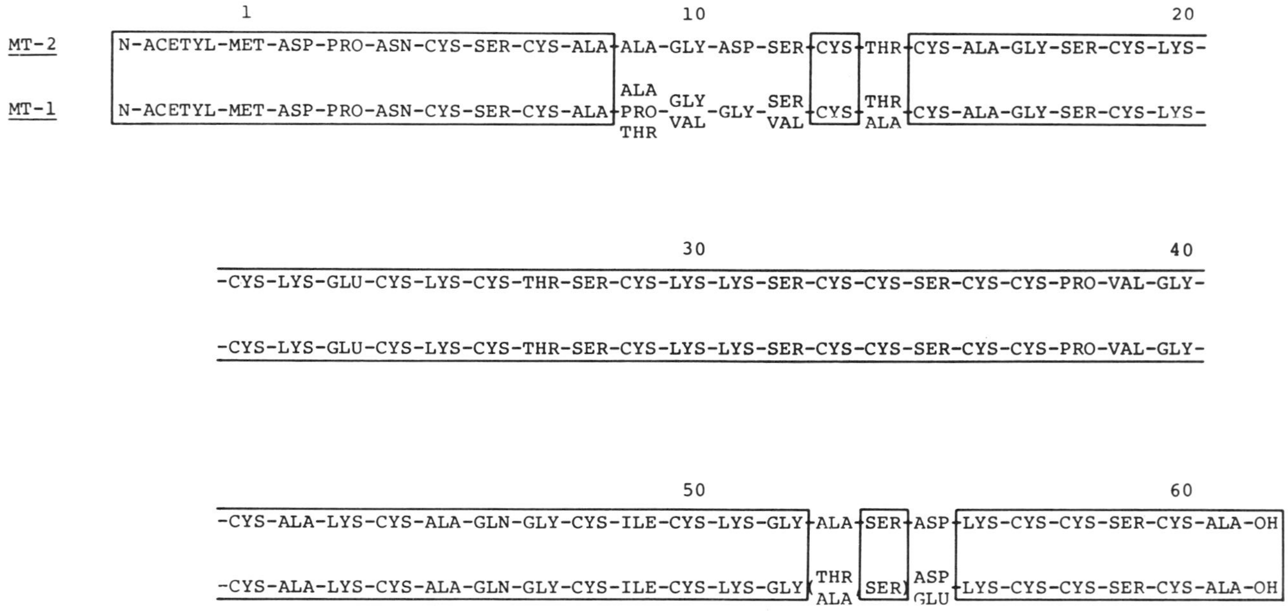

Fig. 2 Primary structures of human hepatic metallothioneins

The amino acid sequences were determined mainly by automated sequence analysis (9) of fragments obtained from the S-pyridylethylated derivatives of the proteins by acid cleavage and by Staphylococcus aureus protease digestion (15,6). The C-terminal sequences were established by digestion of the carbamidomethylated derivatives with carboxypeptidase A and B (9). ( ) Order of residues not as yet established by sequential degradation.

tion in position 11 is likely to be responsible for the greater negative charge of MT-2 (2). An analogous difference was found between the principal equine metallothioneins where Gly-11 in MT-1A is replaced by a glutamyl residue in MT-1B (9). There is in that case, however, no difference in net charge because of a compensatory difference in another position (Glu-Gln replacement in residue 23).

Evolutional aspects

The sequences of the human metallothioneins are very similar, also to those of the two principal equine isometallothioneins (Fig. 3). Large portions of their sequences including all cysteinyl residues and most of the juxtaposed basic amino acids and seryl residues are identical. Pairwise comparison of human MT-2 with the equine isoproteins MT-1A and MT-1B indicates 89% and 90% identity, respectively, and an even closer structural correspondence if conservative substitutions are included also. Hence, judged by standard criteria metallothioneins are rather slowly evolving proteins (10-13).

In spite of their obvious relatedness it is not possible, however, to make unambiguous deductions about the evolution of these mammalian metallothioneins. The difficulty lies in the fact that in the principal human and equine metallothioneins the inter- and intra-species differences are of comparable magnitude and are not discriminating enough to allow a ready identification of orthologous structures, i.e., of distinct sequence types such as the α- and β-chains in the hemoglobins. For this reason it cannot be decided if the human and equine isoproteins arose from the same or from independent gene duplications. The fact that all mammals examined thus far contain two or more forms of metallothionein would seem to favor a single common gene duplication that occurred before the divergence of the euplacental mammals from a common ancestor into the various orders some 75 million years ago. However, as was suggested for the evolution of hemoglobin δ-chains from β-chains (14) it is also conceivable that at about the same time independent duplications occurred in different mammalian lines.

The physiological significance of the duplication of metallothionein gene is also not clear since there is no major difference in the overall structure and in the metal-binding competence of the isoproteins (7). The conservation of all seven chelating oligopeptide sequences is, in fact, a strong indication that the two isometallothioneins serve identical functional roles.

```
                                            1                               10
Human  Metallothionein-2    N-ACETYL-MET-ASP-PRO-ASN-CYS-SER-CYS-ALA-ALA-GLY-ASP-SER-CYS-THR-
                                                                 ⎡ALA⎤
Human  Metallothionein-1                                       " ⎢PRO⎥{GLY} GLY {SER}      {THR}
                                                                 ⎣THR⎦{VAL}     {VAL}      {ALA}
Equine Metallothionein-1A                                      -PRO-THR-GLY- " -SER-      -THR-

Equine Metallothionein-1B                                      -VAL-ALA- " -GLU- "          "


                      20                                 30
-CYS-ALA-GLY-SER-CYS-LYS-CYS-LYS-GLU-CYS-LYS-CYS-THR-SER-CYS-LYS-LYS-SER-CYS-CYS-SER-CYS-CYS-PRO-
                                    "       "       "
                                    "     -ARG-     "
                                  -GLN-     "     -ALA-


            40                                     50                                  60  61
-VAL-GLY-CYS-ALA-LYS-CYS-ALA-GLN-GLY-CYS-ILE-CYS-LYS-GLY-ALA-SER-ASP-LYS-CYS-CYS-SER-CYS-ALA-OH
  "               "                       "         (THR ,SER){ASP}
                                                    (ALA     ){GLU}
-GLY-           -ARG-                   -VAL-          "        "
-VAL-           -LYS-                     "            "        "
```

Fig. 3 Sequence comparison of human metallothioneins with equine forms (7)

ACKNOWLEDGMENT

This work was supported by Schweizerischer Nationalfonds, grants 3.107.73 and 3.531-0.75.

REFERENCES

1. Kägi, J.H.R. (1970) 8th International Congress of Biochemistry, Switzerland, Sept. 1970, Abstracts, pp. 130-131.
2. Bühler, R.H.O. and Kägi, J.H.R. (1974) FEBS Lett. 39, 229-234.
3. Kägi, J.H.R., Himmelhoch, S.R., Whanger, P.D., Bethune, J.L., and Vallee, B.L. (1974) J. Biol. Chem. 249, 3537-3542.
4. Pulido, P., Kägi, J.H.R., and Vallee, B.L. (1966) Biochemistry 5, 1768-1777.
5. Dayhoff, M.O. (1974) Fed. Proc. 33, 2314-2316.
6. Kissling, M.M. and Kägi, J.H.R. (1977) FEBS Lett. 82, 247-250.
7. Kojima, Y., Berger, C., and Kägi, J.H.R. (1979) This volume, pp. 153-161.
8. Bühler, R.H.O. (1974) Ph.D. thesis, Universität Zürich.
9. Kojima, Y., Berger, C., Vallee, B.L., and Kägi, J.H.R. (1976) Proc. Natl. Acad. Sci. USA 73, 3413-3417.
10. Zuckerkandl, E. and Pauling, L. (1965) In: "Evolving Genes and Proteins", V. Bryson and H.J. Vogel eds., Academic Press, New York, pp. 97-166.
11. Margoliash, E. and Smith, E.L. (1965) In "Evolving Genes and Proteins", V. Bryson and H.J. Vogel eds., Academic Press, New York, pp. 221-242.
12. Kimura, M. (1969) Proc. Natl. Acad. Sci. USA 63, 1181-1188.
13. King, J.L. and Jukes, T.H. (1969) Science 164, 788-798.
14. Boyer, S.H., Crosby, E.F., Noyes, A.N., Fuller, G.F., Leslie, S.E., Donaldson, L.J., Vrablik, G.R., Schaefer, E.W., und Thurmon, T.F. (1971) Biochemical Genetics 5, 405-448.
15. Houmard, J. and Drapeau, G.R. (1972) Proc. Natl. Acad. Sci. USA 69, 3506-3509.

Metallothionein
Jeremias H. R. Kägi and Monica Nordberg, eds.
Birkhäuser Verlag, Basel/Boston/Stuttgart, 1979

THE AMINO ACID SEQUENCE OF EQUINE METALLOTHIONEINS

Yutaka Kojima, Christine Berger and Jeremias H.R. Kägi

Biochemisches Institut
Universität Zürich
CH-8028 Zürich, Switzerland

The determination of the amino acid sequences of metallothioneins has been undertaken with the aim of obtaining information on 1) the distribution of metal-liganding cysteinyl residues in the polypeptide chain, 2) the mode of metal-protein interaction, 3) the relationship between chemical structure and metal-ion selectivity, 4) the structural basis of chromatographic polymorphism, 5) the identity or difference of metallothioneins from different organs, and 6) the evolution of metallothioneins. The answers to these questions were also expected to provide some further clues for the understanding of the physiological function of metallothionein.

Since the elucidation of the sequence of equine renal metallothionein-1B (1), our efforts have been extended to a variety of other chromatographic forms of metallothionein of the horse and of other species (2-5). In this paper, the primary structure of two principal forms and of some subforms occurring in equine kidney and liver are compared.

MATERIALS AND METHODS

Equine metallothioneins were purified from renal cortex and from liver by the procedure of Kägi <u>et al.</u> (6). Purified metallothioneins were further resolved into their isoproteins by ion-exchange chromatography on DEAE-cellulose DE-23, using as eluting buffer 0.025 M ammonium acetate adjusted to pH 10.0 with ammonium hydroxide.

Submitted to the "First International Meeting on Metallothionein and Other Low Molecular Weight Metal-binding Proteins", Zürich, July 17 - 22, 1978

The sequence analysis was carried out mainly on S-pyridylethylated polypeptides using a Beckman Sequencer model 890B (updated) (1,2). The main sets of fragments employed were obtained by cleavage of the penultimate acid-labile Asp-Pro bond in acidic conditions, by tryptic digestion at the single arginyl bond (Arg-25) (for metallothionein-1B) after succinylation of the lysyl ε-amino groups, or by cleavage of the bond between Glu-23 and Cys-24 by Staphylococcus aureus protease (for metallothionein-1A). The ninhydrin-negative amino-terminal dipeptide was identified by chemical degradation and by comparison with an authentic sample of N-acetyl-Met-Asp using thin-layer chromatography (1). The sequence of the carboxyl-terminal region was determined on the S-carbamidomethylated derivative of the protein by digestion with bovine pancreatic carboxypeptidases A and B. In some cases amino acid positions were confirmed by manual subtractive Edman degradation and by digestion with carboxypeptidases A, B and C of tryptic peptides of the S-carboxymethylated derivative separated on Dowex AG 1W-X2.

RESULTS AND DISCUSSION

Amino acid and metal compositions

Many mammalian species contain several forms of metallothionein which have the same molecular size but which differ substantially in amino acid composition (7-13). Thus far, in the horse three chromatographic forms have been isolated from liver and two from kidney. The amino acid and metal compositions of these proteins are listed in Table I. The abundance of cysteine comprising about one third of all residues and the absence of aromatic amino acids and of histidine are common to all metallothioneins.

The metal compositions of the hepatic and renal metallothioneins have long been known to be strikingly different (6, 7, 14). The hepatic proteins used in this study contained almost exclusively zinc, while the renal metallothioneins typically had substantial proportions of both zinc and cadmium. There were also some differences in the metal composition of the two forms isolated from kidney.

Amino acid sequence

The amino acid sequences of all forms of equine metallothionein thus far purified from liver and kidney have been established and are shown in Fig. 1. The carboxyl-terminal alanine and the amino-terminal acetylmethionine are found in all mammalian metallothioneins (1,2,15). The methionine could be the initiator for the biosynthesis of this protein since the amino-terminal Met-Asp bond has been reported to be resistant to

TABLE I

Amino Acid and Metal Compositions of Equine Metallothioneins

	Metallothionein-1A including Variant[a] Leu-54		Metallothionein-1A Variant Arg-39	Metallothionein-1B	
	Liver	Kidney	Liver	Liver	Kidney
Amino acid	Residues per molecule[b]				
Cys	18.0 (20)	20.5 (20)	19.0 (20)	17.95(20)	20.70(20)
Ser	6.91 (8/7)[c]	6.76 (8/7)[c]	7.75 (8)	7.17 (8)	7.29 (8)
Lys	5.55 (6)	5.81 (6)	6.18 (6)	6.20 (7)	6.79 (7)
Arg	1.82 (2)	1.82 (2)	2.90 (3)	1.17 (1)	1.16 (1)
Ala	4.8 (5)	5.05 (5)	5.01 (5)	6.62 (7)	7.11 (7)
Gly	6.41 (7)	7.25 (7)	6.13 (6)	5.13 (5)	5.60 (5)
Val	0.98 (1)	0.82 (1)	0.86 (1)	2.76 (3)	2.74 (3)
Asx	2.77	3.01	3.04	2.78	3.05
Asp	(2)	(2)	(2)	(2)	(2)
Asn	(1)	(1)	(1)	(1)	(1)
Glx	2.30	2.46	2.78	2.92	2.87
Glu	(1)	(1)	(1)	(1)	(1)
Gln	(1)	(1)	(1)	(2)	(2)
Pro	2.95 (3)	2.82 (3)	2.80 (3)	2.12 (2)	1.80 (2)
Met	1.0 (1)	0.98 (1)	1.07 (1)	1.0 (1)	0.94 (1)
Thr	2.70 (3)	2.80 (3)	2.91 (3)	1.40 (1)	1.09 (1)
Leu	0.29 (0/1)[c]	0.34 (0/1)[c]			
Total	(61)	(61)	(61)	(61)	(61)
Chain Weight[d]	6082/6108[c]	6082/6108[c]	6181	6109	6109
Metal	Gram atoms per molecule				
Zinc	5.72 (92%)[e]	3.20 (45%)	5.97 (97%)	5.72 (92%)[e]	2.55 (35%)
Cadmium	0.33 (5%)	3.86 (54%)	0.13 (2%)	0.33 (5%)	4.61 (63%)
Copper	0.18 (3%)	0.07 (1%)	0.08 (1%)	0.18 (3%)	0.11 (2%)
Total	6.23(100%)	7.13(100%)	6.18(100%)	6.23(100%)	7.27(100%)

[a]Variant Leu-54 was not separable with standard chromatographic procedures.

[b]Figures in brackets indicate number of amino acid residues per molecule as determined by sequence analysis.

[c]The first figure refers to Metallothionein-1A, the second to its variant Leu-54.

[d]Calculated from sequence data; the values include the amino-terminal acetyl group.

[e]Representative analysis obtained from unresolved mixture of both hepatic metallothioneins (6).

Fig. 1 Amino Acid Sequences of Equine Metallothioneins

```
                              Metal
                    Source  composition
                                               1                                      10
Metallothionein-1A  Liver    Zn      N-ACETYL-MET-ASP-PRO-ASN-CYS-SER-CYS-PRO-THR-GLY-GLY-
        "        "  Kidney   Zn + Cd                                       "   "       "
  Variant Leu-54    Liver    Zn                                            "   "       "
        "        "  Kidney   Zn + Cd                                       "   "       "
  Variant Arg-39    Liver    Zn                                            "   "       "
Metallothionein-1B  Liver    Zn                                          -VAL-ALA-   -GLU-
        "        "  Kidney   Zn + Cd                                       "   "       "


                                  20                                      30
-SER-CYS-THR-CYS-ALA-GLY-SER-CYS-LYS-CYS-LYS-GLU-CYS-ARG-CYS-THR-SER-CYS-LYS-LYS-SER-CYS-CYS-SER-CYS-
                                              "               "
                                              "               "
                                              "               "
                                              "               "
                                            -GLN-           -ALA-
                                              "               "


          39  40                                     50               54                     60   61
-CYS-PRO-GLY-GLY-CYS-ALA-ARG-CYS-ALA-GLN-GLY-CYS-VAL-CYS-LYS-GLY-ALA-SER-ASP-LYS-CYS-CYS-SER-CYS-ALA-OH
          "               "                                           "
          "               "                                          -LEU-
          "               "                                           "
         -ARG-            "                                          -SER-
         -VAL-           -LYS-                                        "
          "               "                                           "
```

the posttranslational cleavage by methionine aminopeptidase (16).

The chain weight of all mammalian metallothioneins deduced from amino acid analysis and sequence data is approximately 6000. When the metal is included, the value for the minimum molecular weight is similar to that found previously by sedimentation equilibrium analysis of native equine renal metallothionein, thereby indicating that metallothionein is a single chain protein (6). The large Stokes radius (16.1 Å) measured by gel filtration suggests a nonglobular ellipsoid shape with an axial ratio of 6 (6). The very similar behaviour of most mammalian metallothioneins on gel filtration implies that their molecular weight and shape are nearly identical.

Mode of metal-binding

It is well established that the metal ions are bound through thiolate (mercaptide) linkages and that all cysteinyl residues participate in metal coordination (17,18). Based on the fact that the ratio of cysteine to bivalent metal is close to 3 (6,7,10,14,17,18), it has been suggested that the metal ions are generally bound to three thiolate ligands forming negatively charged trithiolate (trimercaptide) complexes, *i.e.*, $[Metal^{2+} (Cys^{-})_3]^{-}$ (1,19). The cysteinyl residues in equine metallothioneins are fairly equally distributed along the entire chain. Fourteen out of 20 cysteinyl residues form part of seven -Cys-x-Cys- tripeptide sequences, where x is an amino acid residue other than cysteine. In addition to these sequences whose predominance is unique to metallothionein, there are three -Cys-Cys- and three -Cys-x-y-Cys- sequences, where x and y are amino acid residues other than cysteine. In most metalloproteins of known tertiary structure, two or more of the metal-liganding amino acid residues (L) occur in vicinal positions in the chain forming -L-x-L- or -L-x-y-L- oligopeptide sequences (1,20,21). It is tempting, therefore, to allocate each of the -Cys-x-Cys- sequences to one of the seven metal-binding sites of equine metallothionein. Chelation of the metal ions by these vicinal ligands has been suggested to be the first step in the protein-metal interaction (1). This complex formation would then be followed by the additional binding of the metal to the third cysteinyl residue located in another region of the chain. Such a model of separate metal-binding sites is consistent with earlier spectrophotometric titration data (17) and is supported also by evidence obtained recently by dark-field electron microscopy (22).

Another outstanding general feature of the sequence is the juxtaposition of the basic amino acids and of serine to the cysteinyl residues. Together, these residues form extended clusters separated by stretches of the remaining amino acids. The center portion of the chain (residues 18-37) consists almost exclusively of these residues. Since coordination to only three ligands is uncommon (23), it seems probable that in addition to the three cysteinyl residues each metal-binding site may yet include additional ligands donated either by the peptide backbone, by amino acid side chains, or by the solvent. Such additional ligands would have to be of the uncharged type, however, since at neutral pH the two negative charges of the protein as measured by moving boundary electrophoresis are fully accounted for by the sum of the negative charges of the trithiolate complexes and the positive charges of the polypeptide chain (1). On these grounds, the possibility that an unprotonated lysyl side chain amino group may function as a fourth ligand in an inner sphere complex must be ruled out. It has been suggested, however, that the charged protonated form of the side chain of lysyl and arginyl residues could play a role in metal-binding by neutralizing the negative charge of the trithiolate complex through formation of an ion pair and thus contribute indirectly to the stability of the complex (24).

Isometallothioneins

Comparative sequence studies of the equine metallothioneins reveal remarkable similarities. The positions of the cysteines as well as of the adjacent serines and the basic amino acids are virtually identical in all forms of metallothionein. Amino acid replacements which give rise to the multiple forms of metallothionein are generally located outside of these regions and are explicable mainly by single base changes.

In kidney and liver there are two principal forms of metallothionein (MT-1A and MT-1B) which differ in 7 amino acid positions. This extent of substitution exceeds that to be expected for allelic proteins and implies that they are coded by different cistrons that have evolved by duplication from an ancestral gene.

From the principal form MT-1A there are now also two variants known which differ by a single amino acid substitution and which might be attributed to an allelic cistron arising from a point mutation. The joint occurrence of variant forms and subforms of metallothionein is a cause of difficulties in the purification of this protein and seems responsible for the nonintegral amino acid composition of most preparations. Especially allelic variants resulting from single conservative amino acid replacement are difficult to separate from each other. Thus, it has not been possible as yet to resolve the Ser-54 and the Leu-54 subforms of the isometallothionein-1A (Fig. 1). On the other hand, on account of their charge difference, the (Gly-39 → Arg-39)-variant of metallothionein-1A was readily recognized and obtained in pure form.

Unity of metal-binding structure in spite of wide metal ion specificity

The polypeptide structures of the principal forms of metallothionein isolated from liver and from kidney are completely identical (Fig. 1). This is perhaps surprising in view of the very striking differences which exist between the renal and hepatic metallothioneins with regard to their metal composition (Table I) (6). While the renal proteins generally contain about equal amounts of cadmium and zinc, hepatic metallothioneins are often solely zinc-containing proteins. This gross disparity in metal composition signifies that the metal ion specificity of the polypeptide chain for the different group 2b elements is rather wide and it implies that the selection of the metal must be determined primarily by local physiological circumstances.

Notwithstanding this variability, the conservation of all metal-binding cysteinyl residues and the persistence of seven metal-binding sites in all forms of metallothionein leave little doubt that the metal-binding competence is the central feature of these proteins and constitutes the prerequisite for their biological function. In much the same way as the phenomenon of metallothionein induction by metal ions (9-13,25-28), the elucidation of the primary structure is thus contributing to the understanding of the physiological role of this protein in metal metabolism.

ACKNOWLEDGEMENT

This work was supported by Schweizerischer Nationalfonds, grants 3.107.73 and 3.531-0.75.

REFERENCES

1. Kojima, Y., Berger, C., Vallee, B.L., and Kägi, J.H.R. (1976) Proc. Natl. Acad. Sci. USA 73, 3413-3417.

2. Kissling, M.M., and Kägi, J.H.R. (1977) FEBS Lett. 82, 247-250.

3. Kissling, M.M., and Kägi, J.H.R. (1979) This volume, pp. 145-151.

4. Kissling, M.M., Berger, C., Kägi, J.H.R., Andersen, R.D., and Weser, U. (1979) This volume, pp. 181-185.

5. Lerch, K. (1979) This volume, pp. 173-179.

6. Kägi, J.H.R., Himmelhoch, S.R., Whanger, P.D., Bethune, J.L., and Vallee, B.L. (1974) J. Biol. Chem. 249, 3537-3542.

7. Bühler, R.H.O., and Kägi, J.H.R. (1974) FEBS Lett. 39, 229-234.

8. Bremner, I., and Young, B.W. (1976) Biochem. J. 155, 631-635.

9. Nordberg, G.F., Nordberg, M., Piscator, M., and Vesterberg, O. (1972) Biochem. J. 126, 491-498.

10. Bremner, I., and Davies, N.T. (1975) Biochem. J. 149, 733-738.

11. Bremner, I., and Young, B.W. (1976) Biochem. J. 157, 517-520.

12. Winge, D.R., Premakumar, R., and Rajagopalan, K.V. (1975) Arch. Biochem. Biophys. 170, 242-252.

13. Kimura, M., Otaki, N., Yoshiki, S., Suzuki, M., Horiuchi, N., and Suda, T. (1974) Arch. Biochem. Biophys. 165, 340-348.

14. Pulido, P., Kägi, J.H.R., and Vallee, B.L. (1966) Biochemistry 5, 1768-1777.

15. Kägi, J.H.R., Kojima, Y., Kissling, M.M., and Berger, C. (1977) Fed. Proc. Fed. Am. Soc. Exp. Biol. 36, 699.

16. Schechter, I., Zemell, R., and Burstein, Y. (1978) FEBS Federation of European Biochemical Societies, 11th Meeting, Copenhagen, 1977, Vol. 47, Symposium A6, pp. 103-112, Pergamon Press, Oxford.

17. Kägi, J.H.R., and Vallee, B.L. (1961) J. Biol. Chem. 236, 2435-2442.

18. Sokolowski, G., and Weser, U. (1975) Hoppe-Seyler's Z. Physiol. Chem. 356, 1715-1726.

19. Kägi, J.H.R. (1970) 8th International Congress of Biochemistry, Switzerland, September, 1970, Symposium, Abstracts, pp. 130-131.

20. Liljas, A., and Rossmann, M.G. (1974) Annu. Rev. Biochem. 43, 475-507.

21. Waara, I., Lövgren, S., Liljas, A., Kannan, K.K., and Bergstén, P.-C. (1972) Adv. Exp. Med. Biol. 28, 169-187.

22. Fiskin, A.M., Peterson, G., and Brady, F.O. (1977) Ultramicroscopy 2, 389-395.

23. Cotton, F.A., and Wilkinson, G. (1972) Advanced Inorganic Chemistry, Interscience Publishers, New York, 3rd ed., p. 24.

24. Szwarc, M. (1969) Acc. Chem. Res. 2, 87-96.

25. Piscator, M. (1964) Nord. Hyg. Tidskr. 45, 76-82.

26. Wisniewska-Knypl, J.M., and Jablonska, J. (1970) Bull. Acad. Pol. Sci. Ser. Sci. Biol. 18, 321-327.

27. Shaikh, Z.A., and Lucis, O.J. (1971) Experientia 27, 1024-1025.

28. Webb, M., and Daniel, M. (1975) Chem. Biol. Interact. 10, 269-276.

Metallothionein
Jeremias H. R. Kägi and Monica Nordberg, eds.
Birkhäuser Verlag, Basel/Boston/Stuttgart, 1979

RABBIT LIVER METALLOTHIONEIN

TENTATIVE AMINO ACID SEQUENCE OF METALLOTHIONEIN-B

Masami Kimura, Noriko Otaki and Mieko Imano

Department of Experimental Toxicology, National Institute of Industrial Health, Kawasaki, Kanagawa, Japan 213.

The primary structure of equine renal metallothionein-1B was first reported by Kojima et al. (1). Most recently, Kissling and Kägi (2) have elucidated the amino acid sequence of human hepatic metallothionein-II and compared it to the known structure of the former. Huang et al. (3) have also demonstrated the complete sequence of hepatic metallothionein-I obtained from the liver of mouse administered with cadmium. We report the amino acid sequence of liver metallothionein-B isolated from rabbits exposed to cadmium.

MATERIALS AND METHODS

Rabbit liver metallothionein-A and -B were isolated from rabbits exposed to cadmium (by 21 subcutaneous injections of 1 mg/kg body weight with intervals of 2-3 days). On the second day after the last injection the animals were killed. To one volume of the liver homogenate in 0.02 M TrisHCl buffer, pH 8.6, cold ethanol and chloroform (1.00 : 1.05 : 0.08, by volume) were added with cooling. To the supernatant obtained by centrifugation three volumes of cold ethanol were added subsequently. The solution was allowed to stand at -20° overnight. The produced precipitate was collected by centrifugation. It was dissolved in 0.02 M ammonium carbonate and then it was charged to a Sephadex G-75 column. The elution was carried out by use of 0.02 M ammonium carbonate. The Cd-binding protein fraction of low molecular weight on the gel filtration was further separated into the Cd-binding proteins, metallothionein-fA, a minor component, -A and -B, major components, by ion exchange chromatography on a DEAE Sephadex A-25 column.

Submitted to the "First International Meeting on Metallothionein and Other Low Molecular Weight Metal-binding Proteins", Zürich, July 17 - 22, 1978

Metallothionein was S-pyridylethylated with 4-vinylpyridine at pH 8.0 by using the modified method of Friedman et al. (5) and S-^{14}C-carboxymethylated with iodoacetic acid-2-^{14}C at pH 8.0 by using the modified method of Crestfield et al. (6) after the removal of metals by the treatment of 6 M guanidine HCl in 0.2 N HCl. Acid hydrolysates of the intact protein and the oxidized one with performic acid were analyzed by an amino acid analyzer. S-^{14}C-carboxymethylated metallothionein was digested with trypsin. The tryptic peptides were separated by ion exchange chromatography on a Dowex-50 (x2) column with gradient elution of pyridine-acetate buffer. Further purification of peptide was carried out by use of paper electrophoresis and/or paper chromatography. The fragment used for the automated sequence analysis was obtained by cyanogen bromide cleavage and by selective cleavage of labile bond between Asp-Pro with 70% formic acid. The fragments were sequenced by a JEOL JAS-47K sequence analyzer. Sequence analysis of peptide was carried out by the manual Edman degradation modified by Iwagana et al. (7). PTH-amino acid was identified by gas chromatography, thin layer chromatography and back-hydrolysis to amino acid.

RESULTS AND DISCUSSION

The gel filtration pattern of the precipitate with ethanol from liver homogenate of Cd-exposed rabbits is shown in Fig. 1. And the ion exchange chromatographic elution profile of the Cd-binding protein fraction separated by the gel filtration is shown in Fig. 2. It was very convenient to use the ammonium carbonate solution in the separation system because the desalted preparation could be obtained only by lyophilization. Ammonium carbonate as well as TrisHCl was effective in the separation of protein. Metallothionein-fA, -A and -B, prepared from the livers of two Cd-exposed rabbits, weighed about 50, 210 and 230 mg, respectively.

Amino acid composition of metallothionein is given in Table 1. Rabbit liver metallothionein as well as mouse liver metallothionein contains 61 amino acid residues and no detectable amounts of histidine, tyrosine and phenylalanine. The total number of amino acid residues of rabbit liver metallothionein is identical with that of the other metallothionein reported (1-3, 10). If the oxidized protein was hydrolyzed without removal of metals by the gel filtration on Sephadex G-25, the cysteinyl residues could not be completely recovered as cysteic acid. As the cysteine derived from the hydrolysate of the intact protein was eluted near the position of proline on the ion exchange chromatography for amino acid analysis, the numbers of proline obtained from the hydrolysate of the intact protein might be influenced with it. Neither amino acid compo-

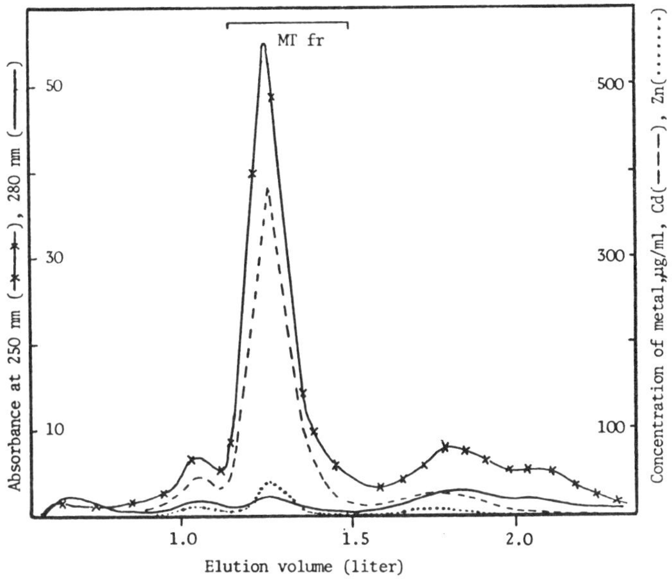

Fig. 1 Gel filtration (Sephadex G-75) of precipitate with ethanol from liver homogenate of Cd-exposed rabbits. The liver (about 200 g) obtained from two Cd-exposed rabbits was used for the preparation of metallothionein. Column size; 5.0 x 90 cm. Eluted with 0.02 M ammonium carbonate at a flow rate of 50 ml/h. Monitored by measuring the absorbances at 250 nm and 280 nm and the concentrations of Cd and Zn.

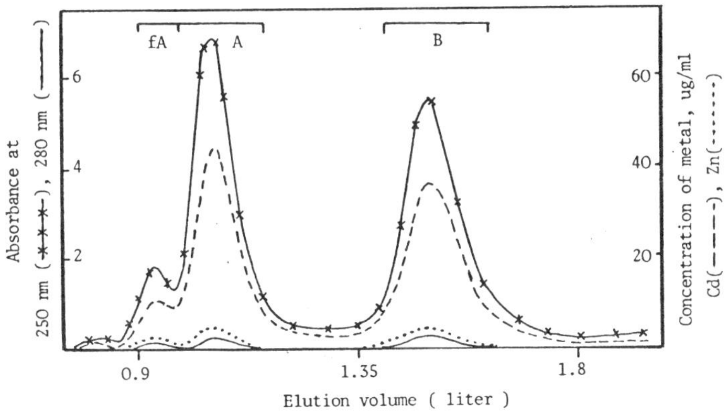

Fig. 2 Ion exchange chromatography (DEAE Sephadex A-25) of Cd-binding protein fraction separated by gel filtration. Column size ; 2.5 x 40 cm. Eluted with the gradient with a mixing chamber containing 1 liter of 0.05 M ammonium carbonate and a reservoir containing 1 liter of 0.20 M ammonium carbonate with a flow rate of 40 ml/h. Monitored by measuring the absorbances at 250 nm and 280 nm and the concentrations of Cd and Zn.

Table 1. Amino acid composition of mouse and rabbit liver metallothionein

Amino acid residues[a)]	Liver metallothionein Mouse I	II	Rabbit A	B
1/2Cys[b)]	19.6(20)	20.0(20)	20.3(20)	19.7(20)
Asx	3.9(4)	4.3(4)	4.1(4)	4.7(4)
Thr	4.7(5)	1.1(1)	3.3(3)	3.0(3)
Ser	8.8(9)	7.9(10)	7.5(8)	7.2(9)
Glx	1.1(1)	3.0(3)	2.4(2)	1.4(1)
Pro	1.8(2)	2.2(2)	1.7(2)	2.4(2)
Gly	5.0(5)	4.0(4)	3.5(4)	3.8(4)
Ala	5.0(5)	6.0(6)	7.0(7)	8.0(8)
Val	1.8(2)	1.2(1)		
Met[c)]	0.9(1)	1.1(1)	1.1(1)	0.8(1)
Ile		1.0(1)	0.9(1)	0.9(1)
Leu			0.7[d)]	
Lys	7.4(7)	7.6(8)	7.8(8)	7.3(8)
Arg			0.9(1)	
Total amino acid residues	(61)	(61)	(61)	(61)

a) Amino acid composition was proposed by the analytical results of oxidized protein hydrolysates (see Text). The number of amino acid residues listed is based on 5.0, 6.0, 7.0 and 8.0 moles of Ala per one mole of mouse liver metallothionein-I, -II, rabbit liver metallothionein-A and -B, respectively. The numbers in parentheses are calculated by the amino acid sequence except those of rabbit liver metallothionein-A. b) Measured as cysteic acid. c) Met (mouse liver metallothionein) was measured as methionine sulfone, while Met (rabbit liver metallothionein) was measured by the amino acid analysis of intact protein. d) Preliminary sequence analysis of rabbit liver metallothionein-A indicated that Leu was partially substituted to Ser (-Gly-Ala-$\frac{\text{Ser}}{\text{Leu}}$-Asp-Lys-).

sition of rabbit liver metallothionein-A nor -B may be therefore identical with those reported by Nordberg et al. (8).

The tentative structure of rabbit liver metallothionein-B, presenting a single band on acrylamide gel electrophoresis, has been deduced from the results of the automated sequence analysis [Asp-2 or Pro-3 to Gly-11] and the manual sequence analysis of tryptic peptides in comparison with the known amino acid sequence of several metallothioneins. The estimated structure of rabbit liver metallothionein-B is indicated in Fig. 3. The sequence of Ser-12 to Lys-20 has not been confirmed. The substitution of Ala to Glx at the position of 23 may be proposed because the peptide of Cys-Lys-Glx-Cys-Lys in a small quantity was found in the tryptic digest. It may be related to the microheterogeneity of metallothionein molecules, as pointed out in the

5 10

N-X-Met-Asp-Pro-Asn-Cys-Ser-Cys-Ala-Ala-Asp-Gly-(Ser, Cys, Thr, Cys, Ala,

20 25 30

Thr, Ser, Cys)-Lys-Cys-Lys-Ala/Glx-Cys-Lys-Cys-Thr-Ser-Cys-Lys-Lys-Ser-Cys-

35 40 45 50

Cys-Ser-Cys-Cys-Pro-Ser-Gly-Cys-Ala-Lys-Cys-Ala-Gln-Gly-Cys-Ile-Cys-Lys-

55 60

Gly-Ala-Ser-Asp-Lys-Cys-Ser-Cys-Cys-Ala-OH

Fig. 3 Tentative chemical structure of rabbit liver metallothionein-B

other report, described by Kimura et al. (9). However, the chemical structure of metallothionein obtained from the animals exposed to cadmium is as a whole similar to that from the intact animal (1) and human being (2), as illustrated by the present report and the other papers described by Huang et al. (3, 10).

REFERENCES

1. Kojima, Y., Berger, C., Vallee, B.L. and Kägi, J.H.R. (1976) Proc. Natl. Acad. Sci. USA, 73, 3413-3417.

2. Kissling, M. and Kägi, J.H.R. (1977) FEBS Lett., 82, 247-250.

3. Huang, I., Yoshida, A., Tsunoo, H. and Nakajima, H. (1977) J. Biol. Chem., 252, 8217-8221.

4. Tsunoo, H., Kino, K., Nakajima, H., Hata, A., Huang, I. and Yoshida, A. (1978) J. Biol. Chem., 253, 4172-4174.

5. Friedman, M., Krull, L.H. and Cavins, J.F. (1970) J. Biol. Chem., 245, 3868-3871.

6. Crestfield, A.M., Moore, S. and Stein, W.H. (1963) J. Biol. Chem., 238, 622-627.

7. Iwanaga, S., Wallen, P., Grondahl, N.J., Henschen, A. and Blomback, B. (1969) Eur. J. Biochem., 8, 189-199.

8. Nordberg, G.F., Nordberg, M., Piscator, M. and Vesterberg, O. (1972) Biochem. J., 126, 491-498.

9. Kimura, M., Otaki, N. and Kakefuda, T. (1979) This volume, pp. 187-196.

10. Huang, I., Tsunoo, H., Kimura, M., Nakajima, H. and Yoshida, A. (1979) This volume, pp. 169-172.

Metallothionein
Jeremias H. R. Kägi and Monica Nordberg, eds.
Birkhäuser Verlag, Basel/Boston/Stuttgart, 1979

PRIMARY STRUCTURE OF MOUSE LIVER METALLOTHIONEIN -I and -II

I-Y. Huang*, H. Tsunoo**, M. Kimura***, H. Nakashima** and A.Yoshida*
* City of Hope Medical Center, Duarte, California, U.S.A.
** Tokyo Medical and Dental University, Tokyo, Japan
*** National Institute of Industrial Health, Kawasaki, Japan

Primary structure of metallothionein from several animal species has been determined or is under investigation in several laboratories. This paper reports isolation, chemical properties and amino acid sequence of mouse liver metallothionein -I and -II.

Isolation of metallothionein

Two forms of liver metallothionein, i.e. metallothionein -I and -II, were prepared from mice injected with Cd Cl_2. The details of the procedure have been reported elsewhere(1). The stepwise purification procedures used are as follows: a) an extraction of the protein from liver homogenate by 0.9% NaCl, b) Gel filtration with Sepnadex G-50, c) column chromatography with DEAE-Sephadex, and d) second gel filtration with Sephadex G-50. Metallothionein -I and -II were completely seprated from each other by the DEAE-Sephadex step. Both metallothionein -I and -II were obtained in homogeneous form as judged by a single protein band on acrylamide gel electrophoresis. From the livers (wet weight 120-140g) of about 100 mice exposed to cadmium, 30-40mg of metallothionein -I and 20-30mg of metallothionein -II were routinely obtained.

Chemical properties

The total metal content consisting of Cd and Zn is close to 6g atom per mole for both metallothioneins. The ratio of Cd to Zn is about 1.9 in metallothionein -I and about 1.6 in metallothionein -II. Metallothionein -I eluted from the DEAE-Sephadex column at lower salt concentrations than metallothionein -II. Metallothionein -II moved faster towards the anode than metallothionein -I. These results indicate that metallothionein -II is more acidic than metallothionein -I. Molecular weight of both metal-free thioneins estimated by agarose gel filtration was about 6,000, which is in agreement with the molecular weight calculated from the amino acid composition, i.e., 6,063 for metallothionein -I and 6,160 for metallothionein -II. Both metallothioneins are composed of 61 amino acid residues. Although their amino acid composition is similar, metallothionein -II contains one isoleucine which is absent in metallothionein -I, and it contains more serine, glutamine, alanine and lysine but less threonine, glycine and valine than metallothionein -I (Table 1).

Submitted to the "First International Meeting on Metallothionein and Other Low Molecular Weight Metal-binding Proteins", Zürich, July 17 - 22, 1978

Table 1

Amino acid composition of mouse liver metallothionein -I and -II.

	MT-I		MT-II	
Amino acid	Molar ratio	Probable no. of residues per mol.	Molar ratio	Probable no. of residues per mol.
Aspartic acid	3.9	4	4.34	4
Threonine	4.7	5	1.07	1
Serine	8.8	9	8.72	10*
Glutamic Acid	1.1	1	3.01	3
Proline	1.8	2	2.21	2
Glycine	5.0	5	4.00	4
Alanine	5.0	5	6.00	6
Half Cystine	19.6	20	20.04	20
Valine	1.8	2	1.17	1
Isoleucine	0	0	1.03	1
Methionine	0.9	1	1.13	1
Lysine	7.4	7	7.64	8
Total		61		61

*Total number of serine residues estimated from analysis of tryptic peptide of the protein.

Amino acid sequence

The determination of amino acid sequence of metallothionein -I has previously been reported in detail(2). A similar method was used for the determination of primary structure of metallothionein -II. An outline of the procedures used is as follows:

1) Cyanogen bromide cleavage of S-pyridylethylated thionein, and tryptic digestion and papain digestion of oxidized thionein, S-carboxymethylated thionein or S-pyridylethylated thionein.

2) Isolation of tryptic and papain peptides by gel filtration, chromatography with SP-Sephadex and high voltage paper electrophoresis.

3) Determination of amino acid sequence of the major cyanogen bromide peptide consisting of 60 amino acid residues by automated Edman degradation.

4) Determination of amino acid sequence of the tryptic and papain peptides by manual Edman degradation.

5) Determination of the COOH-terminal by hydrazinolysis, and determination of the blocked NH_2-terminal by gas chromatographic identification of acetic acid in the hydrolysate of the NH_2-terminal peptide obtained by papaid digestion.

Summary of the sequence determination of metallothionein -I is given in Fig. 1.

Figure 1

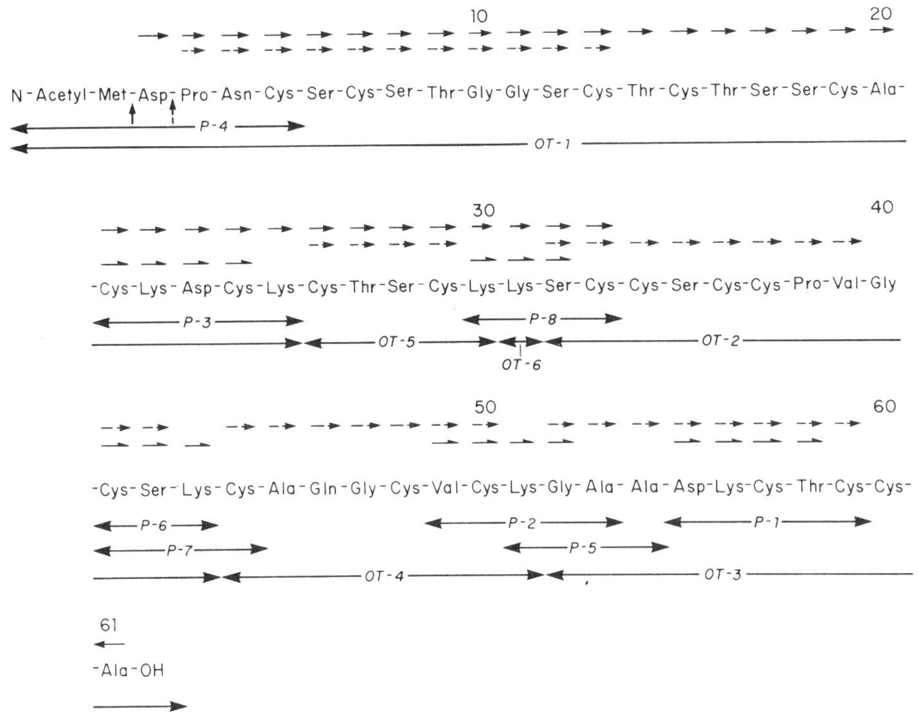

Fig. 1 The complete amino acid sequence of mouse liver metallothionein -I. Continuous arrows (⟶) indicate the automated Edman degradation of S-pyridylethylated, cyanogen bromide-treated protein, broken arrows (--→) indicate manual Edman degradation of individual tryptic peptides of oxidized protein, and half arrows (----⇀) indicate manual degradation of individual papain peptides of S-pyridylethylated protein. Other degradations used: OT, tryptic peptides of oxidized protein; P, papain peptides of S-pyridylethylated protein, ↑, cyanogen bromide cleavage of S-pyridiylethylated protein; ⇡, 70% HCOOH cleavage of peptide OT-1; ←, hydrazinolisis of the protein and peptide OT-3.

The primary structures of the two forms of mouse metallothionein are as follows:

	10
(I)	N-Acetyl-Met-Asp-Pro-Asn-Cys-Ser-Cys-Ser-Thr-Gly-
(II)	N-Acetyl-Met-Asp-Pro-Asn-Cys-Ser-Cys-Ala-Ser-Asp-
	20
(I)	Gly-Ser-Cys-Thr-Cys-Thr-Ser-Ser-Cys-Ala-
(II)	Gly-Ser-Cys-Ser-Cys-Ala-Gly-Ala-Cys-Lys-
	30
(I)	Cys-Lys-Asp-Cys-Lys-Cys-Thr-Ser-Cys-Lys-
(II)	Cys-Lys-Gln-Cys-Lys-Cys-Thr-Ser-Cys-Lys-

	40
(I)	Lys-Ser-Cys-Cys-Ser-Cys-Cys-Pro-Val-Gly-
(II)	Lys-Ser-Cys-Cys-Ser-Cys-Cys-Pro-Val-Gly
	50
(I)	Cys-Ser-Lys-Cys-Ala-Gln-Gly-Cys-Val-Cys-
(II)	Cys-Ala-Lys-Cys-Ser-Gln-Gly-Cys-Ile-Cys-
	60 61
(I)	Lys-Gly-Ala-Ala-Asp-Lys-Cys-Thr-Cys-Cys-Ala-OH
(II)	Lys-Gln-Ala-Ser-Asp-Lys-Cys-Ser-Cys-Cys-Ala-OH

Remarkable structural homology is observed and the amino acid sequence from residue No. 24 to No. 44 is identical in the two types of mouse thionein. However, 15 amino acid substitutions exist between the two proteins. Except for one substitution, i.e. Ala ↔ Lys at position 20, the substitutions can be correlated to single base substitutions. When comparing the amino acid sequences of mouse metallothionein -I and -II with equine and human metallothioneins,(3-5) remarkable structural homology can be observed. Except for one location, i.e. position 59 of the mouse proteins instead of position 58 of equine and human proteins, all cysteine positions are identical in the metallothioneins from these different sources. It should be pointed out that the amino acid sequence of this portion of mouse metallothionein -I and -II was double checked by Edman degradation of the COOH-terminal tryptic peptide (position 52-61) and papain peptide (position 55-59). The sequence of this portion of equine and human thionein was deduced from carboxypeptidase digestion alone (3,4), and further confirmation may be required.

Despite the overall homology of metallothionein structure among these mammals, there are, however, 12 amino acid differences between mouse -I and human liver -2, 10 differences between mouse -II and human liver -2, 12 differences between mouse -I and equine I-A, 13 differences between mouse -II and equine I-A, 13 differences between mouse -I and equine I-B, and 13 differences between mouse -II and equine I-B. The fact suggests that divergence of the two metallothionein components occurred in the early stage of evolutional divergence of mammalia species.

ACKNOWLEDGMENTS

This work was supported by the U.S. Public Health Service grant HL-15125 and Scientific Research grant No. 177086 from the Ministry of Education, Science and Culture of Japan.

REFERENCES

1. Tsunoo, H., Kino, K., Nakajima, H. Hata, A., Huang, I-Y., and Yoshida. (1978) J. Biol. Chem. 253, 4172-4174.
2. Huang, I-Y., Yoshida, A., Tsunoo, H., and Nakajima, H. (1977) J. Biol. Chem. 252, 8217-8221.
3. Kojima, Y., Berger, C., Vallee, B.L., and Kägi, J.H.R. (1976) Proc. Natl. Acad. Sci. U.S.A. 73, 3413-3417.
4. Kissling, M.M., and Kägi, J.H.R. (1977) FEBS Lett. 82, 247-250.
5. Kojima, Y., and Kägi, J.H.R. (1978) Trends Bioch. Sci. April 90-93.

Metallothionein
Jeremias H. R. Kägi and Monica Nordberg, eds.
Birkhäuser Verlag, Basel/Boston/Stuttgart, 1979

AMINO-ACID SEQUENCE OF COPPER-METALLOTHIONEIN FROM *NEUROSPORA CRASSA*

K. Lerch

Biochemisches Institut
Universität Zürich
CH-8028 Zürich, Switzerland

Metallothioneins are low-molecular weight (6000-7000), extremely cysteine and metal-rich proteins containing unusually high amounts of Zn, Cd and/or Cu depending on the source of isolation (1). These proteins have been isolated first from kidney and liver of mammalians but similar proteins have also been described from lower vertebrates (2) and yeast (3,4). The Zn- and Cd-containing metallothioneins have been studied extensively (5,6) and the primary structures of different variants of metallothionein from equine kidney (7), and from equine, mouse and human liver metallothioneins (1,8,9) are now known. Recently, also some copper-containing forms of metallothionein have been isolated from fetal livers of cow and pig (10,11). On the other hand, a number of copper-containing metallothionein-like proteins have been claimed by several authors (12,13) to represent a different class of metal-binding, cysteine-rich proteins. These copper proteins, termed Cu-chelatin (4), cuprodoxin or Cu-thionein (3) have been obtained from a variety of sources previously exposed to the metal (3,4,12,13).

Thus, at the moment, the question whether naturally occurring or metal-induced copper proteins differ indeed from the zinc- and cadmium-metallothioneins is not answered yet.

In this communication, the amino-acid sequence of a copper-containing metallothionein from the ascomycete *Neurospora crassa* is reported and compared to the known primary structures of equine renal, human and mouse liver metallothioneins.

MATERIALS AND METHODS

Neurospora crassa wild-type strain (FGSC No. 320, Arcata, Calif.) was grown on a medium supplemented with 500 µM $CuSO_4$ for 72 hr at 25° according to (14). Copper-containing metallothionein was isolated from the lyophilized mycelium by a heat treatment of the crude extract (10 min. at 60°) and by a combination of gel filtration on Sephadex G-50 and ion-exchange chromatography on DEAE-cellulose. The pure protein was modified by S-pyridylethylation (15), aminoethylation (16) and carboxymethylation with ^{14}C-iodoacetic acid (17) following removal

Submitted to the "First International Meeting on Metallothionein and Other Low Molecular Weight Metal-binding Proteins", Zürich, July 17 - 22, 1978

of the metal by exposure to 50% formic acid and 50 mM EDTA. Copper analyses were performed by atomic absorption spectrometry and colorimetrically with 2,2-bisquinoline (18) on performic acid oxidized, acid hydrolyzed (6 N HCl, 22 hr, 110°) material. Amino acid analyses were made by established procedures (19) on a Durrum D-500 amino acid analyzer. Sequence determinations were carried out with a Beckman Sequencer 890 B (updated) on [^{35}S]-pyridinethylated thionein previously *in vivo* labelled by growing the fungus on a [^{35}S]-sulfate supplemented medium. PTH-amino acids were identified by gas-chromatography (20), by amino acid analysis after reconversion to the free amino acids (21), by thin-layer chromatography (22) and [^{35}S]-PTH-pyridylethylcysteine by liquid scintillation counting.

The aminoethylated protein was also digested with trypsin (treated with tosylphenylalanyl chloromethyl ketone) and the resulting peptides were separated on a Beckman M-72 cationic exchange resin (23) and further purified by paper electrophoresis at pH 1.9 or 6.5. Sequence analysis of the tryptic peptides was performed by manual subtractive Edman degradations and by digestions with carboxypeptidase B and aminopeptidase M.

RESULTS

The amino acid composition of *Neurospora* copper metallothionein is shown in Table 1. Both the high cysteine content (28%) and the complete lack of aromatic amino acids are remarkable features of this protein. In addition, the presence of only seven different amino acids and the high content of serine (28%) and glycine (24%) are also rather striking. The minimal molecular weight as calculated from the amino acid composition was found to be 2200, in reasonable agreement of the value of 2800 obtained from gel filtration of the [^{14}C]-carboxymethyl labelled protein in 6 M guanidine hydrochloride. The molecule binds 6 g-atoms of copper per mole of protein resulting in a copper content of 13%.

In order to determine the amino acid sequence the protein *in vivo* labelled with [^{35}S]-sulfate was alkylated with vinylpyridine and subsequently subjected to automated sequence analysis. The complete primary structure of *Neurospora*-thionein is shown in Fig. 1. All 25 amino acid residues could be unambiguously identified. The positions of the cysteinyl residues were also determined by liquid scintillation counting of the [^{35}S]-PTH-derivative as documented in Fig. 2. The amino acid sequence was confirmed by amino acid analysis and partial sequence analysis of the aminoethylated, tryptic peptides. These results are included in Fig. 1. The amino acid composition obtained from the sequence is in excellent agreement with the

Table 1

Amino-acid composition of Neurospora copper-metallothionein

Residue	Number per molecule	
	Analysis	Sequence
Cys [a)]	6.95	7
Ser [b)]	6.83	7
Gly	5.92	6
Asx	3.01	3
Asp		1
Asn		2
Ala	1.08	1
Lys	0.96	1
Total residues		25
Chain weight		2220
Molecular weight including copper [c)]		2600

a) Determined as aminoethyl-cysteine. b) Extrapolated to zero time of hydrolysis. c) Based on a copper content of 5.8 g-atoms per mole.

Figure 1 Amino-acid sequence of Neurospora copper-metallothionein. T-Fragments obtained from tryptic digestion of S-aminoethylated thionein. → automated Edman degradation; ⇀ subtractive manual Edman degradation; ➝ digestion with aminopeptidase M; ← digestion with carboxypeptidase B.

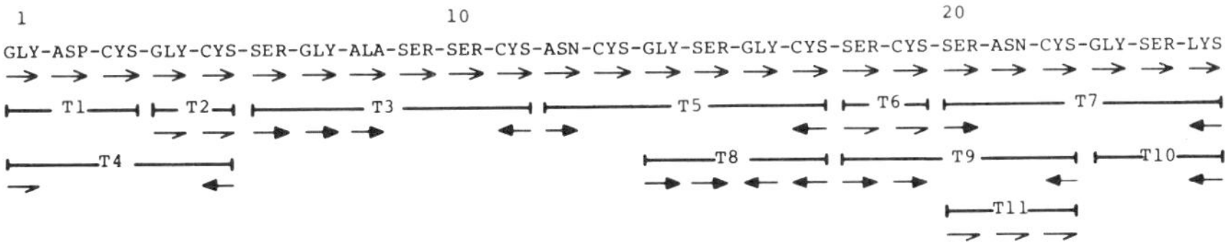

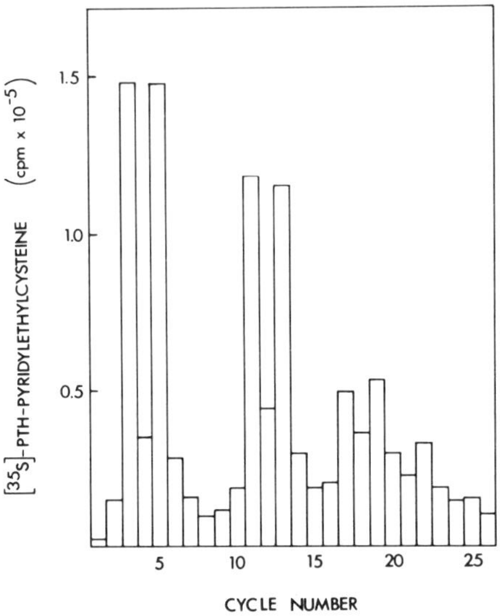

Figure 2

Automated sequence analysis of [^{35}S]-pyridylethylated Neurospora metallothionein. The radioactivity profile of [^{35}S]-PTH-pyridylethylcysteine was obtained by liquid scintillation counting of samples from the aqueous phase.

data determined by amino acid analysis as shown in Table 1. These results establish that Neurospora copper metallothionein is a single chain protein with a molecular weight of 2600 including the bound metal.

DISCUSSION

Low molecular weight, copper-binding proteins have been described from a large variety of sources (for a review see ref.1). At the present time the structural features of these proteins are less well characterized than the zinc- and cadmium-metallothioneins (7).

In an attempt to answer the question whether these proteins binding different metal ions are structurally identical or not, a low molecular weight copper-containing protein was purified from Neurospora crassa and the primary structure subsequently determined. This fungus was found to be especially suitable for such a study because the growth of this organism is hardly influenced by high concentrations of copper and because genetically well-defined, haploid strains are available. In the mycelium of the fungus grown on copper concentrations from 10 μM to 3 mM, the intracellular copper was found to be present in a high molecular weight fraction (mw > 20'000) and a low molecular weight fraction (mw ~ 3000) which contained up to 40% of the total copper content. From the low molecular weight fraction a copper-binding protein was purified to homogeneity as judged by the amino acid composition presented in Table 1. The high cysteine content (28%) and the complete lack of aromatic amino acids which are typical features of the Zn- and Cd-binding metallothioneins (7) would classify this copper protein as a metallothionein.

It is also rather striking that Neurospora copper metallothionein is composed of only 7 different amino acids as compared to the 14 found in zinc- and cadmium-metallothionein. Furthermore, the high content of serine (28%) and glycine (24%) is a rather unusual feature of this protein. The published amino acid compositions from a number of other low molecular weight copper proteins show variable cysteine contents (4.5 - 30%) depending on the source of isolation and the purification procedure. Although these proteins were shown to be homogeneous on polyacrylamide gel electrophoresis in sodium dodecylsulfate, the amino acid compositions suggest a variable degree of heterogeneity. In this study Neurospora copper metallothionein was in addition characterized by amino acid sequence analysis and found to consist of a single polypeptide chain of 25 residues (Fig. 1).

The most prominent feature of this primary structure is the arrangement of 6 cysteine residues in the form of Cys-X-Cys clusters, which have been proposed to constitute the primary chelation sites for the metal ions bound in zinc- and cadmium-metallothioneins (7). The alignment of the Neurospora copper-metallothionein primary structure with those of equine renal, human liver and mouse liver metallothioneins reveals a striking sequence homology of the cysteinyl residues (Fig. 3). All 7 cysteinyl residues of Neurospora copper-metallothionein occupy exactly the same positions as the first 7 cysteinyl residues in the amino-terminal region of the zinc- and cadmium-metallothioneins. The outstanding conservation of the cysteine residues points to an evolutionary relationship of these metallothioneins.

Figure 3 Primary structures of human, equine, mouse and Neurospora metallothioneins. a) Kissling and Kägi (9); b) Kojima et al. (7); c) Huang et al. (8); d) this study. a)-c) For comparative purposes only the amino-terminal sequences are shown.

```
                      1                   10                      20
Human MT-2 a)         Ac-MET-ASP-PRO-ASN-CYS-SER-CYS-ALA-ALA-GLY-ASP-SER-CYS-THR-CYS-ALA-GLY-SER-CYS-LYS-CYS-LYS-GLU-CYS-LYS-CYS-THR-SER-..
Equine MT-1B b)       Ac-MET-ASP-PRO-ASN-CYS-SER-CYS-VAL-ALA-GLY-GLU-SER-CYS-THR-CYS-ALA-GLY-SER-CYS-LYS-CYS-LYS-GLN-CYS-ARG-CYS-ALA-SER-..
Mouse MT-1 c)         Ac-MET-ASP-PRO-ASN-CYS-SER-CYS-SER-THR-GLY-GLY-SER-CYS-THR-CYS-THR-SER-SER-CYS-ALA-CYS-LYS-ASP-CYS-LYS-CYS-THR-SER-..
Neurospora-MT d)                     GLY-ASP-CYS-GLY-CYS-SER-GLY-ALA-SER-SER-CYS-ASN-CYS-GLY-SER-GLY-CYS-SER-CYS-SER-ASN-CYS-GLY-SER-LYS
```

Neurospora copper-metallothionein is the smallest metallothionein found so far. The binding capacity of 6 copper ions per molecule protein makes it a very efficient chelator for this metal. Since copper is regarded as a fairly toxic element for most organisms, it has been speculated that the protein might serve a detoxicating function analogous to that discussed for metallothionein with respect to cadmium (5). However, recent studies on the biosynthesis of Neurospora copper-metallothionein (K. Lerch, unpublished) indicate that this metallothionein may play an important role in copper homeostasis as well.

ACKNOWLEDGMENTS

The author is indebted to Miss Doris Ammer for excellent technical assistance, to Mr. René Budry for the isolation of the protein, and to Mr. Edwin Schwander for running the sequenator.

I would like to thank Dr. J.H.R. Kägi and Dr. Y. Kojima for helpful suggestions. This work was supported by Schweizerischer Nationalfonds, Grants 3.409-0.74 and 3.018.76.

REFERENCES

1. Kojima, Y. and Kägi, J.H.R. (1978) Trends in Biochemical Sciences 3, 90-93.
2. Marafante, E. (1976) Experientia 32, 149-150.
3. Prinz, R. and Weser, U. (1975) Hoppe-Seyler's Z. Physiol. Chem. 356, 767-776.
4. Premakumar, R., Winge, D.R., Wiley, R.D., and Rajagopalan, K.V. (1975) Arch. Biochem. Biophys. 170, 278-288.
5. Kägi, J.H.R., Himmelhoch, S.R., Whanger, P.D., Bethune, J.L., and Vallee, B.L. (1974) J. Biol. Chem. 249, 3537-3542.
6. Bühler, R.H.O. and Kägi, J.H.R. (1974) FEBS Lett. 39, 229-234.
7. Kojima, Y., Berger, C., Vallee, B.L., and Kägi, J.H.R. (1976) Proc. Natl. Acad. Sci. USA 73, 3413-3417.
8. Huang, I.-Y., Yoshida, A., Tsunoo, H., and Nakajima, H. (1977) J. Biol. Chem. 252, 8217-8221.
9. Kissling, M. and Kägi, J.H.R. (1977) FEBS Lett. 82, 247-250.

10. Hartmann, H.-J. and Weser, U. (1977) Biochim. Biophys. Acta 491, 211-222.

11. Bremner, I. and Young, B.W. (1976) Biochem. J. 155, 631-635.

12. Winge, D.R., Premakumar, R., Wiley, R., and Rajagopalan, K.V. (1975) Arch. Biochem. Biophys. 170, 253-266.

13. Riordan, J.R. and Gover, I. (1975) Biochem. Biophys. Res. Commun. 66, 678-686.

14. Vogel, N.J. (1956) Microbial Genet. Bull. 13, 42-43.

15. Friedman, M., Krull, L.H., and Cavins, J.F. (1970) J. Biol. Chem. 245, 3868-3871.

16. Cole, D.R. (1967) in Methods of Enzymology, ed. Hirs, C.H.W. (Academic Press, New York), Vol. 11, pp. 315-317.

17. Gurd, F.R.N. (1967) in Methods of Enzymology, ed. Hirs, C.H.W. (Academic Press, New York), Vol. 11, pp. 532-541.

18. Brumby, P.E. and Massey, V. (1967) in Methods of Enzymology, eds. Estabrook, R.W. and Pullmann, M.R. (Academic Press, New York), Vol. 10, pp. 463-474.

19. Spackman, D.H., Stein, W.H., and Moore, S. (1958) Anal. Chem. 30, 1190-1206.

20. Pisano, J.J., Bonzert, T.J. and Brewer, H.B., Jr. (1972) Anal. Biochem. 45, 43-59.

21. Mendez, E. and Lai, C.Y. (1975) Anal. Biochem. 68, 47-53.

22. Hermodson, M.A., Ericsson, L.H., Titani, K., Neurath, H., and Walsh, K.A. (1972) Biochemistry 11, 4493-4502.

23. Baumann, H., Wilson, K.J., Chen, P.S., and Humbel, R.E. (1975) Eur. J. Biochem. 52, 521-529.

Metallothionein
Jeremias H. R. Kägi and Monica Nordberg, eds.
Birkhäuser Verlag, Basel/Boston/Stuttgart, 1979

THE AMINO-TERMINAL SEQUENCE OF A RAT LIVER METALLOTHIONEIN (MT-2)

M.M. Kissling, C. Berger and J.H.R. Kägi
Biochemisches Institut, Universität Zürich
CH-8028 Zürich, Switzerland

and

R.D. Andersen and U. Weser
Anorganische Biochemie
Physiologisch-chemisches Institut der Universität Tübingen
D-7400 Tübingen, Germany

In 1976 Kojima *et al.* (1) reported the first amino acid sequence of a metallothionein, equine renal MT-1B. Since then, the primary structures have been determined also for the form MT-1A of equine kidney (2,3) as well as for some metallothioneins of the liver of horse (3), man (4,5), rabbit (6), and mouse (7,8). From comparison of the various equine and human sequences we suggested tentatively that cells from euplacental mammals contain two different cistrons coding for metallothionein which may have evolved by duplication from an ancestral gene (5). This view has now received support also from the recently completed sequence studies of the two mouse isometallothioneins (7,8). In this paper, we report the N-terminal amino acid sequence of another rodent isometallothionein, rat MT-2, and document its very close relationship to mouse-MT-II.

Materials and Methods

Rat liver metallothionein MT-2 was isolated as described by Andersen *et al.* (9). The protein was modified by S-pyridylethylation after removing of the metal (1). Sequence determination was carried out with a Beckman Sequencer 890 B (updated) on the polypeptide obtained after cleavage of the acid-labile penultimate Asp-Pro bond in 70% formic acid. The PTH-amino acids were identified by high pressure liquid chromatography, by amino acid analysis after reconversion to the free amino acids (4), and by thin-layer chromatography (4). The structure of the blocked N-terminal dipeptide was determined as described previously (1).

Results and Discussion

The amino acid composition of rat liver MT-2 is listed in Table 1. The analysis of the material employed for sequencing compares fairly well with that reported previously by Winge *et al.* (10) and by Andersen *et al.* (9). Besides the extremely high cysteine content and the single methionine it shows the complete

Submitted to the "First International Meeting on Metallothionein and Other Low Molecular Weight Metal-binding Proteins", Zürich, July 17 - 22, 1978

lack of aromatic amino acids and histidine typical of all metallothioneins.

TABLE 1 Amino acid composition of rat metallothionein-2

	Residues per molecule			Whole residues (Nearest integer)
Residue	This work[a]	(10)[b]	(9)[c]	(9)
Lys	7.2	8.7	8.1	8
Cys[d]	18.8	15.8	20.9	21
Asp	4.3	5.2	4.1	4
Met[d]	1.1	0.7	0.9	1
Thr	2.4	2.5	2.0	2
Ser	8.1	9.7	10.3	10
Glu	4.2	4.5	3.1	3
Pro	2.0	2.3	2.2	2
Gly	4.9	4.5	4.2	4
Ala	4.7	5.4	5.0	5
Val	1.6	0.9	1.2	1
Ile	1.2	0.4	0.8	1
Leu	0.5	0.4	-	-

[a] Based on a total of 61 amino acid residues

[b] Values from metallothionein, form B, of Table II, by Winge *et al.* recalculated to a total of 61 amino acid residues

[c] Based on integral value for alanine of 5

[d] Determined after performic acid oxidation

The results of sequential Edman degradation of the polypeptide chain obtained thus far are shown in Fig. 1. The positions from Pro-3 to Lys-31 (AP-2) were identified by automated sequencing of the chain following removal of the blocked N-terminal fragment (AP-1) by acid cleavage (see method section). The positions from residue 32 to residue 38 were determined by sequence analysis of a fragment (AP-3) resulting from adventitious cleavage of the bond between Asp-10 and Gly-11 during the incubation in 70% formic acid employed for generating a free N-terminus of the polypeptide chain. The analysis of AP-3 also confirmed the sequence from Gly-11 to Lys-31. The structure of the blocked N-terminal dipeptide (AP-1) was deduced from its amino acid composition and its behaviour on thin-layer chromatography (1). It was found to be indistinguishable from N-acetyl-Met-Asp identified previously in equine (1) and human (4) metallothionein.

Fig. 1 N-terminal amino acid sequence of rat metallothionein-2

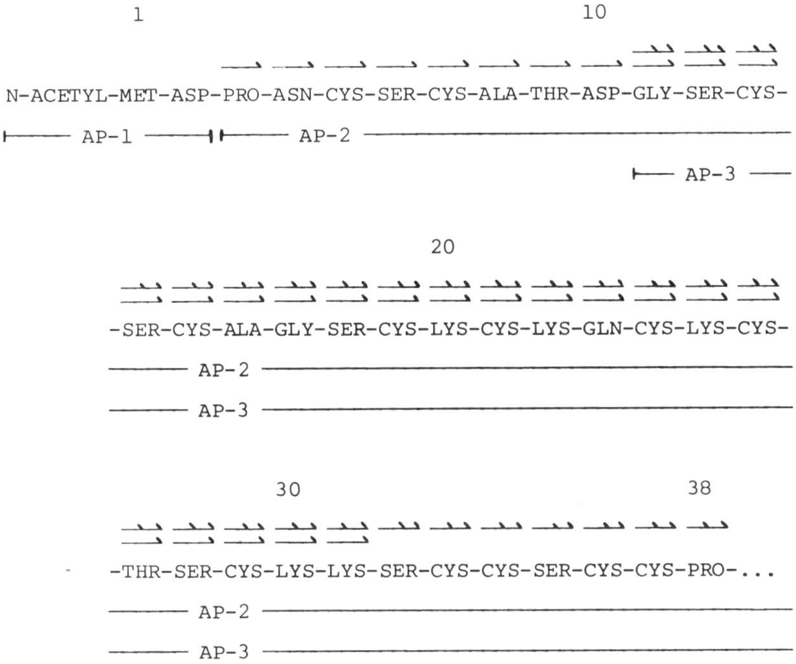

The results obtained with rat MT-2 show a high degree of homology with the corresponding segment of the sequences of other mammalian metallothioneins (Fig. 2). The positions of 26 out of 38 residues of the segment are invariant in all chains. Of all sequences rat MT-2 is most similar to mouse MT-II determined by Huang <u>et al.</u> (8). There are only 2 substitutions located at residue 9 (Ser_{mouse} - Thr_{rat}) and residue 18 (Ala_{mouse} - Ser_{rat}), while all other corresponding N-terminal segments differ from rat MT-2 in 5 or more positions. The smaller difference between rat MT-2 and mouse MT-II fits in well with the observation that the amino acid compositions of the two proteins are at variance only in the threonine and alanine content. The fact that this strong interspecies similarity is not seen between rat MT-2 and mouse MT-I is an unambiguous indication that the separation of

Fig. 2 Sequence comparison of rat metallothionein-2 with other mammalian metallothioneins

```
                                  1                                  10
1)  Rat MT-2:           N-ACETYL-MET-ASP-PRO-ASN-CYS-SER-CYS-ALA-THR-ASP-GLY-SER-CYS-SER-CYS-
2)  Mouse MT-II:                                              " -SER- "   "           "
3)  Mouse MT-I:                                             -SER-THR-GLY- "         -THR-
4)  Equine MT-1B:                                           -VAL-ALA- " -GLU-         "
5)  Equine MT-1A:                                           -PRO-THR- " -GLY-         "
6)  Human MT-2:                                             -ALA-ALA- " -ASP-         "


                     20                                      30                              38
1)  -ALA-GLY-SER-CYS-LYS-CYS-LYS-GLN-CYS-LYS-CYS-THR-SER-CYS-LYS-LYS-SER-CYS-CYS-SER-CYS-CYS-PRO-...
2)    "   " -ALA-     "         -GLN-     "       "
3)  -THR-SER-SER-   -ALA-       -ASP-     "       "
4)  -ALA-GLY- "     -LYS-       -GLN-   -ARG-   -ALA-
5)    "   "   "       "         -GLU-     "     -THR-
6)    "   "   "       "           "     -LYS-     "
```

the genes for the two principal isometallothioneins has occurred much earlier than the divergence of the order of rodents into the different murine genera.

ACKNOWLEDGMENT

This work was supported by Schweizerischer Nationalfonds, Grant 3.125-0.77. Parts of this study were aided by a grant from the Deutsche Forschungsgemeinschaft (We 401/14).

REFERENCES

1. Kojima, Y., Berger, C., Vallee, B.L., and Kägi, J.H.R. (1976) Proc. Natl. Acad. Sci. USA 73, 3413-3417.
2. Kojima, Y. and Kägi, J.H.R. (1978) Trends Biochem. Sci. 3, 90-93.
3. Kojima, Y., Berger, C., and Kägi, J.H.R. (1979) This volume, pp. 153-161.
4. Kissling, M.M. and Kägi, J.H.R. (1977) FEBS Lett. 82, 247-250.
5. Kissling, M.M. and Kägi, J.H.R. (1979) This volume, pp. 145-151.
6. Kimura, M., Otaki, N., and Imano, M. (1979) This volume, pp. 163-168.
7. Huang, I-Y., Yoshida, A., Tsunoo, H., and Nakajima, H. (1977) J. Biol. Chem. 252, 8217-8221.
8. Huang, I-Y., Tsunoo, H., Kimura, M., Nakajima, H., and Yoshida, A. (1979) This volume, pp. 169-172.
9. Andersen, R.D., Winter, W.P., Maher, J.J., and Bernstein, I.A. (1978) Biochem. J. 174, 327-338.
10. Winge, D.R., Premakumar, R., and Rajagopalan, K.V. (1975) Arch. Biochem. Biophys. 170, 242-252.

Metallothionein
Jeremias H. R. Kägi and Monica Nordberg, eds.
Birkhäuser Verlag, Basel/Boston/Stuttgart, 1979

CHARACTERIZATION OF CADMIUM-INDUCED METALLOTHIONEIN IN AFRICAN GREEN MONKEY KIDNEY CELLS *IN VITRO*

Masami Kimura*, Noriko Otaki* and Tsuyoshi Kakefuda**

*Department of Experimental Toxicology, National Institute of Industrial Health, Kawasaki, Japan 213 and **Chemistry Branch, National Cancer Institute, National Institutes of Health, Bethesda, Maryland, USA 20014.

The induced synthesis of metallothionein in animals has been recently studied by several investigators (1-8). The synthesis of Cd-binding protein in cultured human cells was first reported by Lucis *et al.* (9). Webb and his colleagues (10-11) have shown that the cells derived from pig liver as well as kidney induce metallothionein in response to Cd. Also, Hidalgo *et al.* (12) have investigated the induction of Cd-thionein in the isolated rat liver cells. Rugstad and Norseth (13) have observed the induction in human skin epithelial cells and a correlation between the levels of metallothionein and the cells' resistance to Cd.

This report describes the biochemical properties of Cd-induced metallothionein in an established line (VERO) and secondary culture of African green monkey kidney cells in culture.

MATERIALS AND METHODS

An established cell line of African green monkey kidney cells (VERO) was obtained from the American Type Culture Collection, Rockville, Maryland. Secondary culture cells were prepared by subculturing primary cells which were obtained from Flow Laboratories, Bethesda, Maryland. The cells in Falcon petri dishes (15 cm in diameter) in Eagle's minimum essential medium (10 ml) supplemented with 10% fetal calf serum were incubated at 37°C with 5% CO_2. When the cultured cells became 90% confluent (2×10^6 cells), the medium was changed one hour prior to the administration of Cd. For the preparation of ^{109}Cd labelled cells, ^{109}Cd (2 mCi/ml, < 0.1 mg/ml; as $CdCl_2$ in 0.5 N HCl, carrier-

Submitted to the "First International Meeting on Metallothionein and Other Low Molecular Weight Metal-binding Proteins", Zürich, July 17 - 22, 1978

free; New England Nuclear) was diluted ten times by the addition of phosphate buffered saline (PBS). This is referred to as $^{109}CdCl_2$ solution in the present paper. One hundred microliters of the $^{109}CdCl_2$ solution were added to 10 ml of the culture medium. The cells were incubated for 0.5, 1, 2, 4 and 20 hrs. After the incubation the culture medium was removed and the cells were washed several times with cold PBS. The cells scraped from the dishes were suspended in 10 ml of the solution containing 0.32 M sucrose, 0.002 M $MgCl_2$, 0.001 M NaCl, 0.001 M potassium phosphate buffer, pH 6.8. The cell suspension was further washed twice by centrifugation in the same volume of the solution. The cells pooled from several dishes were stored at -20°C. The cells were then homogenized in 0.02 M TrisHCl buffer, pH 8.6. In order to measure the induced ^{109}Cd-thionein the supernatant obtained from the cell homogenate was fractionated by using a Sephadex G-75 column (1.5 x 80 cm).

For large scale experiments one hundred dishes of the secondary culture cells exposed to $CdCl_2$ (2 μg/ml of Cd) were used for the preparation of metallothionein. This was then mixed with ^{109}Cd labelled cells prepared separately as described above.

The cells were homogenized in 20 ml of 0.02 M TrisHCl buffer, pH 8.6, in a Waring blender for 60 seconds. The suspension was centrifuged at 100,000 x g for one hr. The supernatant (approx. 40 ml) was charged to a Sephadex G-75 column (5 x 90 cm) equilibrated with 0.02 M TrisHCl buffer, pH 8.6. It was eluted with the same buffer at a flow rate of 50 ml/hr. The absorbance at 250 nm and the radioactivity of ^{109}Cd were measured. A main radioactive fraction corresponded to the metallothionein position. It was desalted by passing through a Sephadex G-25 column (2.5 x 40 cm) with 0.02 M ammonium carbonate as eluent. The protein fraction was further separated by ion exchange chromatography on a DEAE Sephadex A-25 column (2.5 x 40 cm). Gradient elution was carried out with a mixing chamber of 1 liter of 0.05 TrisHCl buffer, pH 8.6, and a reservoir of 1 liter of 0.2 M TrisHCl buffer, pH 8.6. This was followed by gradient elution with a mixing chamber of 500 ml of 0.2 M TrisHCl buffer, pH 8.6, and a reservoir of 500 ml of 0.25 M TrisHCl buffer, pH 8.6. The last fraction was eluted with 0.25 M TrisHCl buffer, pH 8.6, containing 0.2 M NaCl. The radioactivity counts were monitored. The pooled metallothionein fractions (the bracket region in Fig. 2) were desalted by passing through a Sephadex G-25 column.

One third of the protein sample was hydrolyzed in 6 N HCl at 110°C for 24 hrs and another third was hydrolyzed after performic acid treatment. The rest of the protein sample was subjected to metal analysis using an atomic absorption spectrophotometer (Hitachi Model 308). Amino acid analysis of the hydrolysate was performed by an amino acid analyzer (JEOL-JLC-6AH). Radioactivity was measured in a Packard TriCarb liquid scintillation spectrophotometer.

RESULTS AND DISCUSSION

Induction of metallothionein in cultured monkey kidney cells. In the gel filtration of the supernatant obtained from the cell homogenate which was exposed to $^{109}CdCl_2$, the radioactivity was found in two protein fractions, one of high molecular weight eluted at or near the void volume and the other of low molecular weight eluted at a similar position to rabbit liver metallothionein (14). Metallothionein appeared in the cytosol as early as 30 minutes after incubation. Its radioactivity was 31% of the maximum metallothionein induction which was about 2 to 4 hrs after the administration of $^{109}CdCl_2$. The supernatant after 20 hrs incubation contained approximately 20% of the added ^{109}Cd and the metallothionein fraction was 57% of the total radioactivity recovered in the cytosol of the established cell line VERO. The secondary culture of the kidney cells behaved identical to the established cell line except that the rate of the induction of metallothionein in the secondary culture was significantly higher than the established cell (Fig. 1). More than 90% of the total ^{109}Cd count was found in the low molecular weight region. The details of the inducibility of metallothionein in terms of cell aging will be described elsewhere.

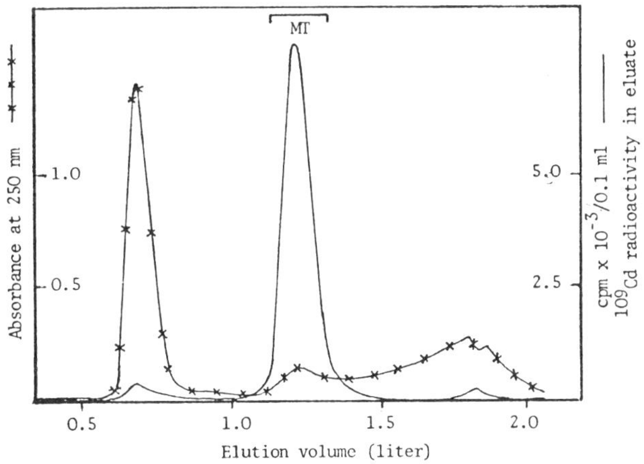

Fig. 1 Gel filtration profile of supernatant obtained from Cd-incubated monkey kidney cells.

Absorbance at 250 nm and radioactivity of ^{109}Cd in each 8 ml fraction were measured. Metallothionein is indicated by the bracket. See the METHODS.

Bryan and Hidalgo (8) have shown that metallothionein is hardly found in rat liver one hr after injection of Cd but more than 90% of Cd is bound to metallothionein in 24 hrs. Cempel and Webb (1) have indicated that the onset of Cd-thionein synthesis in rat liver occurs between 3 and 4 hrs and the inducible synthesis appears to be complete within 8 hrs. Squibb et al. (2) have described that the binding of Zn to thionein in rat liver is measurable by 2.5 hrs and reaches a plateau by 18 hrs after injection of Zn. The significantly faster appearance of metallothionein in our tissue culture system using African green monkey kidney cells may be due to the fact that uptake of Cd into the cells is by simple diffusion (T. Kakefuda, M. Kimura et al., publication in preparation), thus providing efficient induction and subsequent binding of Cd to thionein.

Condition of incubation for preparation of metallothionein. Webb and Daniel (10) have reported that incubation with 0.5 μg/ml of Cd for 7 days gives no detectable change in pig kidney cells. In our preliminary experiments the VERO cells incubated with $CdCl_2$ (2 μg/ml of Cd) supplemented with $^{109}CdCl_2$ (2 μC/ml of ^{109}Cd) for 20 hrs showed no detectable morphological change while incubation with above 10 μg/ml of Cd supplemented with $^{109}CdCl_2$ resulted in a marked shrinkage and apparent death of the cells. Administration of more than 10 μg/ml of Cd was highly toxic as described above. However, the metallothionein recovered from these dead cells 20 hrs after incubation was identical in amount (radioactivity) to that of 2 hrs after incubation at which time the cells showed no marked morphological changes. About 90% of human skin epithelial cells died in the culture medium containing Cd at the concentration of 2.24 μg/ml for 2-3 days (3). Taking this information into consideration our experimental condition was set to induce the cells for 20 hrs with 2 μg/ml of Cd for the preparation of metallothionein.

Preparation and properties of metallothionein. The cell homogenates obtained from 100 dishes of the secondary culture cells mixed with ^{109}Cd labelled cells (see METHODS) were centrifuged and the supernatant was charged to a Sephadex G-75 column. The elution profile is shown in Fig. 1. Most radioactivities were recovered in the low molecular weight protein eluted at the similar position to rabbit liver metallothionein (14). The Cd-binding protein fraction isolated in the Sephadex G-75 was further separated on a DEAE Sephadex A-25 column as shown in Fig. 2. Seven radioactive peaks were found in ion exchange chromatography. The elution profile obtained from the present experiment is different from the chromatographic profile of metallothionein obtained from the isolated rat liver cells reported which represents two major fractions (12). The results of amino acid analysis and metal determination of the protein

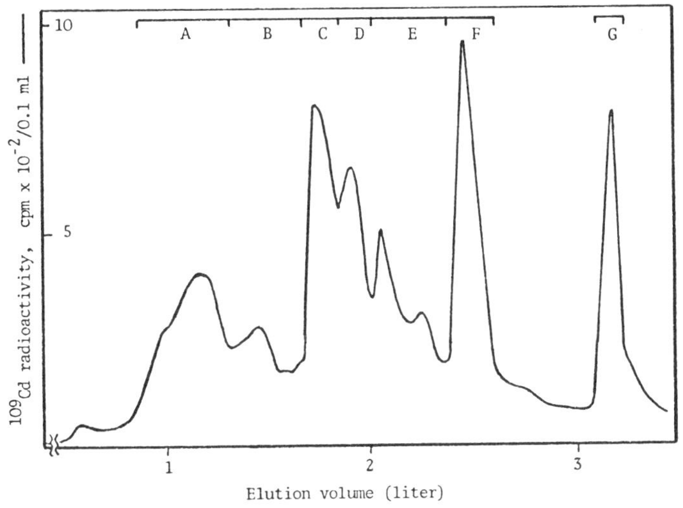

Fig. 2 Ion exchange chromatographic profile of metallothionein fraction obtained by gel filtration.

Radioactivity of ^{109}Cd in each 10 ml fraction was measured. Metallothionein is indicated by the bracket. See the METHODS.

(A) to (G) are shown in Table 1. These proteins except (G) contain many cysteinyl residues, determined as cysteic acid, and no aromatic amino acid residues. These amino acid compositions are similar to those of metallothionein previously reported (15-19). The protein (A) to (F) could be recognized as metallothionein but the protein (G) did not coincide with the definition of metallothionein because of the differences in its chromatographic behavior and amino acid composition. Metal analysis presents the various contents of Cd and Zn in each protein. The more basic protein seems to contain more Cd than Zn. Assuming the chain molecular weight of 6,000, the Cd content appears to vary among these proteins. However, the metal contents of these proteins seem to be lower than those of other metallothionein reported (15-19) except the protein (C). Though it was difficult to investigate further the biochemical properties of metallothionein molecules prepared by the cell culture owing to the limited quantities of the proteins recovered in the present experiment (2.4 to 13.5 nmoles), the differences in structural properties among these isolated metallothionein molecules were nevertheless evident. The microheterogeneity of induced protein molecules may be due to the different phenotypic expression of the cells in culture against cadmium ions which are assumed to be far more rapidly introduced into the cell that occurs in <u>in</u>

Table 1. Amino acid composition of metallothionein obtained from Cd-incubated monkey kidney cells

Amino acid**	Metallothionein (Ratio of amino acid)*						
	A	B	C	D	E	F	G
$CySO_3H$	(24.6)	(18.7)	(16.8)	(19.3)	(22.8)	(17.6)	(6.7)
Asx	4.3	6.2	4.6	5.4	6.3	7.1	7.8
Thr	3.7	5.0	4.8	3.2	3.3	3.4	4.1
Ser	7.6(8.0)	7.8(7.6)	7.9(6.7)	7.8(7.8)	6.8(7.7)	7.3(7.6)	6.2(5.7)
Glx	4.4(5.2)	6.9(8.5)	3.9(5.6)	3.9(4.7)	4.4(5.3)	3.3(5.3)	6.2(6.2)
Pro	3.8(1.7)	5.3(2.0)	2.0(+)	1.6(+)	2.6(+)	3.5(+)	1.3(3.6)
Gly	8.2	8.9	7.5	8.6	7.7	8.2	6.1
Ala	8.0	8.0	8.0	8.0	8.0	8.0	8.0
Val	3.6	4.7	+ (5.8)	3.7	3.4	3.6	2.9
Met	0.6	0.3		0.9	0.6		1.1
Ile	0.5	2.0		+	0.9		1.4
Leu	0.5	2.3		+	0.8		2.4
His	0.3	0.4	0.2		0.3		0.6
Lys	8.1	9.8	5.9	7.6	7.7	7.3	4.2
Arg	0.4	0.9		+			1.3
Tyr							1.4
Phe							1.7
Yield,nmole***	13.5	7.5	2.4	4.5	10.2	2.4	43.5
Cd/Zn,mole/mole****	20.7	not	10.3	6.5	5.7	1.5	4.3
Cd/MT,mole/mole*****	2.3	2.3	9.9	1.8	3.4	3.8	0.3

* Calculated on the basis of 8 moles of alanine / one mole of metallothionein (MT).

** Values are obtained from acid hydrolysates and values in parentheses are obtained from acid hydrolysates of oxidized samples.

*** Total yield of MT was calculated by analytical result of amino acid composition.

**** Measured by atomic absorption spectrophotometer.

***** Based on metallothionein molecular weight 6,000.

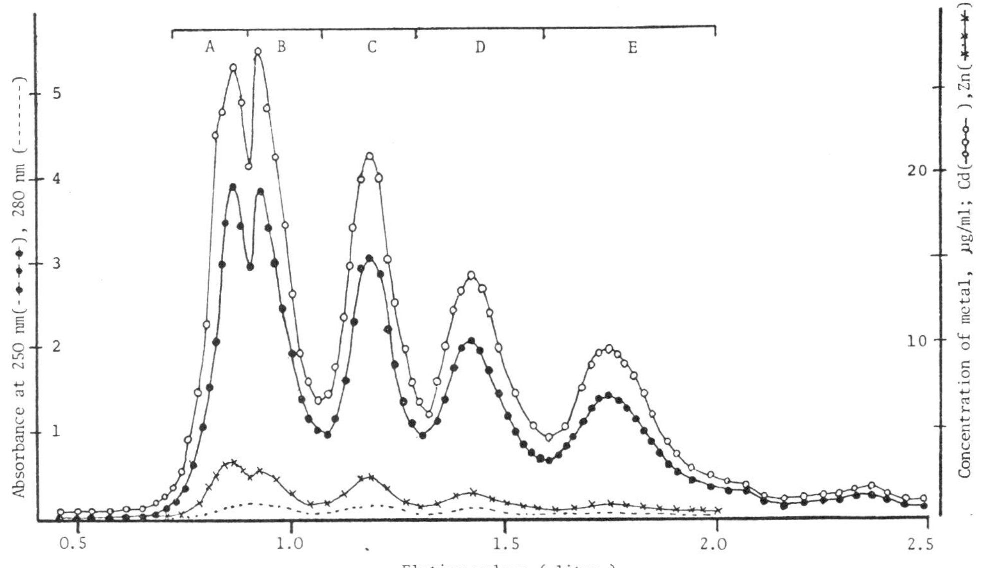

Fig.3 Ion exchange chromatography of Cd-binding protein fraction separated by gel filtration of ethanol-precipitate from liver of Cd-exposed monkey. Column size of DEAE Sephadex A-25 ; 2.8 x 45 cm. Eluted with the gradient of ammonium carbonate (0.05M → 0.20 M and following 0.20 M → 0.25 M). The volume of each ammonium carbonate was 1 liter. Monitored by measuring the absorbances at 250 nm and 280 nm and the concentrations of Cd and Zn.

Table 2 Amino acid composition of metallothionein obtained from Cd-exposed monkey kidney.[a)] a) The values are presented as % of total number of residues in molecule. b) The data are obtained from the analyses of hydrolysates of intact and oxidized metallothionein for 24 hrs. c) ½Cys was measured as cysteic acid.

Amino acid[b)]	Monkey kidney metallothionein A	B	C	D	E
½Cys[c)]	28.3	28.7	30.3	29.9	29.4
Asx	5.3	4.6	8.5	7.4	7.0
Thr	5.1	4.7	3.6	3.6	3.9
Ser	11.4	11.1	10.8	13.0	11.1
Glx	6.0	5.8	4.4	4.3	5.7
Pro	3.7	3.4	3.7	4.5	3.7
Gly	8.3	9.6	8.8	7.9	7.4
Ala	10.2	10.6	9.5	9.1	10.6
Val	4.5	4.0	4.0	3.0	3.8
Met	2.1	2.1	2.0	1.9	1.3
Ile	0.5	0.4	1.5	1.5	1.0
Leu	0.9	0.6	0.6	0.6	1.0
Lys	13.2	13.9	13.1	13.8	13.8
Arg	1.1	0.4	0.2	0.2	+
Cd µmole/mg	0.695	0.596	0.680	0.762	0.570
Zn µmole/mg	0.226	0.113	0.132	0.152	0.059

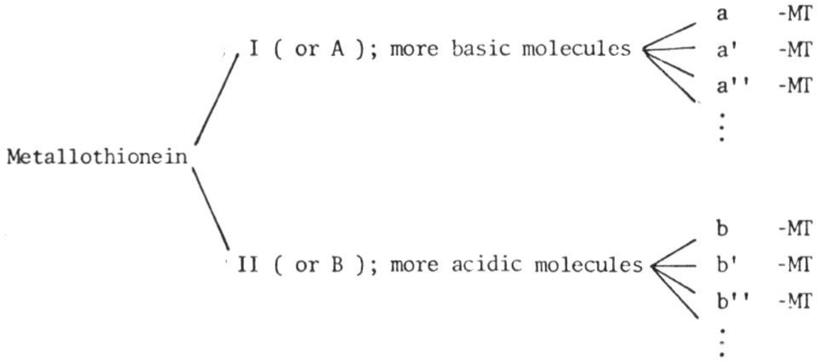

Fig. 4 Scheme for microheterogeneity of metallothionein molecule

vivo system. About 240 μg of apoprotein were obtained from the cultured monkey kidney cells of approximate 2 x 10^8 exposed to $CdCl_2$ (2 μg/ml of Cd) for 24 hrs. The total amount of metallothionein was calculated to be 40.5 nmoles. This information may be useful to study the inducible synthesis of metallothionein.

SUPPLEMENT

Metallothionein was prepared from the Cd-exposed rhesus monkey (Macaca mulatta) by using the method described in the other report (14). Subcutaneously, 3 mg/kg of Cd was 14 times injected into the animal (about 9 kg) every other day.

A crude fraction of the Cd-binding protein was prepared from the liver homogenate by ethanol precipitation and was passed over a Sephadex G-75 column. It was further fractionated by ion exchange chromatography on a DEAE Sephadex A-25 column. The chromatographic pattern is shown in Fig. 3. Five chromatographic forms of liver metallothionein (A to E) were obtained from the Cd-exposed monkey. As shown in Table 2, their amino acid compositions are slightly different from one another. Cadmium induces thus both in vivo and in vitro several variants of metallothionein.

Based on their chromatographic behavior and amino acid composition these metallothionein variants could be divided into a more basic and a more acidic group. The basic forms were metallothionein-A and -B, and the more acidic ones metallothionein -C, -D and -E. A scheme depicting this microheterogeneity of Macaca mulatta metallothionein is shown in Fig. 4.

REFERENCES

1. Cempel, M. and Webb, M. (1976) Biochem. Pharmacol., 25, 2067-2071.

2. Squibb, K.S., Cousins, R.J. and Feldman, S.L. (1977) Biochem. J., 164, 223-228.

3. Piotrowski, J.K., Trojanowska, B., Wisniewska-Knypl, J.M. and Bolanowska, W. (1974) Toxicol. Appl. Pharmacol., 27, 11-19.

4. Shaikh, Z.A. and Smith, J.C. (1977) Chem.-Biol. Interactions, 19, 161-171.

5. Nordberg, G.F., Piscator, M. and Lind, B. (1971) Acta pharmacol. et toxicol., 29, 456-470.

6. Probst, G.S., Bousquet, W.F. and Miya, T.S. (1977) Toxicol. Appl. Pharmacol., 39, 51-60.

7. Shaikh, Z.A. and Smith, J.C. (1976) Chem.-Biol. Interactions, 15, 327-336.

8. Bryan, S.E. and Hidalgo, H.A. (1976) Biochem. Biophys. Res. Comm., 68, 858-866.

9. Lucis, O.J., Shaikh, Z.A. and Embil Jr., J.A. (1970) Experientia, 26, 1109-1110.

10. Webb, M. and Daniel, M.R. (1975) Chem.-Biol. Interactions, 10, 267-276.

11. Daniel, M.R., Webb, M. and Cempel, M. (1977) Chem.-Biol. Interactions 16, 101-106.

12. Hidalgo, H.A., Koppa, V. and Bryan S.E. (1978) Biochem. J., 170, 219-225.

13. Rugstad, H.E. and Norseth, T. (1975) Nature 257, 136-137.

14. Kimura, M., Otaki, N. and Imano, M. (1979) This volume, pp. 163-168.

15. Nordberg, G.F., Nordberg, M., Piscator, M. and Vesterberg, O. (1972) Biochem. J. 126, 491-498.

16. Kojima, Y., Berger, C., Vallee, B.L. and Kägi, J.H.R. (1976) Proc. Natl. Acad. Sci. USA, 73, 3413-3417.

17. Kissling, M.M. and Kägi, J.H.R. (1977) FEBS Letters 82, 247-250.

18. Huang, I., Yoshida, A., Tsunoo, H. and Nakajima, H. (1977) J. Biol. Chem., 252, 8217-8821.

19. Bühler, R.H.O. and Kägi, J.H.R. (1974) FEBS Letters, 39, 229-234.

Metallothionein
Jeremias H. R. Kägi and Monica Nordberg, eds.
Birkhäuser Verlag, Basel/Boston/Stuttgart, 1979

PHYSIOLOGICAL AND CHEMICAL CHARACTERIZATION OF INVERTEBRATE METALLOTHIONEIN-LIKE PROTEINS

R.W. Olafson, R.G. Sim and A. Kearns

Australian Institute of Marine Science
Townsville, Qld. 4810, Australia

ABSTRACT

Metallothionein-like proteins have been isolated from marine invertebrates. Three crustaceans, *Scyllus serratus*, *Cancer magister* and *Acetes sibogae* together with a mollusc, *Cryptochiton stelleri*, have been investigated. *S. serratus* hepatopancreas was shown to contain cadmium and zinc inducible metallothionein-like proteins with an amino acid analysis very similar to vertebrate metallothioneins. The molecular weight, ultra-violet spectrum and isoelectric points of *S. serratus* metallothioneins are also comparable to the vertebrate proteins, suggesting a fundamental biological function conserved throughout evolution.

INTRODUCTION

Metallothionein research in marine science has developed out of a need to investigate the mechanisms underlying heavy metal tolerance in organisms. Heavy metals are especially toxic to aquatic invertebrates (1, 2). However, it has become apparent through unreported field observations in heavy metal polluted environments, as well as through experimental observations, that marine organisms can develop tolerance to known toxic levels of metals (3, 4, 5). Induction of metallothionein-like proteins provides a plausible explanation for these observations.

In the twenty years since the discovery of metallothionein by Margoshes and Vallee, investigators have directed

Submitted to the "First International Meeting on Metallothionein and Other Low Molecular Weight Metal-binding Proteins", Zürich, July 17 - 22, 1978

much of their efforts towards the question of the physiological role of this group of proteins (6). The problem is still unresolved. However, the well known characteristic of induction by certain metals could be of considerable practical significance to marine biochemists should metallothioneins be found in the marine environment. This work was based on the hypothesis that metallothioneins have a fundamental role in biology and that as a result they are ubiquitously distributed throughout the phyla.

METHODS AND MATERIALS

Hepatopancreas tissue of the crabs *Cancer magister* and *Scyllus serratus* were homogenized at 0° and fractionated on Sephadex G-75 as described previously (7). To avoid extensive proteolytic digestion the 27,000 x g supernatant was chromatographed directly. The chiton *Cryptochiton stelleri*, digestive gland and the shrimp *Acetes sibogae*, were similarly processed. Ten microcuries of cadmium-109 were routinely introduced during homogenization. Metal determinations were carried out by atomic absorption spectrophotometry. Differential pulse polarographic analysis for sulphydryl groups was conducted using the Brdička reaction as described by Paleček and Pechan (8) and Kehr (9). Gel filtration fractions were used directly in all analytical procedures.

Lyophilized fractions were applied to a 2.6 x 30 cm DEAE cellulose column as described by Kägi *et al* (10), except the limit buffer was 0.5M Tris/HCl pH 8.6.

Isoelectric focusing was carried out at 5° using a 110 ml LKB column with the following carrier ampholytes: in the dense solution, 2.0 ml of pH 3.5-5; in the light solution 0.7 ml of pH 3.5-5. Focusing took 24 hours at 500-700 volts (3-2mA).

RESULTS AND DISCUSSION

Figure 1 depicts typical gel permeation profiles of *S. serratus* hepatopancreas homogenate. Marked similarity with elution profiles of known vertebrate systems can be observed (7, 11, 12).

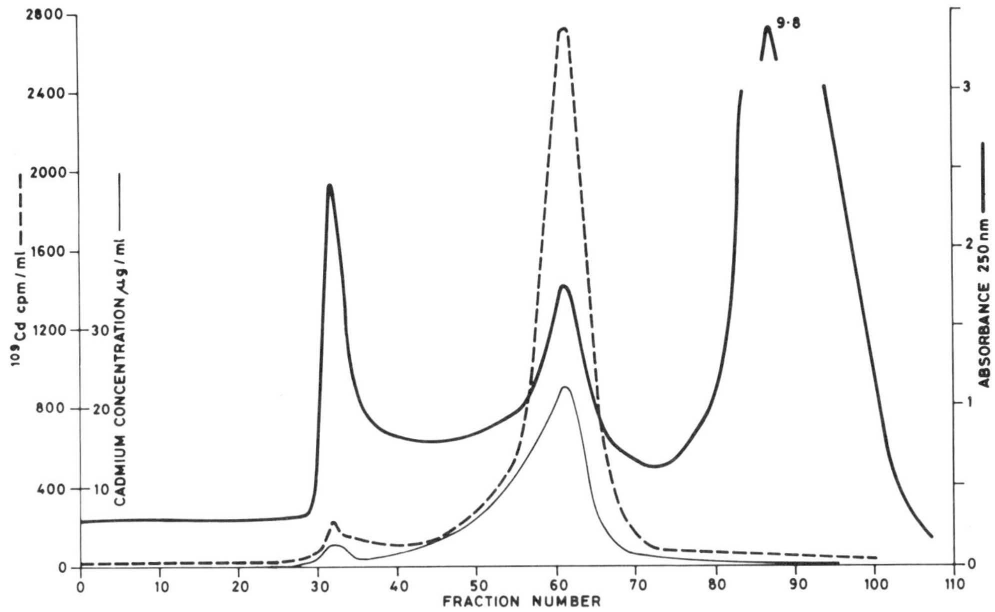

Fig. 1 Sephadex G-75 fractionation of cadmium exposed S. serratus hepatopancreas homogenate. Cadmium -109 ----; Non-radioactive cadmium—— ; Absorbance 250 nm ▬▬.

Metallothionein-like protein elutes with the 10,000 molecular weight peak (fractions 55-65) as indicated by the specific *in vitro* cadmium-109 and in *in vivo* non-radioactive cadmium binding. These results were obtained from cadmium exposed animals using heat denatured hepatopancreas homogenates. Identical profiles, with respect to the 10,000 molecular weight peak, were observed with undenatured and ethanol precipitated material.

Tissue homogenates from three other invertebrates have similarly been fractionated on Sephadex G-75. The apparent molecular weight of the high sulphydryl content heavy metal binding peak from these gel filtrations are listed in Table 1.

TABLE 1

APPARENT MOLECULAR WEIGHT OF HEAVY METAL BINDING FRACTIONS ON CALIBRATED SEPHADEX G-75 COLUMNS

Species		Determinations
S.Serratus (hepatopancreas)	9,600	4
C.magister (hepatopancreas)	11,500	2
A.sibogae (whole animal)	9,500	2
C.stelleri (digestive gland)	9,800	1

Further purification of S. serratus metallothionein was accomplished by DEAE cellulose ion exchange chromatography. Two well-resolved, 250 nm absorbing peaks were demonstrated to coincide with elution of cadmium-109 (Fig. 2).

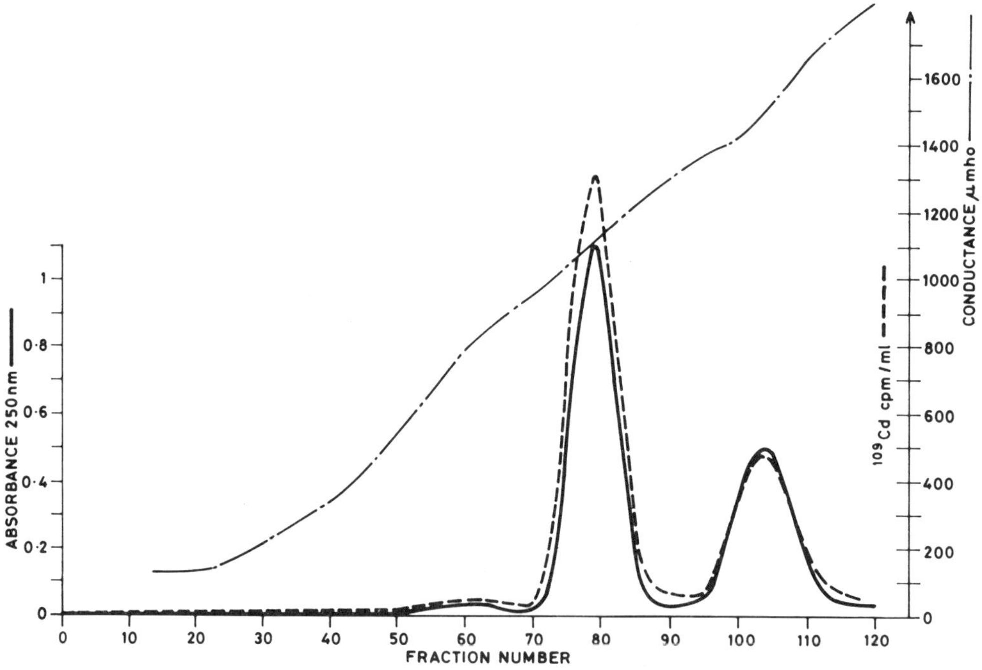

Fig. 2 DEAE cellulose chromatography of S. serratus hepatopancreas homogenate 10,000 molecular weight fraction isolated by gel filtration. Cadmium-109 --- ; Absorbance 250 nm ——.

Isoelectric focusing of Sephadex G-75 fractionated protein afforded results quite similar to those reported by Nordberg _et al._ (13) for rabbit hepatic metallothionein. Isoelectric points were 4.3 ± 0.2 and 4.6 ± 0.1, within 0.4 and 0.1 units respectively of those reported by the former workers. However, a further peak with pI 4.9 ± 0.1 was consistently observed on isoelectric focusing columns. Whether this is a third isoprotein or not is under investigation at this time.

S. serratus hepatopancreas metallothionein was also observed to have the characteristic high 250/280 nm absorbance ratio for cadmium bound material. The low aromatic residue composition inferred by low 280 nm absorbance was corroborated by amino acid analysis of isoprotein MT-I (Table 2).

TABLE 2

AMINO ACID COMPOSITION OF SCYLLUS SERRATUS METALLOTHIONEIN ISOPROTEIN MT-I

Amino Acid Residue	Residues per Molecule (d)	Nearest Integral Number
Half-Cystine (b)	14.2	14
Aspartic Acid	3.3	3
Threonine (c)	3.2	3
Serine (c)	6.5	7
Glutamic Acid	5.9	6
Glycine	5.1	5
Alanine	2.2	2
Valine (a)	1.0	1
Methionine (b)	-	-
Isoleucine (a)	0.1	-
Leucine (a)	0.3	-
Tyrosine	-	-
Phenylalanine	-	-
Histidine	-	-
Lysine	7.5	8
Arginine	1.0	1
Proline	3.9	4

a Taken from 72 hours hydrolysates
b Taken from hydrolysates of performic acid-oxidized protein
c Extrapolated to zero time
d Calculated on the basis of 4 residues of gly/molecule

Evidence that crustacean metallothionein synthesis is induced in the presence of cadmium is indicated on comparison of gel filtration profiles of cadmium exposed animals with a control unexposed animal (Fig. 4). Injection of cadmium or zinc salts results in marked increases in *in vivo* metal incorporation and absorbance at 250 nm (Fig. 1) as well as sulphydryl group concentration and incorporation of tritiated lysine in the 10,000 molecular weight fraction (Fig. 3). Increase in sulphydryl group concentration was followed by differential pulse polarography and confirmed by Ellmans reagent (DTNB).

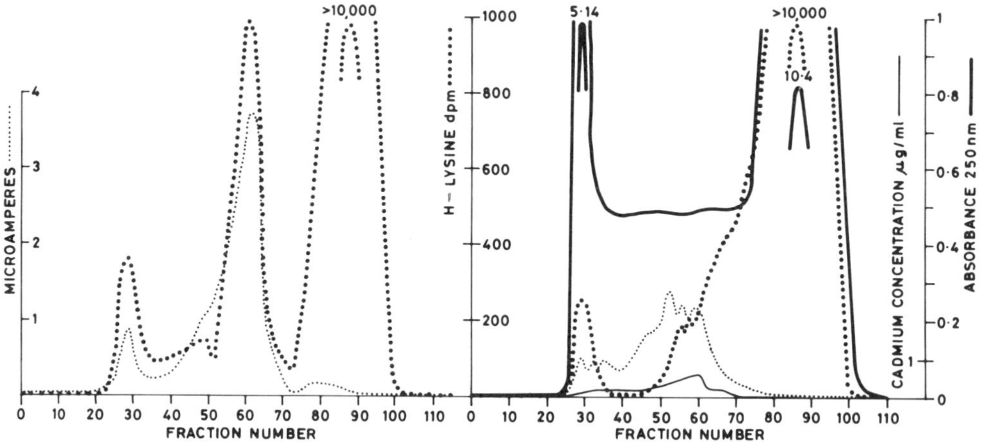

Fig.3 Sephadex G-75 fractionation of cadmium chloride exposed *S. serratus* hepatopancreas homogenate. Polarographic response ...... ; ^{3}H-Lysine incorporation •••• .

Fig.4 Sephadex G-75 fractionation of control *S. serratus* hepatopancreas homogenate. Non-radioactive cadmium ——; Absorbance 250 nm ▬▬; Polarographic response...... ; ^{3}H-Lysine incorporation •••••.

No indication of induction of metallothionein by mercury or copper could be obtained for *S. Serratus*.

It should be noted however, that decapod blood contains the cuproprotein hemocyanin, perhaps necessitating unique controls on the mobility of large divalent cations. Why cadmium and zinc should be exceptional is not apparent. Further studies with S. serratus as well as non-hemocyanin containing invertebrates are necessary to help elucidate this problem. The small sergestid shrimp Acetes sibogae is such an organism and has already provided evidence for induction when exposed to cadmium chloride treated sea water. Studies of exposure to other metals will be conducted shortly.

Isolation of crustacean metallothionein has proven to be in every respect similar to that reported for vertebrate metallothioneins, supporting the hypothesis that metallothioneins, like many other proteins, have been structurally conserved throughout evolution. The phylogenetic distribution of metallothionein-like proteins is broad. The existence of low molecular weight heavy metal binding proteins in invertebrates has been demonstrated in earlier studies on molluscs (14, 15, 16). This report gives unequivocal evidence for the presence of metallothionein-like protein in an invertebrate and is the first demonstration of the existence of this protein in crustaceans. Others have shown the presence of low molecular weight heavy metal binding species in single celled organisms, including a blue-green alga (Anacystis nidulans) (Maclean et al, (17), and a yeast (Candida utilis) (Failla et al. (18)). (The yeast Saccharomyces cerevisiae is reported to contain a copper binding protein distinct from metallothionein (Premakumar et al (19)). Work is proceeding to elucidate more of the chemical characteristics and phylogenetic distribution of this protein as part of a program investigating heavy metal detoxification and tolerance in marine organisms.

ACKNOWLEDGEMENTS

The valuable advice of Dr. K.G. Boto is gratefully acknowledged.

REFERENCES

1. Brown, B. and Ahsanullah, M. (1971) Mar.Poll.Bull. 2, 182-6.
2. Conner, P.M. (1972) Mar.Poll.Bull. 3, 190.
3. Bryan, G.W. and Hummerstone, L.G. (1971) J.Mar.Biol.Ass. 51, 845-63.
4. Braek, G.S. and Jensen, A. (1976) J.Exp.Mar.Biol.Ecol. 25, 37-50.
5. Saliba, L.J. and Ahsanullah, M.(1973) Mar.Biol.23,297-302.
6. Margoshes, M. and Vallee, B.L. (1957) J.A.Chem.Soc. 79,4813.
7. Olafson, R.W. and Thompson, J.A.J. (1974) Mar.Biol. 83-86.
8. Paleček, E. and Pechan, J. (1971) J.Anal. Biochem. 42, 59-65.
9. Kehr, P. (1973) Pulse Polarographic Studies of Cadmium-induced Proteins, Thesis, Purdue University, Indiana.
10. Kägi, J.H.R., Himmelhoch, S.R., Whanger, P.D., Bethune,J.L. and Vallee, B.L. (1974) J.Biol.Chem. 249,3537-3542.
11. Shaikh, Z.A. and Lucis, O.J. (1971) Experientia 27, 1024-1025.
12. Bouquegneau, J.M., Gerday, C.H. and Disteche, A. (1975) FEBS Lett. 55, 173-177.
13. Nordberg, G.F., Nordberg, M., Piscator, M. and Vesterberg, O. (1972) Biochem.J. 126,491-498.
14. Howard, A.G. and Nickless, G. (1975) J.Chromatog. 104, 457-459.
15. Casterline, Jr. J.L. and Yip, G. (1975) Arch.Environ. Contam.Toxicol. 3,319-329.
16. Noel-Lambot, F. (1976) Experientia 32, 324-325.
17. Maclean, F.I., Lucis, O.J., Shaikh, Z.A. and Jansz, E.R. (1972) Fed.Proc. 31,699a.
18. Failla, M.L., Benedict, D.C. and Weinberg, E.D. (1976) J.Gen.Microbiol. 94,23-26.
19. Premakumar, R., Winge, D.R., Wiley, R.D. and Rajagopalan, K.V. (1975) Arch.Biochem.Biophys. 170,278-288.

PHYSICOCHEMICAL PROPERTIES OF METALLOTHIONEIN

Metallothionein
Jeremias H. R. Kägi and Monica Nordberg, eds.
Birkhäuser Verlag, Basel/Boston/Stuttgart, 1979

DETERMINATION OF THE CHARGE OF HORSE KIDNEY METALLOTHIONEIN BY FREE BOUNDARY ELECTROPHORESIS

J.L. Bethune, A.J. Budreau, J.H.R. Kägi and B.L. Vallee

Biophysics Research Laboratory, Department of Biological Chemistry, Harvard Medical School, Boston, Massachusetts, USA, and Biochemisches Institut der Universität Zürich, Switzerland, and the Department of Radiology, Harvard Medical School, Boston, Massachusetts, USA

INTRODUCTION:

Traditionally, the charge of a protein molecule as determined by electrophoresis has been compared to that revealed by pH titration, and any lack of coincidence has been ascribed to ion binding, and the two results have been brought into agreement by adjustment of binding parameters (1). Metallothionein allows a unique opportunity to examine the validity of the electrophoretic approach, since the amino acid sequence and metal atom binding studies allow the absolute charge of the molecule to be computed (2). This then can be compared to the charge determined from electrophoretic mobility measurements to allow evaluation of the theories relating charge to mobility.

EXPERIMENTAL:

Metallothionein was prepared and characterized as described (3). Buffers for electrophoresis, all of an ionic strength of 0.1, were prepared according to Creeth and Nichol (4). All chemicals were of reagent grade and the methods for purification of water and cleaning of glassware have been described (5).

Electrophoresis was performed in a Tiselius moving boundary apparatus (Spinco Model H) at a temperature of 1° and at protein concentrations of approximately 0.15 mg/ml and fields of 4-5 V/cm. Rayleigh fringe patterns were analyzed according to Longsworth (6). Conductivities were measured with a Radiometer Copenhagen conductivity meter and pH with a model 22 Radiometer.

Submitted to the "First International Meeting on Metallothionein and Other Low Molecular Weight Metal-binding Proteins", Zürich, July 17 - 22, 1978

RESULTS AND DISCUSSION:

Over the pH range 7.41 to 9.61, horse kidney metallothionein exhibits a single boundary with a mobility of $-2.28 \pm 0.02 \times 10^{-5}$ cm^2/volt (see Table I).

TABLE I

pH Dependence of Metallothionein Mobility

pH	Buffer	Mobility X 10^5
7.41	veronal	-2.28
7.90	veronal	-2.27
7.93	veronal	-2.24
8.94	tetraborate	-2.33
9.61	tetraborate	-2.27

This remarkable insensitivity of mobility to pH changes is completely consistent with the amino acid sequence of the protein (2), which reveals no groups capable of undergoing changes in ionization over this range. Indeed, the amino acid sequence, together with knowledge of the metal binding capacity and accompanying hydrogen ion release, demonstrates a charge of minus 2 per metallothionein molecule.

While in conventional situations the calculation of protein charge from mobility measurements is fraught with perils (7,8,9), metallothionein constitutes a unique system to evaluate the now existent criteria which allow discrimination of the ranges of validity of the different equations which have been proposed to relate mobility and charge. Specifically, the two functions

$$E = 6 \pi \eta e \mu / DkT$$

and

$$q_o = (8\pi e^2 IN/1000\ DkT)^{\frac{1}{2}} r$$

where η = solution viscosity

e = electron charge, 4.804×10^{-10} esu.

μ = electrophoretic mobility

k = Boltzmann constant, 1.3805×10^{-16} deg -1

T = Temperature

I = Ionic strength

N = Avogadro's number, 6.0226×10^{23}

r = Radius of the molecule studied

have been elaborated to allow such discrimination (7,9).

For metallothionein, with $\eta = 0.01731$ poise and $\mu = -2.28 \pm 0.02 \times 10^{-5}$ cm^2/v-sec, or -6.84×10^{-3} cm^2/stat-v-sec, $D = 87.6$, $T = 274.16^{\circ}$, at an ionic strength of 0.1 and $r = 1.6 \times 10^{-7}$ cm (2),

$$E = -0.324 \text{ and } q_o = 1.65$$

These indicate (see Figures 2 and 8 of Ref. 7) that Henry's equation is appropriate to calculate the charge. This equation is:

$$Q = 6\pi\ r\ \eta\ \mu\ (1 + q_o)/f(q_o)$$

where Q is the charge (in esu) and $f(q_o)$ is a series given in Ref. 8. After an obvious correction, for $q_o = 1.65$ $f(q_o)$ has a value of 1.052. Substituting the appropriate values, $Q = 9.08 \pm 0.08 \times 10^{-10}$ esu, equivalent to a particle charge, Z, of -1.89 ± 0.02 elementary charges/mole.

This extremely close correspondence to the theoretical charge of -2, a difference of 5%, then, represents the deviation of theory and experiment and indeed may have been anticipated. Thus, while the theory applied is in itself rigorous, it is based on certain assumptions that do not necessarily reflect reality. Among these are, that the molecule of interest is a sphere (metallothionein has an axial ratio of 6 (2)) and that no relaxation of the ion atmosphere occurs during motion in the electric field. While it appears that these effects are of the order of only a few percent under the circumstances here pertaining (7,9), they could account for the small discrepancy. We conclude therefore, that for metallothionein the charge calculated from amino acid sequence and metal binding data explain the electrophoretic behavior of the molecule over the pH range 7.5-9.5, in accord with Henry's equation.

REFERENCES

1. Aoki, K. and Foster, J.F. (1957) J. Am. Chem. Soc. 79, 3388.
2. Kojima, Y., Berger, C., Vallee, B.L. and Kägi, J.H.R. (1976) Proc. Nat. Acad. Sci., 73, 3413.
3. Kägi, J.H.R. and Vallee, B.L. (1961) J. Biol. Chem., 236, 2435.
4. Creeth, M. and Nichol, L.W. (1960) Biochem. J., 77, 230.
5. Vallee, B.L., Coombs, T.L. and Williams, R.J.P. (1958) J. Am. Chem. Soc., 80, 397.
6. Longsworth, L.G., In Electrophoresis, M. Bier, Ed. Academic Press, N.Y. (1950) 161.
7. Wiersema, P.H., Loeb, A.L. and Overbeek, J. Th. G. (1966) J. Colloid and Interface Sci., 22, 78.
8. Henry, D.E. (1931) Proc. Roy. Soc. London A133, 106.
9. Overbeek, J. Th. G. and Wiersema, P.H., In Electrophoresis, II, M. Bier, Ed., Academic Press, N.Y. (1967) 1.

Metallothionein
Jeremias H. R. Kägi and Monica Nordberg, eds.
Birkhäuser Verlag, Basel/Boston/Stuttgart, 1979

SPECTROSCOPIC PROPERTIES OF ZINC-METALLOTHIONEIN

Rolf H.O. Bühler and Jeremias H.R. Kägi

Biochemisches Institut
Universität Zürich
CH-8028 Zürich, Switzerland

Metallothioneins exhibit spectral features which have long been known to be characteristic for their mode of metal binding. In fact, studies of the absorptive and chiroptical properties of both cadmium- and zinc-containing metallothioneins of equine and human kidney have given the earliest clues that the metal ions are bound to the protein through mercaptide bonds (1,2,3). These results have since been confirmed on metallothioneins from other sources (4,5,6) and have also been extended to their copper- (6,7), silver- (5), mercury- (5,8), and lead- (9) containing derivatives. However, there are no comparable studies on record on the solely zinc-containing forms of metallothionein which occur naturally in mammalian liver. The present report gives a description and tentative interpretations of the absorption and circular dichroism features of metallothionein-2 (MT-2), one of the major zinc proteins from human liver (10).

Human hepatic MT-2, which contained 7 zinc ions and 20 cysteinyl residues per chain weight of 6085 daltons and which was essentially free of cadmium and copper, was prepared by the method of Bühler and Kägi (10). From pH-titration studies it was found that all cysteine mercapto groups participate in mercaptide bonding with zinc. Absorption spectra and circular dichroism spectra were recorded on Cary 15 and Cary 61 instruments, respectively. Protein concentration was determined by quantitative amino acid analysis.

Difference circular dichroism was measured using the difference circular dichroism attachment of the spectropolarimeter. For measurements below 200 nm, the monochromators were purged with nitrogen. Molecular ellipticity, $[\theta]$, is expressed in units of degree $cm^2 dmol^{-1}$ and was calculated from the relationship $[\theta] = (\theta \times 100)/(c \times d)$, where θ = measured ellipticity in degrees, c = concentration in moles per liter, and d = path length in cm. The molar absorptivity ε and the molecular ellipticity $[\theta]$ refer to peptide chain concentration. The mean residue absorptivity, ε_{MRW}, and the mean residue ellipticity, $[\theta]_{MRW}$, refer to amino acid residue concentration as based on a mean residue

Submitted to the "First International Meeting on Metallothionein and Other Low Molecular Weight Metal-binding Proteins", Zürich, July 17 - 22, 1978

weight (MRW) of 99 (11). Molar difference absorptivity and difference ellipticity are based on the concentration of zinc in the sample and are designated as $\Delta\varepsilon_{Zn}$ and $\Delta[\theta]_{Zn}$, respectively.

Absorption spectrum

The ultraviolet absorption spectrum of native MT-2 is remarkably plain (Fig. 1). Because of the complete lack of aromatic amino acids and of cystine, there is no protein absorption above 260 nm. Its sole characteristics are a large absorption band with maximum at 190 nm and a shoulder at about 220 nm. On removal of the metal by exposure to low pH, the 190 nm band is reduced in intensity and the shoulder is abolished (Fig. 1, Table I). The remaining absorption can be attributed to the amide absorption of the polypeptide and to the absorption of the side chains of the sulphur-containing amino acids. As documented in Table II, there is excellent agreement between the measured molar absorptivity at 190 nm and the arithmetic sum of the absorptivities of the contributing chromophores, calculated from ε_{190}-values of low-molecular weight model compounds (12) and the known

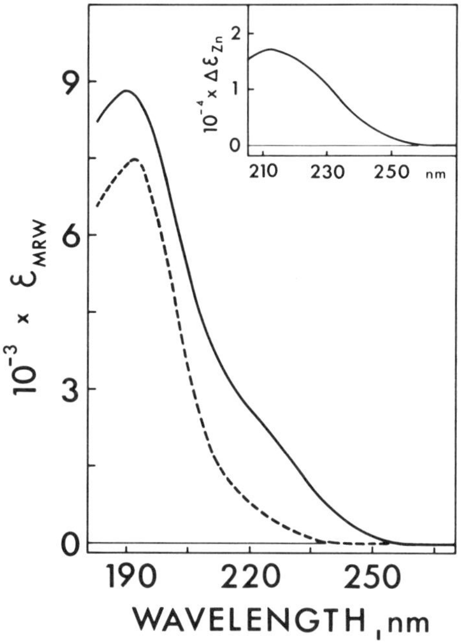

FIG. 1. Absorption spectra of human hepatic metallothionein (MT-2) in 0.05 M sodium phosphate buffer, pH 7.0 (——) and in 0.02 M perchloric acid, pH 1.7 (----); ordinate: mean residue absorptivity, ε_{MRW}. Insert: pH-difference spectrum; ordinate: difference molar absorptivity referred to zinc concentration, $\Delta\varepsilon_{Zn}$.

structure of the polypeptide chain (11). The pH-difference absorption spectrum (Fig. 1, insert) resembles in shape and in position of the maximum (212.5 nm) very closely that of the zinc mercaptide complex formed from zinc ions in the presence of excess β-mercaptoethanol (1). Thus, the difference spectrum may be considered as an accurate representation of the absorption spectrum of the protein's zinc mercaptide chromophores. The slight

TABLE I. Representative molar absorptivities ε and mean molar residue absorptivities ε_{MRW} of human hepatic metallothionein (MT-2) at neutral and acid pH. (Conditions as in Fig. 1)

	pH	190 nm	220 nm
ε	7.0	544'000	159'000
	1.7	458'000	47'300
ε_{MRW}	7.0	8'800	2'600
	1.7	7'400	770

shoulder discernible near 220 nm is an indication of the composite nature of the spectrum. The maximum molar difference absorptivity $\Delta\varepsilon_{Zn}$ based on the metal content is 17'000 $cm^{-1}M^{-1}$, which is in fair agreement with early data obtained in metal substitution studies of equine renal metallothionein (1). It is also comparable to the average maximum difference absorptivity $\Delta\varepsilon_{245}$ = approx. 16'500 $cm^{-1}M^{-1}$ of the lowest energy cadmium mercaptide band as measured in solely cadmium-containing forms of metallothionein (1).

The exact nature of these mercaptide absorption bands is still conjectural. The spectral location as well as the intensity of the band and their broad and featureless shape leave no doubt, however, that they have their origin in transitions from orbitals of the sulphur ligands. Very similar structureless bands are observed in alkaline solutions of aliphatic mercaptans such as cysteine (λ_{max} = 237 nm (13)) and 2-mercaptoethanol (λ_{max} = 233 nm (14)) and have been attributed to the lowest energy transitions from valence shell orbitals of the deprotonated ligand to excited states, in which both the 4s and the 3d atomic orbitals of sulphur and the σ* molecular orbital of the C-S bond (intraligand charge transfer) are thought to play a role (15,16).

TABLE II. Origin of far-ultraviolet absorption of metal-free human hepatic metallothionein. Additivity of chromophore absorbance in the amide region (190 nm).

Chromophoric group contributing to absorbance at 190 nm	190[a]	No. of Groups	$\Sigma\varepsilon_{190}$
Secondary amide	6500	61	397'000
Primary amide	5600	2	11'200
Cysteinyl side chain	2070	20	41'430
Methionyl side chain	1930	1	1'930
Total sum of group absorptivities			451'560
Measured molar absorptivity, ε_{190}, of polypeptide chain[b]			458'000

[a] taken from McDiarmid (ref. 12) for poly-L-lysine

[b] conditions: in 0.02 M perchloric acid, pH 1.7

The large separation of the absorption maximum of the zinc trimercaptide complex (λ_{max} = 212.5 nm) from that of the cadmium complex (λ_{max} = 245 nm) exemplifies the influence of the central metal ion on the energy of the metal mercaptide transitions (1,4,5,6,7,8,9). An analogous difference is seen in inorganic sulfides, where the absorption profile of colorless ZnS is situated also at much lower wavelengths than that of yellow CdS (17). Based on the large extent of these shifts and their direction it was suggested that the bands are of the charge-transfer type (1). Comparison with a more recent study of the absorption spectra of halide complexes of zinc and cadmium, which show similar differences with respect to the relative positions of the major bands, confirm this view (18). These models imply that in post-transition-metal sulfide complexes, the lowest energy bands arise from the transition of a sulphur electron to a σ^*-molecular orbital of the sulphur-metal bond (charge transfer) which includes the lowest empty s-orbital of the metal. Because of the lower optical electronegativity of Zn^{2+}, as compared to Cd^{2+}, such a band would be located at lower wavelength in complexes containing zinc (19).

Charge-transfer bands constitute new absorption bands and are usually independent of the existence of other ligand-localized or metal-localized transitions. However, if the charge-transfer transitions are rather close to those of mercaptide ions in aqueous solution (vide supra) as in the zinc and cadmium trimercaptide complexes, it becomes likely that the orbitals will mix and give rise to new transitions of intermediate character. That this may be the case in zinc and cadmium metallothionein is indicated by the observation that the absorption of the mercaptide ions as such does not persist in the difference absorption spectrum of the zinc trimercaptide complex (Fig. 1). However, that the ligand-localized transitions contribute in a major way to the intensity of the zinc and cadmium trimercaptide bands is suggested by a comparison of the magnitudes of the absorptions. Both in the zinc and cadmium complexes the maximum difference absorptivity of the low energy bands (vide supra) corresponds fairly well to the sum of the known maximum difference absorptivities of the three cysteine mercaptide ligands (ε_{237} = 5-6000 $cm^{-1}M^{-1}$) participating in the complex.

<u>Circular dichroism</u>

The circular dichroism spectrum of hepatic metallothionein is also exceptional (Fig. 2). Comparison of the spectra of metal-containing and metal-free protein indicates large ellipticity contributions from both the polypeptide structure and the metal-polypeptide interaction. The contribution of the former dominates the region below 210 nm, manifesting itself by a strong negative band located at or near 200 nm. The prominence of this band, which is also seen in the circular dichroism spectra of random-coil polyamino acids and fully denatured proteins, indicates the lack of substantial secondary structure in metallothionein (20,21). A similar inference can be made from the absorption spectrum of the metal-free protein (Fig. 1), where the absence of hypochromism as indicated by the additivity of chromophore absorbance (Table II) and amide band splitting excludes the occurrence of substantial α-helix conformation in the chain (22). The latter conclusion has also been reached on the basis of NMR-studies on both the metal-containing and metal-free metallothionein from chicken (23) and from horse (24).

These experimental results also fit well with the theoretical predictions made from the amino acid sequence of the peptide chain (11) using the method of Chou and Fasman (25), in which α-helical conformation is forbidden and β-structure is only marginally possible. However, this lack of regular polypeptide conformation should not be construed as indicating that metallothionein is devoid of a fixed three-dimensional structure. Judged from the slow exchange of many peptide hydrogens of metallothionein with the solvent (24,26) and from its stability to tryptic digestion (27,28) it is in fact highly probable that the protein exists as a tight and well defined tertiary structure. This is also implied by the astonishing preservation of the primary structure of the metallothioneins during evolution (29).

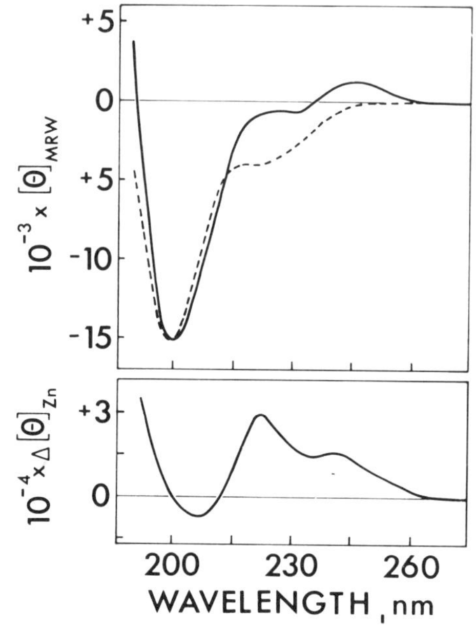

FIG. 2. Above: Circular dichroism spectrum of human hepatic metallothionein (MT-2) in 0.05 M sodium phosphate buffer, pH 7.0 (———) and in 0.02 M perchloric acid, pH 1.7 (----); ordinate: mean residue ellipticity, $[\theta]_{MRW}$.

Below: pH difference circular dichroism spectrum evaluated from the data above; ordinate: molar difference ellipticity referred to zinc concentration, $\Delta[\theta]_{Zn}$.

Since the ellipticity of the amide transition does not seem to be affected appreciably by metal-binding, it is admissible to attribute the large changes occurring on acidification mainly to the loss of the zinc mercaptide chromophore. Like the pH-difference absorption spectrum (Fig. 1, insert), the pH-difference circular dichroism spectrum (Fig. 2, bottom) may be visualized as representing the actual circular dichroism spectrum of the zinc mercaptide chromophore. Based on this assumption, the difference circular dichroism spectrum implies the existence of at least four physically meaningful ellipticity bands associated with zinc-binding. Of these, the nearest positive difference band ($\Delta[\theta]_{Zn,239}$ = +15'000) has also been observed in zinc-containing liver metallothionein obtained from rats pretreated by injection of large doses of zinc sulphate (30) as well as in zinc substituted chicken liver metallothionein (6). It is noteworthy that its spectral location at 239 nm does not coincide with a known difference absorption band of the zinc mercaptide complexes, thus indicating the existence of a low-intensity transition hidden in the descending portion of the zinc-mercaptide difference absorption spectrum. The larger positive difference ellipticity band at 222 nm ($\Delta[\theta]_{Zn,222}$ = +29'000) and the negative difference ellipticity band at 207 nm ($\Delta[\theta]_{Zn,207}$ = -7'400) also do not correspond to distinct difference absorption bands but are positioned above and below the 212 nm maximum of the zinc-mercaptide absorption envelope. This relationship between circular dichroism and absorption spectrum is analogous to the situation in solely cadmium containing metallothionein, where a larger positive ellipticity band occurs at the long wavelength side ($\Delta[\theta]_{Cd,257}$ = +36'000) and a smaller ellipticity band at the short wavelength side ($\Delta[\theta]_{Cd,240}$ = -12'000) of the 245 nm cadmium-mercaptide absorption maximum (6,30,31). It thus appears that an association of these transitions with both a positive and a negative circular dichroism band, i.e. biphasic circular dichroism, is an inherent and typical feature of zinc and cadmium complexes in metallothionein.

The molecular origin of the metal-induced optical activity of metallothionein is also still largely unknown. From our proposal that the absorption of the metal-mercaptide complex is to a large extent a property of the sulphur ligands it follows that the chiroptical features must also be associated with these ligands and thus reflect the dissymmetric interaction with their immediate environment. The most likely sources of rotatory power of these transitions, besides the configurational asymmetry of each cysteinyl residue, are the chiral structures of the multidentate metal-mercaptide complexes. This latter origin of optical activity in asymmetric coordination was already suggested in our early studies on the optical rotatory dispersion properties of cadmium-containing metallothionein (2,3). This interpretation gains additional support from the present observation that the metal-mercaptide absorption bands are associated with biphasic or conservative circular dichroism (32). This feature, which is well known from the circular dichroism spectra of regularly folded polypeptides and polynucleotides, is often a manifestation of rotatory power generated by exciton interaction, i.e. by electrostatic coupling of the corresponding electric transition dipole moments in pairs of identical but dissymmetrically positioned chromophoric groups (33). The highly conspicuous occurrence of such conservative circular dichroism in cadmium and zinc metallothionein permits the inference that the metal polymercaptide complexes constitute such exciton systems. According to this model, the biphasic circular dichroism of the metal polymercaptide complexes may be visualized as arising from the dipol-dipol interaction of pairs of dissymmetrically positioned sulphur ligands within each metal complexing site of metallothionein. Examples in which strong biphasic circular dichroism is generated by such exciton coupling of degenerate ligand transitions are well known for a number of resolved enantiomers of hexa-coordinated complexes of heavy metals with strongly absorbing ligands such as 1,10 phenanthroline and 2,2'-bipyridyl (34).

Conclusion

The study of zinc metallothionein suggests that the chromophoric features of complexes of thionein with non-transition elements arise from a combination of ligand-localized sulphur transitions and a charge-transfer transition to a sulphur-metal molecular orbital. The circular dichroism of the polypeptide chain indicates little or no secondary structure in metallothionein and does not seem to be affected appreciably by metal binding. The strong circular dichroism associated with metal binding is ascribed tentatively to an exciton interaction between pairs of asymmetrically positioned sulphur ligands within the coordination sphere of each metal complexing site.

ACKNOWLEDGMENT

The authors are indebted to Dr. W. Ludwig, Zürich, and Dr. H.A.O. Hill, Oxford, for helpful suggestions. This work was supported by Schweizerischer Nationalfonds grants 3.107.73 and 3.531-0.75.

REFERENCES

1. Kägi, J.H.R. and Vallee, B.L. (1961) J. Biol. Chem. 236, 2435-2442.
2. Ulmer, D.D., Kägi, J.H.R., and Vallee, B.L. (1962) Biochem. Biophys. Res. Comm. 8, 327-330.
3. Pulido, P., Kägi, J.H.R., and Vallee, B.L. (1966) Biochemistry 5, 1768-1777.
4. Weser, U., Rupp, H., Donay, F., Linneman, F., Voelter, W., Voetsch, W., and Jung, G. (1973) Eur. J. Biochem. 39, 127-140.
5. Winge, D.R., Premakumar, R., and Rajagopalan, K.V. (1975) Arch. Biochem. Biophys. 170, 242-252.
6. Rupp, H. and Weser, U. (1978) Biochim. Biophys. Acta 533, 209-226.
7. Rupp, H., Voelter, W., and Weser, U. (1975) Hoppe Seyler's Z. Physiol. Chem. 356, 755-765.
8. Sokolowski, G., Pilz, W., and Weser, U. (1974) FEBS-Lett. 48, 222-225.

9. Ulmer, D.D. and Vallee, B.L. (1969) Trace Substances in Environmental Health-II. Proc. Univ. Missouri Ann. Conf. Trace Substances Environ. Health, 2nd, ed. D.D. Hemphill, 7. Columbia, Univ. Missouri

10. Bühler, R.H.O. and Kägi, J.H.R. (1974) FEBS-Lett. 39, 229-234.

11. Kissling, M.M. and Kägi, J.H.R. (1977) FEBS-Lett. 82, 247-250.

12. McDiarmid, R.S. (1965) Ph.D. Thesis, Harvard University, cited by Gratzer, W.B. (1967) in: Poly-α-Amino acids, G.D. Fasman ed., M. Dekker, Inc., New York, pp. 177-238.

13. Benesch, R.E. and Benesch, R. (1955) J. Am. Chem. Soc. 77, 5877-5881.

14. Donovan, J.W. (1969) in: Physical Principles and Techniques of Protein Chemistry, Part A (Leach, S.J., ed.) Academic Press, New York, pp. 101-170.

15. Thompson, S.D., Carroll, D.G., Watson, F., O'Donell, M., and McGlynn, S.P. (1966) J. Chem. Phys. 45, 1367-1379.

16. Clark, L.B. and Simpson, W.T. (1965) J. Chem. Phys. 43, 3666-3672.

17. Landolt-Börnstein (1955), 6th Ed., Atom- und Molekularphysik, Part 4, Springer Publishing House, Berlin, p. 885.

18. Day, P. and Seal, R.M. (1972) J. Chem. Soc. 2054-2058.

19. Jørgensen, C.K. (1970) Progr. Inorg. Chem. 12, 101.

20. Quadrifoglio, F. and Urry, D.W. (1968) J. Am. Chem. Soc. 90, 2760-2765.

21. Fasman, G.D., Hoving, H., and Timasheff, S.N. (1970) Biochemistry 9, 3316-3323.

22. Rosenheck, K. and Doty, P. (1961) Proc. Natl. Acad. Sci. USA 47, 1775-1785.

23. Rupp, H., Voelter, W., and Weser, U. (1974) FEBS-Lett. 40, 176-179.

24. Galdes, A., Vasak, M., Hill, H.A.O., and Kägi, J.H.R. (1978) FEBS Lett. 92, 17-21.

25. Chou, P.Y. and Fasman, G.D. (1974) Biochemistry 13, 222-245.

26. Ulmer, D.D. and Vallee, B.L. (1971) Adv. Chem. Ser. 100, 187.

27. Kägi, J.H.R. (1970) 8th International Congress of Biochemistry, Switzerland, Sept. 1970, Abstracts, pp. 130-131.

28. Webb, M. (1972) Biochem. Pharmacol. 21, 2751-2765.

29. Kissling, M.M. and Kägi, J.H.R. (1979) This volume, pp. 145-151.

30. Sokolowski, G. and Weser, U. (1975) Hoppe Seyler's Z. Physiol. Chem. 356, 1715-1726.

31. Kägi, J.H.R., unpublished observations.

32. Schellman, J.A. (1968) Accounts Chem. Res. 1, 144-151.

33. Tinoco, I. and Cantor, Ch.R. (1970) Meth. Biochem. Anal. 18, 81-203.

34. Bosnich, B. (1969) Accounts Chem. Res. 2, 266-273.

Metallothionein
Jeremias H. R. Kägi and Monica Nordberg, eds.
Birkhäuser Verlag, Basel/Boston/Stuttgart, 1979

COPPER-THIONEIN AND OTHER METAL-SULPHUR-PROTEINS

Ulrich Weser and Heinz Rupp
Anorganische Biochemie
Physiologisch-chemisches Institut
der Universität Tübingen, D-7400 Tübingen
Hoppe-Seyler-Str. 1

It is attempted to show the relationship of metallothionein to other metal-sulphur coordinated proteins. Special emphasis is placed to characterize the metal-sulphur binding in yeast copper-thionein employing optical methods and x-ray photoelectron spectrometry. The latter technique is particular suited to detect the charge of both copper and sulphur. In addition the appearance of distinct satellites in the Cu 2p-region allow the assignment of the 3 d^9 electronic state.

COMMON ASPECTS OF METAL-SULPHUR-COORDINATION

Proteins containing metal sulphur chromophores are widely distributed in biochemical systems. Originally, metal-sulphur binding was thought to be present only in the iron-sulphur proteins containing inorganic labile sulphide. By way of contrast, a great number of metals other than iron have to be considered when metal-sulphur coordination is extended to all sulphur species including thiolate sulphur, the C=S and/or methionine sulphur. We shall not extend our discussion on those metal proteins but rather restrict ourselves to the metal-thiolate coordination. For example, liver alcohol dehydrogenase contains two tetrahedrally coordinated zinc ions. One could be called a "rubredoxin-type" chromophore attributable to the exclusive coordination with four thiolate sulphurs in a distorted geometry very similar to that found in rubredoxin.

Submitted to the "First International Meeting on Metallothionein and Other Low Molecular Weight Metal-binding Proteins", Zürich, July 17 - 22, 1978

The second zinc known to be the catalytically active metal is still bound to two thiolate sulphurs, one histidine nitrogen and the fourth binding site being reserved for the substrate. As in the case of the "rubredoxin type" zinc thiolate cluster, the distorted tetrahedral overall structure of the catalytically active zinc ion is maintained.

It is very tempting to regard the metal chromophore of tetrahedrally surrounded sulphurs as a general unit in biological systems. The most familiar species for tetrahedral coordination is present both in the 2Fe-ferredoxins and in adrenodoxin. Four distorted iron-sulphur tetrahedra form the cubane like iron cluster of the microbial ferredoxins (Fig.1) (1,2).

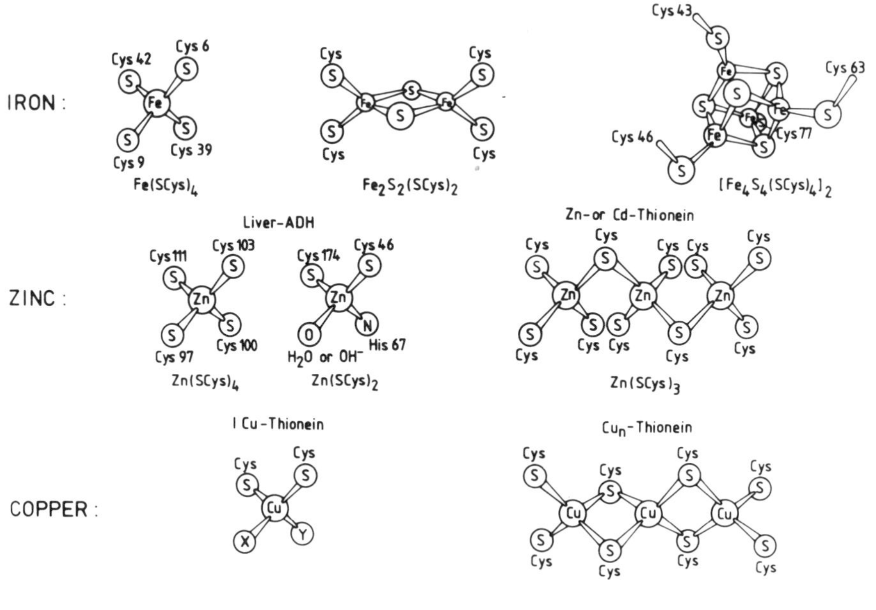

Figure 1: Metal Sulphur-Coordination

Unlike the clearcut crystallographic data published for liver alcohol dehydrogenase and the iron sulphur proteins direct proof for the existence of tetrahedrally arranged copper-thiolate and/or zinc-thiolate clusters in the metallothioneins is not available. At the moment our knowledge is restricted to some chemical and physicochemical data including the elemental ratio of metal and sulphur being 1:3 in the case of Zn-and Cd-thioneins (3) and 1:2 in the Cu-thioneins (4). The zinc-thiolate clusters in alcohol dehydrogenase and the existence of model metal chelates of tetrahedral coordination including Zn_2(dimethyldithiocarbamate)$_4$ (5) and Cu(I)(thiourea)$_3$ (6) encouraged us to speculate a tetrahedral structure for Cu and Zn in those ubiquitous metallothioneins. Two three-coordinated sulphur atoms each should be considered in the nCu-thioneins in a way similar as observed for the linear chain of [Cu(I)-(thiourea)$_3$]$_n$ -tetrahedra where one sulphur serves as a bridging ligand between two Cu(I). Several other complex structures including one or more cubic units similar to the Cu_4S_6 -adamantane core (7) are also conceivable. The basis for this structural proposal was obtained from electron absorption spectrometry, circular dichroism, high resolution NMR and X-ray photoelectron spectrometry.

Cu-THIONEIN

The class of Cu-thioneins have been thought to be substantially different from the metallothioneins. We succeeded to convert chicken Cd,Zn-thionein into Cu-thionein in 1974 (8). Titration of Cd,Zn-thionein with Cu(I) yielded Cu-thioneins having a 1:2 stoichiometry with regard to copper and sulphur (9). Shortly thereafter the successful isolation of a homogeneous Cu-thionein from yeast was reported (10,11). The copper to sulphur ratio was also 1:2 . Premakumar et al. (12) isolated the same type of protein and called it copper chelatin. Due to the different copper and sulphur content in the Cu-chelatins initially, a direct relationship to our Cu-thionein was not seen. Nevertheless, the electron absorption profile of either Cu-thionein or Cu-chelatin was similar. Surprisingly, identical ϵ-values were obtained for all Cu-thioneins isolated from animal sources provided the absorption was expressed per mole of copper. Care had to be taken to avoid oxidation of the Cu-thioneins leading to disturbing disulphides. Significantly higher ϵ_{Cu} -values were measured when microbial Cu-thionein was examined (Table 1).

Table 1: Electron Absorption of Different Cu-Thioneins

Source	250nm	280nm	300nm
		ϵ_{Cu} [$M^{-1}cm^{-1}$]	
Rat (13)	3170	2220	1030
Bovine fetus	3320	2870	1210
Chicken (in vitro)	3580	2240	1390
Saccharomyces cerevisiae	5210	3650	2040

Conformational studies on the copper chromophores using circular dichroism suggested a similar chemical environment of the copper (Table 2). Although the magnitude of the Cotton extrema varies, their position remained essentially at the same wavelength regardless whatever Cu-thionein was employed.

Table 2: Circular Dichroism of Cu-Thioneins, Position of Cotton Bands.

Origin	Wavelength (nm)					
Chicken liver Cu(I)-titrated	350(+)	310(-)		290(-)	260(+)	220(s-)
Fetal bovine liver	370(+)			275(-) 285(s-)	255	220(s-)
Saccharomyces cerev.	360(+)	330(+)	300(s-)	280(-)	250(+)	215(s-)

Attributable to the high quality and distinct resolution of the circular dichroism spectra of the different yeast Cu-thioneins theses proteins were selected for more detailed investigations (Fig. 2). The magnitude of the observed Cotton extrema strongly parallels the copper and cysteine-sulphur content of the proteins. It was suggested that these Cotton bands can be attributed to the coordination of copper with the two cysteine-sulphurs in conjunction with the overall protein conformation. Similar to the ultraviolet absorption a linear dependence of the molecular ellipticity-values from the copper concentration can be drawn (Fig. 3).

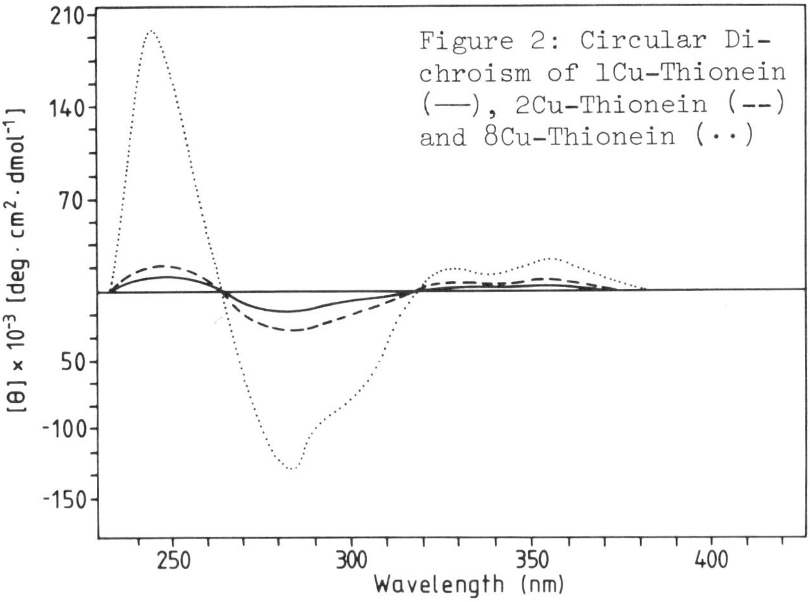

Figure 2: Circular Dichroism of 1Cu-Thionein (—), 2Cu-Thionein (--) and 8Cu-Thionein (··)

This spectrochemical dependency of copper content on the chiroptical and ultraviolet data has its counterpart in the variable composition of cysteine. The cysteine content was found to rise concomitantly with the copper concentration. A 1: 2 stoichiometry between copper and cysteine sulphur is clearly detectable. More detailed analyses of the amino acid composition can be taken from (14). It was intriguing to note that the sum of the 5 different amino acid residues including alanine,cysteine,isoleucine,leucine and valine is always close to 23.4±2.3%. Upon plotting the %-amino acid residues versus the copper content a linear dependency is seen (Fig. 4). This phenomenon seems to be restricted to the microbial Cu-thioneins and a decision cannot be made whether this is attributable to anabolic or catabolic processes.

The use of SH-Sephadex chromatography to isolate human fetal liver Cu-thionein was elegant to recover all cysteine residues. This technique ascertains the most gentle treatment of the protein portion, however, some of the metals probably will be bound to the thiolate rich Sephadex gel (15).

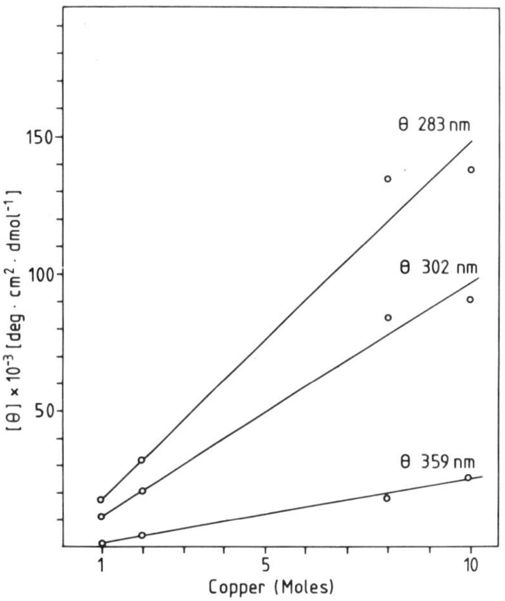

Figure 3: Correlation of Copper content and Molecular Ellipticity

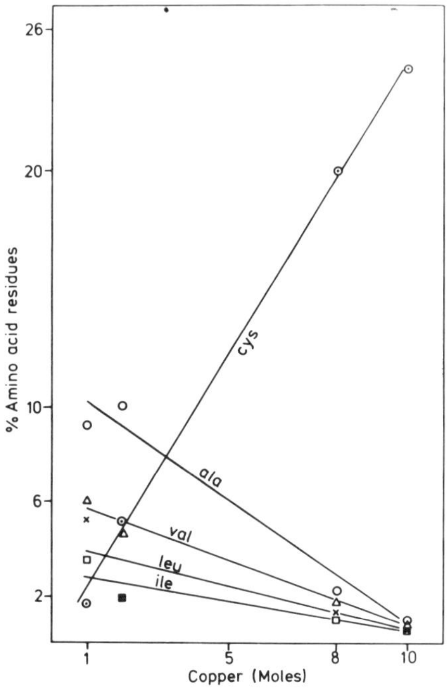

Figure 4: Correlation of Amino Acid Residues and Copper Content

THE Cu(THIOLATE)$_2$-CHROMOPHORE

The 1:2 stoichiometry of copper to sulphur has been repeatedly shown including elemental analyses, the composition of amino acid residues and correlation of Θ and ∈-values with the copper content. Supporting evidence of both the 1:2 stoichiometry and the coordination of copper with thiolate residues was obtained from titration with p-chloromercuribenzoate. The addition of 2.2 moles of p-chloromercuribenzoate caused the complete disappearance of the characteristic Cotton extrema (Fig. 5).

Displacement of zinc and cadmium by protons using Zn-,Cd-or the mixed Cd,Zn-thioneins confirmed the earlier ultraviolet spectrometric titration studies of Kägi and Vallee (16). Upon titration of Cu-thionein an extraordinary higher stability of the metal chromophore is seen (Fig. 6,7). Essentially no changes were detectable up to a pH-value of one. Within half a pH-unit all copper has been displaced. It was of interest to note that the Cotton extrema remained at the same position throughout which was paralleled by a constant decrease of the molecular ellipticity values.

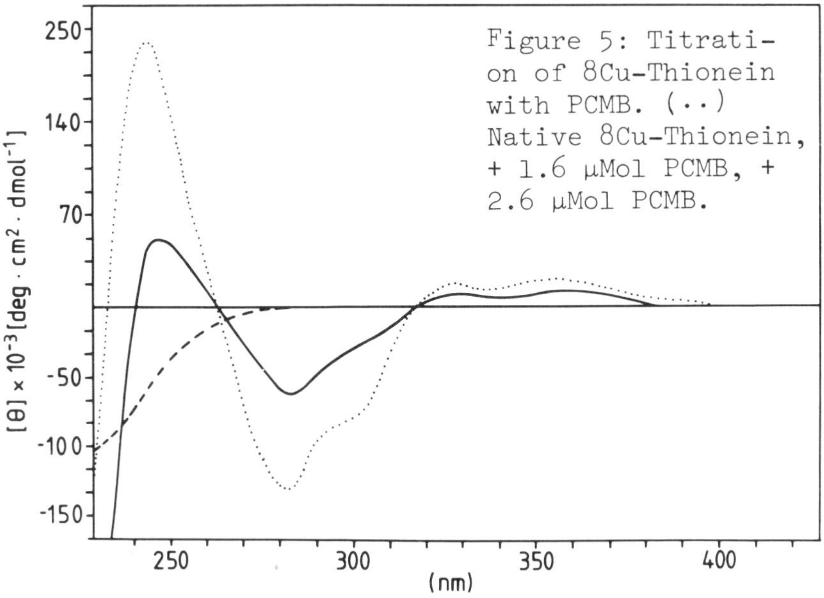

Figure 5: Titration of 8Cu-Thionein with PCMB. (··) Native 8Cu-Thionein, + 1.6 μMol PCMB, + 2.6 μMol PCMB.

REDOX STATE OF COPPER

X-ray photoelectron spectrometry is a powerful tool to examine the genuine redox state of an element. The binding energy of the core electrons reflect thereby the involvement of the binding electrons of the outer sphere in a chemical bonding (For details see 17,18).

X-ray photoelectron spectrometric studies on copper complexes of biochemical significance (17-19) allowed a clean-cut decision between $3d^9$ -[Cu(III)] and $3d^{10}$[Cu(I)]. The spectra of Cu(I)-complexes show one homogeneous $Cu2p_{3/2}$ signal around 932 eV. A blue shift of the main signal by approximately 2 eV is indicative for a higher charge on the copper nucleus. The appearance of characteristic more or less split satellites from 940±2 eV are seen when Cu is present in the $3d^9$ state. Owing to the undisturbed appearance of these Cu 2p-satellites even antiferromagnetically coupled $3d^9$[Cu(II)] can be seen (17). When native 8Cu-thionein is measured only one large and homogeneous $Cu2p_{3/2}$ signal appears at 932.3 eV attributable to a reduced charge on the copper, the electronic state being $3d^{10}$. As stated above antiferromagnetic coupling of Cu(II) is excluded by this method. Exactly the same numerical value of the main signal has been recorded when Cu(I)-cysteine was measured. Oxidation of the Cu(thiolate)$_2$-chromophore by H_2O_2 caused the broadening and shifting of the main signal to 934 eV and two satellites at 940 and 942.4 eV appeared. In addition to the formation of $3d^9$[Cu(II)] a dramatic shift of the sulphur 2p signal from 162.7 to 167.9 eV was seen indicating that all sulphur was oxidized to RSO_3^- . Intermediate signals appearing at 163.7 eV and 166.6 eV were assigned to RSSR and RSO_2^- (Fig. 8).

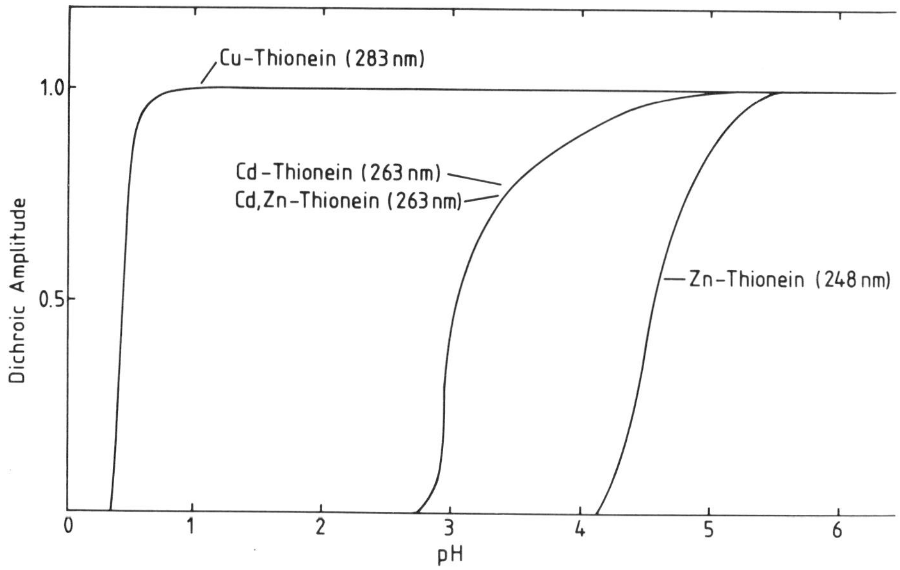

Figure 6: Dependency of Molecular Ellipticity and pH

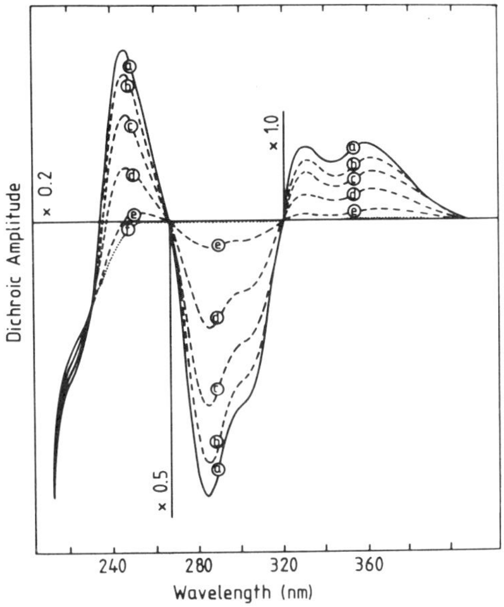

Figure 7: Circular dichroism of Cu-thionein at different pH values: a)6.90, b)0.51, c)0.46, d)0.41, e)0.37, f)0.34

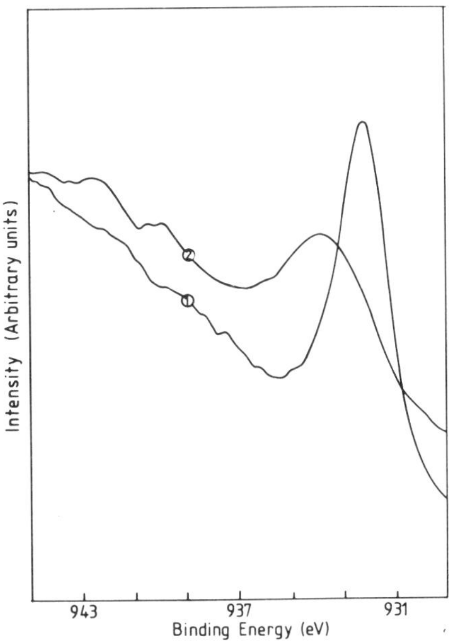

Figure 8: $Cu2p_{3/2}$levels of 8Cu-Thionein before ① and after H_2O_2 treatement ②

REFERENCES

1. Palmer, G. (1973) in "Iron Sulphur Proteins", W.Lovenberg ed., Academic Press, New York 2,285-325.
2. Brändén, C.I., Jörnvall, H., Eklund, H. and Furugren, B. (1975) in "The Enzymes", P.D.Boyer ed., Academic Press, New York 11,103-190.
3. Kägi, J.H.R., Himmelhoch, S.R., Whanger, P.D., Bethune, J. L. and Vallee, B.L. (1974) J.Biol.Chem. 249,3537-3542.
4. Rupp, H. and Weser, U. (1978) Biochim.Biophys.Acta 533, 209-226.
5. Klug, H.P. (1966) Acta Cryst. 21,536-546.
6. Okaya, Y. and Knobler, C.B. (1964) Acta Cryst. 17,928-930.
7. Griffith, E.H., Hunt, G.W. and Amma, E.L.(1976) J.Chem.Soc.

Chem.Comm. 432-433.
8. Rupp, H. and Weser, U. (1974) FEBS Lett. 44,293-297.
9. Rupp, H., Voelter, W. and Weser, U. (1975) Z.Physiol.Chem. 356,755-765.
10. Prinz, R. and Weser, U. (1975) FEBS Lett. 54,224-229.
11. Prinz, R. and Weser, U. (1975) Z.Physiol.Chem. 356,767-776
12. Premakumar, R., Winge, D.R., Wiley, R.D. and Rajagopalan, K.V. (1975) Arch.Biochem.Biophys. 170,278-288.
13. Winge, D.R., Premakumar, R., Wiley, R.D. and Rajagopalan, K.V. (1975) Arch.Biochem.Biophys. 170,253-266.
14. Weser, U., Hartmann, H.-J., Fretzdorff, A. and Strobel, G.J. (1977) Biochim.Biophys.Acta 493,465-477.
15. Rydén, L. and Deutsch, H.F., (1978)J.Biol.Chem.253,519-524
16. Kägi, J.H.R. and Vallee, B.L. (1961) J.Biol.Chem. 236, 2435-2442.
17. Rupp, H. and Weser, U. (1976) Biochim.Biophys.Acta 446, 151-165.
18. Rupp, H. and Weser, U. (1976) Bioinorg.Chemistry 6,45-59.
19. Thomas, W.H. and Weser, U. (1976) Z.Physiol.Chem. 358, 47-52.
20. Supported by the Deutsche Forschungsgemeinschaft, grant No. We 401/14.

Metallothionein
Jeremias H. R. Kägi and Monica Nordberg, eds.
Birkhäuser Verlag, Basel/Boston/Stuttgart, 1979

STRUCTURAL ASPECTS AND REDUCTION OXIDATION REACTIONS OF METALLOTHIONEIN

Heinz Rupp and Ulrich Weser
Anorganische Biochemie
Physiologisch-chemisches Institut
der Universität Tübingen, D-7400 Tübingen
Hoppe-Seyler-Str. 1

In an approach to elucidate the conformation of metallothioneins, far-ultraviolet circular dichroism was measured. Using 270 MHz proton magnetic resonance spectroscopy it was expected to characterize further the structure described by circular dichroism as unordered. In addition, the oxidation-reduction properties of yeast Cu-thionein were studied by electron paramagnetic resonance spectroscopy. The biological implications of the oxidation state of copper in Cu-thionein are discussed.

FAR-ULTRAVIOLET CIRCULAR DICHROISM OF CHICKEN Cd, Zn-THIONEIN

Intrinsic Cotton effects are influenced by the asymmetry of the polypeptide backbone and are characteristic for hydrogen bonded parallel and anti-parallel sheets and helical structure. In the absence of long-range order, circular dichroism is less sensitive for detecting secondary and tertiary structural changes (1). In the far-ultraviolet region, both Cd- and Zn-thionein exhibit a negative Cotton effect at 200 nm and a shoulder near 225 nm. Based on the circular dichroic properties of poly-amino acids measured as a film (2) we can conclude that the contribution of α-helix or pleated sheet structure to the conformation is less than 5%. The unordered structure is maintained inspite of drastic changes in the net charge of Cd,Zn-thionein due to dissociation of the

Submitted to the "First International Meeting on Metallothionein and Other Low Molecular Weight Metal-binding Proteins", Zürich, July 17 - 22, 1978

metal ions at low pH-value or to deprotonated ε-amino groups of lysine at alkaline pH-values. It is noteworthy that even for cysteine-thionein no significant changes were observed at alkaline pH or on heating to 50°C for 20 min despite the high content (10%) of lysine residues. For detailed circular dichroic data of the metal sulphur chromophore see (3,4).

270 MHz PROTON NMR OF CHICKEN Cd,Zn-THIONEIN

As a first approximation the NMR spectrum of a protein may be constructed from the proton resonances of the constituent amino acids taking into account the slow motion of the protein. The chemical shift of individual proton resonances in a protein can be influenced by the conformation, since the folding of the protein chain may result in many perturbations due to the influence of proximate groups. The most important factors which contribute to a shift of a resonance in a protein relative to the free amino acid include ring current shifts, hydrogen bonding, local electrostatic effects and para-magnetic interactions. In most cases the calculated spectra are good representations only for proteins which have been denatured using heat treatment, pH-changes or chaotropic agents (5). In the denatured protein, protons of the same residue, but located in different environments become equivalent. Furthermore, intramolecular interactions are minimized. It was shown that the smallest conformational change detectable by peptide circular dichroism would be in the order of one half turn of α-helix or its equivalent of the pleated sheet structure (1). However, even considerable changes in the unordered conformation would not result in distinct shifts of the Cotton effects and would therefore not be accessible to a meaningful interpretation in terms of different structural elements. On the other hand, NMR can not only follow changes in the secondary and tertiary structure but may also monitor the conformational state of a single residue. In the case of metallothionein NMR should prove most valuable in getting further insight into the structure, described by circular dichroism as unordered. Emphasis will be placed on the question whether or not metallothionein contains regions of considerable rigidity and must therefore be highly ordered. The lack of aromatic amino acid residues in metallothionein proves disadvantageous for this analysis, since ring current effects which are main perturbants of protein resonances are missing. Nevertheless, the existence of hydrogen bonding and electrostatic effects should result in significant shifts of the proton resonances. In accordance with the amino acid analysis no resonances of phenylalanine, tyrosine or tryptophane can be detected in the low-field region of the 270 MHz NMR spectrum of Cd,Zn-thionein (6). In the high-field region the proton resonances are well resolved due to the low-molecular-weight of the protein (Fig. 1). The simulated spectrum is based on chemical shifts of amino acids and peptides (5) using the amino acid composition of chicken metallothionein (7,8). In fact most signals in the actual spectrum of Cd,Zn-thionein can be assigned to resonances of the constituent amino acids. Thus, the signals near 3.02 ppm can be attributed to cysteine β-CH_2 and lysine ϵ-CH_2 protons, the signals near 1.69 ppm to leucine β-CH_2 + γ-CH, lysine δ-CH_2 + β-CH_2 and arginine γ-CH_2 protons and the signals near 1.40 ppm to lysine γ-CH_2 and alanine CH_3 protons. However, there are resonances which cannot be accounted for by the simulated spectrum, the most prominent example being the split resonance at 1.53 ppm, 1.51 ppm which has to be attributed to structural elements in the protein which are closely related to the metal clusters. This conclusion is supported by the pH-dependent NMR spectra of metallothionein. Thus, proton signals which are influenced by the binding of Zn (2.78 ppm, 2.20 ppm, 1.92 ppm resonances disappear) and Cd (1.51 ppm, 1.53 ppm, 1.04 ppm, 1.01 ppm, 0.91 ppm, 0.83 ppm resonances disappear, new signals at 0.92 ppm, 0.94 ppm, 0.96 ppm) could be assigned using the displacement of the metal ions with DCl. The cysteine β-CH_2 resonance at 2.95 ppm is shifted by 0.07 ppm to lower field upon binding of Cd. This downfield shift induced by the binding of diamagnetic metal ions was corroborated by ^{13}C NMR spectra of Cd (D-penicillamine) chelates. Downfield shifts were observed for D-penicillamine COOH (6.5 ppm), β-C (3.8 ppm) and the nonequivalent methyl groups (4.0 ppm, 2.4 ppm) in Cd(D-penicillamine) of 1:1 molar ratio (pH 7.1). We can therefore consider the downfield shift of the cysteine β-CH_2 protons as further direct proof of the coordination of the metal ions to cysteine residues in metallothionein. Furthermore, we can assume that the protonation of the cysteine residues involved in the binding of Zn in Cd,Zn-thionein (pH 3.9) does not result in a considerable change of the

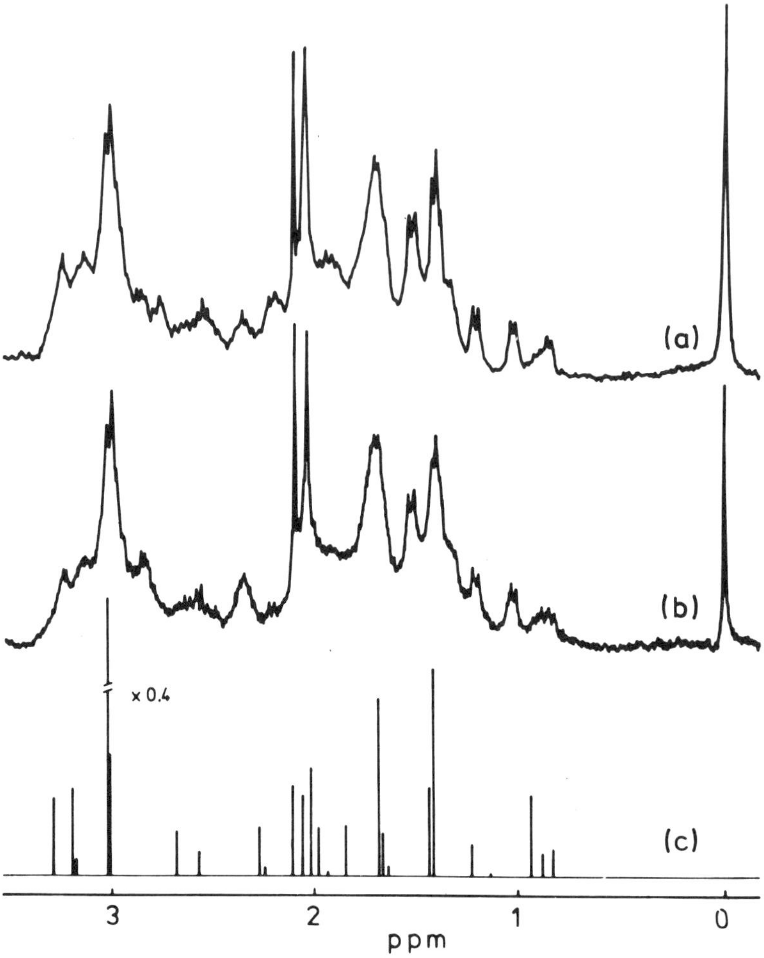

Figure 1: ^{1}H NMR spectra at 270 MHz of Cd,Zn-thionein in D_2O (30 mg/ml) at 25°C. pH-values are uncorrected for deuterium isotope effects and were adjusted with 6 N DCl. (a) pH. 6.7, (b) pH 3.9. Chemical shifts are measured with respect to sodium 2,2,3,3-tetradeutero-3(trimethylsilyl) propionate as internal standard. (c) simulated spectrum based on reported chemical shifts of amino acids and peptides (5) using the amino acid composition of chicken metallothionein (7,8). The cysteine resonance at 3.05 ppm is shown in a reduced scale.

protein structure. From the marked broadening of the 3.02 ppm signal of metallothionein at pH 3.9 we can conclude that the regid structure of the metal clusters is retained despite the dissociation of Zn.

OXIDATION-REDUCTION PROPERTIES OF Cu-THIONEIN

Spectral properties of simple complexes involving copper-sulphur coordination

Cu-thionein belongs to a class of proteins which cannot be attributed either to type I copper proteins (blue copper), type II (non-blue copper) or to type III as jugded from the known examples containing diamagnetic copper (9). In the blue copper proteins a cysteine residue, a methionine and two histidine residues are involved in the coordination of copper (10). Since Cu-thionein contains no histidine residues, the copper binding site is expected to be fundamentally different. Based on the apparant stoichiometey of 1 Cu to 2 cysteine residues, a tetrahedral site involving bridging cysteine sulphur was proposed (4). Both copper and the cysteine sulphur ligand atoms can participate in redox reactions. However, only part of the possible reactions may be involved in the biological activity. Some reactions may be found as intermediate states in the breakdown of the binding site during unfolding of the protein:

$$
[(RS^-)_n Cu(I)] \underset{+e}{\overset{-e}{\rightleftharpoons}} \left[\begin{array}{c} [(RS^-)_n Cu(II)] \\ \downarrow\uparrow \\ [(RS^\cdot)_n Cu(I)] \\ \downarrow\uparrow \\ [(RSSR)_{n/2}\ Cu(I)] \end{array} \right] \underset{+e}{\overset{-e}{\rightleftharpoons}} \left[\begin{array}{c} [(RS^\cdot)_n Cu(II)] \\ \downarrow\uparrow \\ [(RSSR)_{n/2} Cu(II)] \end{array} \right]
$$

fully reduced (a) — half oxidized (b) — fully oxidized (c)

Several possible oxidation states of an isolated copper-sulphur group of the binding site have to be considered. Although the geometry of the metal-sulphur cluster can be decisive in controlling the redox state, as has been shown for ferredoxins

(11), an evaluation of prominent spectral features of simple copper-sulphur complexes might help in understanding the redox reactions of Cu-thionein.

In the fully reduced form of Cu-thionein, both copper and sulphur are reduced, i.e.$(RS^-)_n Cu(I)$ (scheme 1a). Electron absorption and circular dichroism are not expected in the visible region, unless they are due to charge-transfer transitions. An absorption arising from d-d transitions is obviously not possible. Furthermore, the fully reduced state is diamagnetic. Copper can, therefore, not be detected by EPR.

The primary step of the oxidation of $(RS^-)_n Cu(I)$ may involve either the metal ions or the ligand atoms. The redox-active unit in Fe-S proteins is Fe(II)/Fe(III). Oxidation of the ligands RS^- or S^{2-} results in a breakdown of the metal cluster. According to this mechanism, oxidation of Cu-thionein might lead to Cu(II) coordinated to thiolate sulphur. The magnetic properties of $(RS^-)_n Cu(II)$ depend on the presence of antiferromagnetic coupling between the Cu(II) ions. For Fe-S centres in soluble Fe-S proteins (12), in mitochondria (13,14) or in chloroplasts (15) there is now strong evidence that a magnetic interaction between the centres can occur. Oxidation of RS^- would lead to a sulphur radical bound to Cu(I). Although $RS^{\cdot}$ occurs only as a transition state during the metal--catalyzed oxidation of simple thiols, it might be stabilized by a specific ligand geometry. In this state the unpaired electron would reside on the sulphur atom. Based on the assumption that the absence of a satellite structure in the XPS spectrum (Cu $2p_{1/2}$) indicates the presence of Cu (I) (16), i.e. 3 d^{10}, also in the case of non-innocent ligands, a $(RS^{\cdot})_n Cu(I)$ configuration was proposed for the violet chromophore of Cu--D-penicillamine (17,18). In addition, a $(RS^-)_n Cu(I)$ might also be present, but would be difficult to differentiate from $(RS^{\cdot})_n Cu(I)$ by XPS. The absence of dipolar coupling or exchange interaction in $(RS^{\cdot})_n Cu(I)$ would lead to an EPR spectrum which is characteristic for a sulphur radical with additional copper hyperfine interaction. A high spin density at the sulphur atom would result in a hyperfine splitting which is different to that of non-blue copper. It is probable that the hyperfine splitting constant is much smaller than 0.01 cm^{-1}. Dimerization of $(RS^{\cdot})_n$ of $(RS^{\cdot})_n Cu(I)$ to $(RSSR)_{n/2}$ would result in a diamagnetic $(RSSR)_{n/2} Cu(I)$ unit (19). Obviously, this step would not be detectable by EPR. By further oxidation the stable $(RSSR)_{n/2}Cu(II)$ is formed which might exhibit super exchange interaction of Cu(II) via the disulphide group(20).

Reduction-oxidation reactions of yeast Cu-thionein

As already pointed out, the EPR non-detectable Cu (>90% of total copper) in Cu-thionein could be attributed either to antiferromagnetically coupled Cu(II) ions or to diamagnetic Cu(I). A way for probing the oxidation state is by releasing copper with acid anaerobically as was shown for tyrosinase (21). Cu-thionein in 3N HCl under argon did not show an EPR signal due to released cupric copper(Fig.2). Upon aeration, a signal appeared corresponding to 95% of total copper. An oxidation of thiolate to disulphide according to the reaction $[(RS^-)_nCu(II)] \rightarrow n/2\ RSSR + Cu(I)]$ is not involved in the release of copper as was inferred from the UV-absorption of the apoprotein. We assign therefore the EPR silent copper to Cu(I) (22,23). This conclusion is supported by employing p-chloromercuribenzoate for displacing copper and blocking the thiolate group. The effect of oxygen is attributed to an oxidation of cuprous copper released by HCl or p-chloromercuribenzoate.

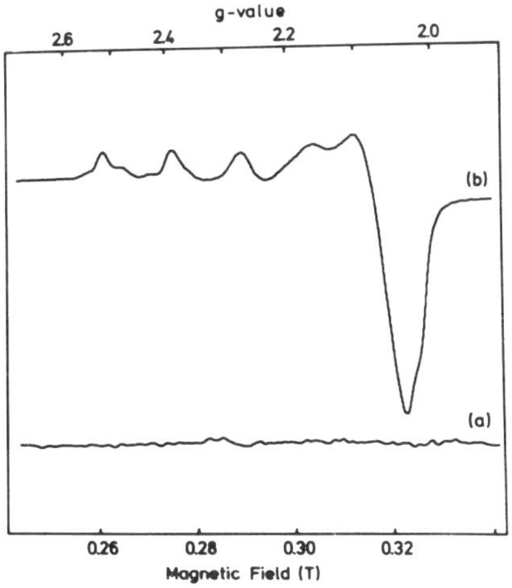

Figure 2: EPR spectra of Cu-thionein in 3N HCl under argon (a) and following admission of oxygen (b). Spectrum (b) corresponds to 95% of total copper present. Temperature 77°K:modulation 1 mT:microwave frequency 9.15 GHz:microwave power 20 mW.

Oxidation of Cu-thionein can result in different products, as shown for simple thiol compounds in scheme 1. Oxidation-reduction potential titrations proved to be most valuable. Cu-thionein was poised at a particular oxidation-reduction potential using a potentiometric procedure(23). Following transfer into quartz EPR tubes, the samples were frozen in liquid nitrogen. All steps were carried out under strictly anaerobic conditions. The amplitude or the integrated area of the EPR signal was determined as a function of the oxidation-reduction potential. Potentiometric titrations using hexachloroiridate (IV) or hexacyanoferrate (III) as oxidizing agents produced several different Cu(II) EPR signals at various stages of oxidation. The former oxidizing agent required a lower redox potential (+350 mV) to oxidize the copper, than the latter (+410 mV). However, the titrations were not fully reversible. The EPR signal from Cu(II) oxidized by hexachloroiridate(IV) resembled that produced by p-chloromercuribenzoate in air, suggesting that the copper was released from its thiolate ligands.

We can conclude that in the fully oxidized samples, both the metal ions and the ligand atoms are oxidized, i.e. $(RSSR)_nCu(II)$. It should be pointed out that EPR on its own does not allow to detect certain possible states, such as $(RS\text{-}SR)_nCu(I)$. Circular dichroism should prove an appropriate method for detecting disulphide groups (see also Rupp and Weser (4)). At present it is not clear whether the oxidation to $(RS\text{-}SR)_nCu(II)$ has any biological significance. It might well be that under physiiological conditions, oxidation reactions are restricted to the ligand atoms $((RS^-)_nCu(I) = (RSSR)_{n/2}Cu(I)$. Nevertheless, oxidation-reduction reactions of Cu-thionein may represent an effective means for controlling metal exchange. The high stability of the $(RS^-)_nCu(I)$ units (half of the metal ions dissociate at pH 0.4) makes it highly probable that metal exchange occurs via a second thiol compound, e.g. glutathione. Alternatively, oxidation of $(RS^-)_nCu(I)$, probably to $(RS^-)_n$-Cu(II) or $(RSSR)_{n/2}Cu(I)$ might provide a less stable bond, thus allowing exchange of the copper with compounds which have a lower affinity for Cu(I), such as free amino acids. Although Cu-thionein might turn out to be a redox enzyme, its high affinity for copper must be considered to be a decisive factor in the homeostasis of copper.

The oxidation-reduction properties of Cu-thionein are interesting also regarding the evolution of metalloproteins. According to Österberg (24) copper proteins could not evolve until the redox potential on earth was high enough to permit the existence of Cu(II). However, owing to the high stability of Cu in Cu-thionein arising from $(RS^-)_nCu(I)$-clusters, the sulphur-rich thionein is a strong chelator also for Cu(I). It is assumed that cysteine-rich polypeptides have reacted both with Fe-thiolate clusters, and with Cu(I), forming structures similar to Cu-thionein. In analogy to the rubredoxins and ferredoxins, which are thought to be the earliest representatives of the known iron-sulphur proteins, microbial Cu-thionein might be regarded as an evolutionary forerunner of the copper proteins.

REFERENCES

1. Bayley,P.M.(1973) in Progress in Biophysics and Molecular Biology(Butler,J.A.V. and Noble,D.,eds.), Vol.27,pp.3-76, Pergamon Press, Oxford.
2. Fasman, G.D., Hoving,H. and Timasheff, S.N.(1970) Biochemistry 9,3316-3324.
3. Rupp, H., Voelter, W. and Weser, U.(1975) Z.Physiol.Chem. 356, 755-765.
4. Rupp.H. and Weser, U.(1978) Biochim.Biophys.Acta 533,209-226
5. McDonald, C.C. and Phillips, W.D.(1969) J.Am.Chem.Soc.91, 1513-1521.
6. Rupp,H., Voleter,W. and Weser,U.(1974) FEBS Lett.40,176-179.
7. Weser, U., Donay,F. and Rupp.H.(1973) FEBS Lett.32,171-173.
8. Weser,U., Rupp,H., Donay,F., Linnemann,F., Voelter,W., Voetsch,W. and Jung,G. (1973) Eur.J.Biochem. 39, 127-140.
9. Malkin,R. and Malmström,B.G. (1970) in Advances in Enzymology (Nord,F.F.,ed.)33, pp.177-244, Interscience Publ., New York.
10. Colman, P.M., Freeman,H.C., Guss,J.H.,Murata,M., Norris, V. A., Ramshaw, F.A.M. and Venkatappa,M.P.(1978) Nature 272, 319-324.
11. Carter,C.W.(1977) in Iron-Sulphur Proteins (Lovenberg,W., ed.) Vol.3, pp.157-204, Academic Press, New York.
12. Rupp,H., Rao,K.K., Hall,D.O. and Cammack,R.(1978) Biochim. Biophys.Acta, in press.

13. Moore,A.L. and Rupp,H.(1978) FEBS Lett.in press.
14. Rupp,H. and Moore,A.L.,Biochim.Biophys.Acta, submitted.
15. Rupp.H., de La Torre,A.and Hall,D.O., Biochim.Biophys.Acta submitted.
16. Rupp,H. and Weser,U.(1976) Bioinorganic Chem.6, 45-59.
17. Rupp,H. and Weser,U.(1976) Biochem.Biophys.Res.Commun.72, 223-229.
18. Rupp,H. and Weser,U.(1976) Biochim.Biophys.Acta 446,151-165.
19. Ottersen,T., Warner,L.G. and Seff, K.(1974) Inorg.Chem. 13, 1904-1911.
20. Kroneck,P. (1975) J.Am.Chem.Soc. 97,3839-3841.
21. Makino,N., McMahill,P., Mason,H.S. and Moss,T.H. (1974) J. Biol.Chem.249, 6062-6066.
22. Rupp,H. and Weser, U. (1974) FEBS Lett. 44,293-297.
23. Rupp,H., Cammack,R., Hartmann,H.J. and Weser,U. Biochem.J. submitted.
24. Österberg, R. (1974) Nature 249, 382-383.
25. Supported by grants from the Deutsche Forschungsgemeinschaft (We401/14) and by NATO Res.Programme (No.1256).

Metallothionein
Jeremias H. R. Kägi and Monica Nordberg, eds.
Birkhäuser Verlag, Basel/Boston/Stuttgart, 1979

THE ^{1}H N.M.R. SPECTRA OF METALLOTHIONEINS

Alphonse Galdes and H.Allen O.Hill
Inorganic Chemistry Laboratory, Oxford, England.
Jeremias H.R. Kägi and Milan Vašák
Institute of Biochemistry, University of Zurich, Switzerland.
Ian Bremner and Brian W.Young
The Rowett Research Institute, Blacksburn, Aberdeen, Scotland.

Nuclear magnetic resonance (n.m.r) spectroscopy is one of the few techniques with the inherent ability to sense each and every part of a protein molecule since magnetic nuclei (^{1}H, ^{13}C, ^{14}N and ^{15}N) are obligatory constituents thereof. The resonance of a given nucleus in a protein is sensitive to both short- and long range effects and consequently can yield information on the primary, secondary, and tertiary structures of the protein, particularly the first and last of these. In this paper we consider the results derived from the investigation of the ^{1}H n.m.r. spectra (1,2) of metallothioneins isolated from equine liver and equine kidney (3,4) and sheep liver (5).

COMPARISON OF APOTHIONEIN AND METALLOTHIONEINS

General features

The ^{1}H n.m.r. spectra of equine liver metallothionein is shown in Figure 1. The holo-protein contains 98% zinc. The most striking observation is that the spectrum of apothionein is practically identical to that of the amino-acid mixture which means that each residue must have the same environment as the corresponding free amino-acid. This is consistent with apothionein having a random-coil configuration but does not exclude other configurations having short-range order yet with each residue essentially fully solvated. The holo-protein gives rise to a related but different spectrum. For example whilst the valinyl and threonyl residues are essentially unperturbed, implying that they may still be exposed to solvent in the holo-protein i.e. on the 'outside' of the protein, there is consider-

Submitted to the "First International Meeting on Metallothionein and Other Low Molecular Weight Metal-binding Proteins", Zürich, July 17 - 22, 1978

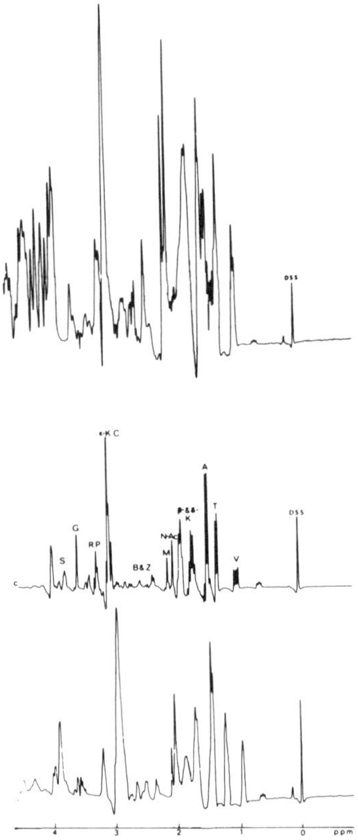

Fig.1. The ^1H n.m.r. spectra at 270 MHz in the high-field region of equine thioneins in 20mM phosphate buffer at pH*7.5 in ^2H_2O. The spectra are of: Apothionein (lower trace); a mixture of amino acids of the same composition as the protein; metallothionein.

able perturbation of the alanyl residues. The ^1H n.m.r. spectra of most native proteins are quite different from the unfolded forms. This difference is caused primarily by the ring currents of aromatic residues. As expected the resonances corresponding to aromatic residues are conspicuously absent as shown in Figure 2. The resonances of the peptide and amide N<u>H</u> protons, expected in this region, are also absent having exchanged with the solvent deuterons. They are visible in the spectrum of the apothionein dissolved in water. The appearance of more than one resonance implies the presence of a remaining short-range order which may e.g. be due to intramolecular hydrogen bonding

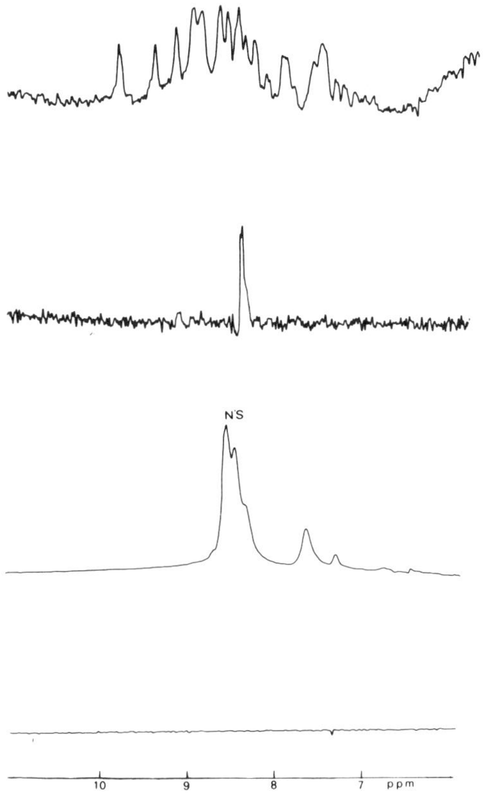

Fig.2. The ^{1}H n.m.r. spectra of equine liver thionein (conditions as in Fig.1.) in the low-field region. The spectra are of, in ascending order: Apothionein in $^{2}H_2O$; Apothionein in $^{1}H_2O$; metallothionein in $^{2}H_2O$; metallothionein in $^{1}H_2O$.

of the peptide group and neighbouring serinyl or cysteinyl residues. This short range order is lost at higher temperatures. Figure 3 shows the ^{1}H n.m.r. spectra of another sample of equine liver apothionein at different temperatures. With increasing temperature the resonances coalesce reflecting a temperature-dependent exchange process, which is reversible. This illustrates the significant thermal stability of the protein.

The spectrum of the holoprotein in $^{2}H_2O$ shows only one remaining resonance which means that part of the protein is inaccessible to solvent. The most striking observation is that of the large number of resonances observed in the spectrum of metallothionein measured in $^{1}H_2O$. This is only consistent with the existence of a well-defined tertiary structure for the holoprotein.

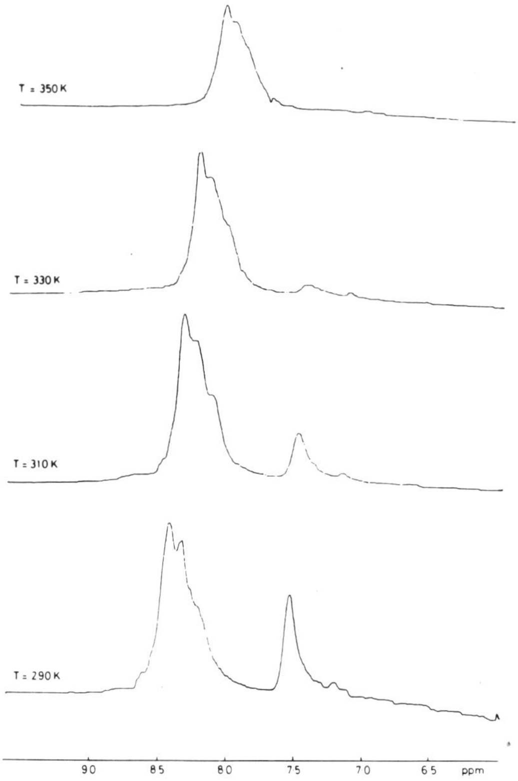

Fig.3. The ^1H n.m.r. spectrum (low-field region) of equine liver apothionein (conditions as in Fig.1) at different temperatures.

Structural variation

The ^1H n.m.r. spectra reveal differences in structure,dependent on source and metal-content of the metallothionein. For example, a protein, isolated from equine kidney and containing 55% cadmium and 45% zinc has (Figure 4) a ^1H n.m.r. spectrum in the high-field region consistent with its amino-acid composition. The low-field spectra in ^2H_2O and ^1H_2O are consistent with the conclusions reached above, though resonances of a greater number of unexchanged peptide N<u>H</u>'s are apparent. This exchange process is time-dependent, as shown by the ^1H n.m.r. spectra of the metallothionein from sheep liver. There are obviously even more slowly-exchanging peptide NH's in this protein, some being particularly inaccessible to solvent (Figure 5).

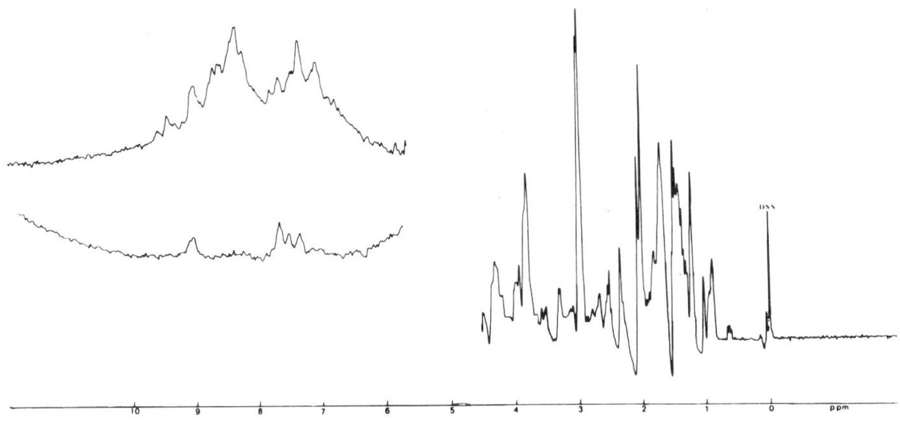

Fig.4. The ^{1}H n.m.r. spectra at 270 MHz of thioneins isolated from equine kidney (conditions as in Fig.1.). The spectra are of the holoprotein in $^{2}H_2O$ and, upper trace, in $^{1}H_2O$.

Metal-binding

It is generally agreed that the cysteinyl residues act as ligands to the constituent metal ions. Confirmation (1) of this role is strikingly illustrated in Figure 6. Whilst most of the resonances are not markedly perturbed, that associated with the cysteinyl residues is much affected. Though our analyses of these spectra are not yet complete they are consistent with a freely-rotating C_α-C_β bond in the cysteinyl residues of the apo-thionein but with this rotation strongly hindered in the metallo-thionein, as expected if the cysteinyl residues act as ligands.

Preliminary ^{1}H n.m.r. experiments reveal the loss of metal ions from the protein on decreasing the pH. The chemical shift of one of the peptide NH resonances has the pH-dependence shown in Figure 7. This behaviour, which is completely reversible, would be that expected for a group having a pK_a of approx. 4. A similar phenomenon is found on the pH-behaviour of sheep metallothionein. We interpret this as reflecting the effect of the titration of the carboxyl of the C-terminyl alanyl residue on the neighbouring peptide nitrogen i.e. C(O)N$\underline{H}$-CH(CH_3)COOH. Our analysis of the effect of pH on the spectrum of equine kidney metallothionein (59% Cd, 41% Zn) shows that approx. 90% of the zinc ions and less than 5% of the cadmium ions are released by pH4; all of the cadmium is released by pH2. The ^{1}H n.m.r. spectra gradually change from that typical of the holoprotein to that of the apo-protein as the pH* is lowered below 6. When the pH is again increased from pH*<2 the spectrum of the form of the holoprotein reappears. The metal ions are at all times in slow exchange (by n.m.r. criteria). The spectrum of the holoprotein formed by lowering, then raising the pH* is similar, but <u>not</u>

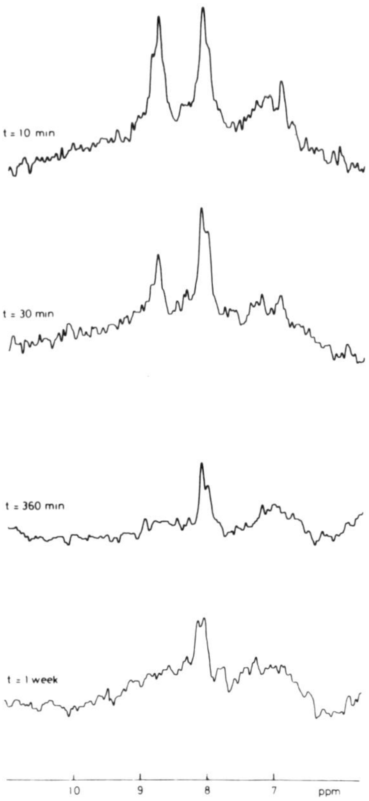

Fig.5. The ^{1}H n.m.r. spectra at 270 MHz of sheep liver metallothionein in 20mM phosphate buffer at pH*7.4 and 300°K. The time after dissolution of the protein is indicated.

identical, to that of the original protein. It is possible, therefore, that the metal ions, Cd^{2+} and Zn^{2+}, bind randomly to the apothionein and that there may not be 'cadmium-preferred' and 'zinc-preferred' sites.

In conclusion, we have shown that whereas apothionein might have a 'random-coil' structure there is every indication that the metallothioneins have well-defined tertiary structures in which the cysteinyl residues act as ligands. The proteins are probably compact with at least part of the interior relatively inaccessible to solvent. The structures of metallothioneins from different sources and/or containing different metal ions, whilst similar, show subtle differences.

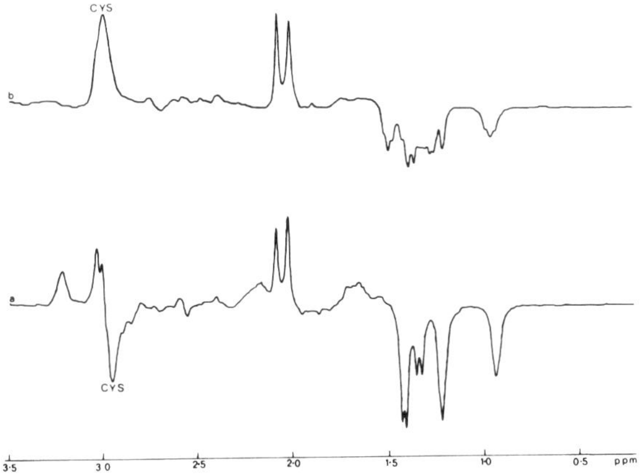

Fig.6. The spin-echo spectrum (1) of equine liver metallothionein measured using a pulse sequence 90-τ-180-τ [collect], with τ=71ms; (a) apothionein (b) metallothionein.

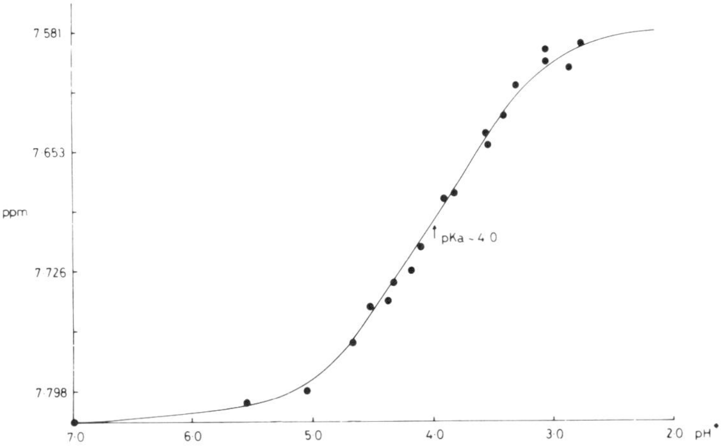

Fig.7. The pH-dependence of a resonance corresponding to a peptide NH in the ^{1}H n.m.r. spectrum at 270MHz of equine kidney metallothionein (59% Cd; 41% Zn) in $^{1}H_2O$.

ACKNOWLEDGEMENTS

We thank E.M.B.O. for a short term Fellowship to M.V. and the Rhodes Trustees for a Scholarship to A.G. H.A.O.H. is a member of the Oxford Enzyme Group.

REFERENCES

1. Galdes, A., Vašák, M., Hill, H.A.O., and Kägi, J.H.R., (1978) FEBS Lett. 92, 17-21.
2. Galdes, A., Hill, H.A.O., Bremner, I., and Young, B.W., to be submitted
3. Kägi, J.H.R., Himmelhoch, S.R., Whanger, P.D., Bethune, J.L., and Vallee, B.L., (1974) J.Biol.Chem., 249, 3537-3542.
4. Kojima, Y., Berger, C., Vallee, B.L., and Kägi, J.H.R., (1976), Proc.Natl.Acad.Sci. U.S.A., 73, 3413-3417.
5. Bremner, I., and Young, B.W., (1976) Biochem.J., 155, 631-635.

Metallothionein
Jeremias H. R. Kägi and Monica Nordberg, eds.
Birkhäuser Verlag, Basel/Boston/Stuttgart, 1979

STRUCTURAL CHARACTERIZATION OF METALLOTHIONEIN BY MULTINUCLEAR NMR

James D. Otvos and Ian M. Armitage
Department of Molecular Biophysics and Biochemistry
Yale University
New Haven, Connecticut 06510

The application of ^{1}H, ^{13}C, and ^{113}Cd Fourier transform NMR to the study of Cd,Zn metallothionein offers the promise of providing the detailed structural information concerning the chemical nature of the metal binding sites which has heretofore been lacking. Among the several factors which make the study of this protein by multinuclear NMR particularly attractive are its ready availability in relatively large amounts from animals injected with cadmium and its small molecular weight, which should allow reasonably well-resolved high field natural abundance ^{13}C and ^{1}H spectra to be obtained. ^{1}H NMR spectra of metallothionein at 270 MHz have been reported and show several spectral changes to occur in the aliphatic region induced by metal ion removal at low pH (1). Unfortunately, assignment of these spectral perturbations to structural events occurring at specific sites in the protein is difficult due to the inability to resolve individual resonances from the large number of overlapping signals confined to the relatively narrow ^{1}H chemical shift range. The prospects for resolution and assignment of individual ^{13}C resonances are much better, however, due to the wide chemical shift range of the ^{13}C nucleus and the absence of spin-spin splitting when ^{1}H decoupling is employed. Recent work at high magnetic field strengths has demonstrated that even in the aliphatic region, numerous resolved single carbon resonances are obtained in ^{13}C NMR spectra of proteins larger than metallothionein (2,3). The problem of the much lower sensitivity of ^{13}C compared to ^{1}H is not a serious one for metallothionein since concentrations $>$ 10 mM can be routinely obtained.

The direct NMR observation of the metal nucleus itself potentially offers

Submitted to the "First International Meeting on Metallothionein and Other Low Molecular Weight Metal-binding Proteins", Zürich, July 17 - 22, 1978

the best probe of the nature of the metal ion binding sites in metallothionein. The ^{113}Cd nucleus (12.3% natural abundance) has spin 1/2 and a sensitivity which has allowed its detection in model complexes in aqueous solution as well as in several large metalloenzymes where the $^{113}Cd^{2+}$ ion was substituted for the native Zn^{2+} (4-7). The chemical shift range of over 700 ppm for common cadmium compounds makes this nucleus an extremely sensitive probe of the nature and coordination geometry of its ligands. Furthermore, dynamic information (ligand exchange rates, metal binding constants, solvent access) not obtainable by other methods should be accessible by ^{113}Cd NMR.

In this report, natural abundance ^{13}C and ^{113}Cd NMR spectra of rabbit liver metallothionein-II are presented. These results are preliminary and serve primarily to illustrate the potential of the multinuclear NMR approach to the study of metallothionein.

Materials and Methods

Rabbit liver metallothioneins were isolated from 6 rabbits which were injected with $CdCl_2$ over a 6 week period (10 mg $CdCl_2$/kg body wt.) and killed 2 weeks after the final injection. The isolation procedure of Kägi et al.(8) was used with the final separation of metallothioneins I and II being carried out by ion-exchange chromatography on DEAE-cellulose using a gradient of 0.02M to 0.15M Tris, pH 8.6. Pure metallothionein-II was concentrated for NMR in an Amicon ultrafilter using UM-2 membranes. 20% D_2O was added to provide the field-frequency lock. The Cd^{2+} and Zn^{2+} concentrations of the final solution determined by atomic absorption were 75.0 and 17.5 mM, respectively. The metallothionein concentration was therefore 13.3 mM assuming 7 eq Me^{2+} are bound to the protein. Proton decoupled ^{13}C NMR spectra were obtained at 67.9 MHz on a Bruker HX-270 spectrometer using 10 mm sample tubes containing 1.3 ml of sample. Chemical shifts are relative to external TMS. ^{113}Cd spectra were obtained on an extensively modified Bruker HFX 90 spectrometer operating at 19.96 MHz (7). Chemical shifts are relative to the resonance position of 0.1M $CdClO_4$.

Results and Discussion

^{13}C NMR spectrum

The ^{13}C NMR spectrum of rabbit liver metallothionein-II consists of 2 groupings of resonances. The region between 170 and 183 ppm contains the carboxyl and carbonyl group resonances while the region between 10 and 70 ppm contains all the aliphatic carbon signals. As expected, the aromatic region of the spectrum between 90 and 170 ppm is devoid of resonances except for several small peaks near 130 ppm which correspond to the presence of a phenylalanine residue, despite the fact that the protein is not reported to contain this amino acid (9). According to the tentative amino acid sequence of rabbit liver metallothionein-II (9), the carbonyl region of the spectrum should contain resonances corresponding to 61 backbone carbonyl carbons, 2 amide carbonyls, and 5 carboxyl carbons. As seen in Figure 1, the majority of these resonances are at least partially resolved. Of particular interest are the 5 resonances downfield of 178 ppm which are separated from the bulk of the carbonyl signals. From studies on other proteins and peptides (10,11), backbone carbonyl carbons are always located upfield of 177 ppm, while the downfield region is occupied by the carboxyl carbons of aspartate (177.0-178.4 ppm), glutamate (181.6-182.8 ppm), and C-terminal residues (180.0-181.5 ppm) as well as the carbonyl carbon of glutamine (178.7-178.9 ppm). Thus, from chemical shift considerations alone, the 3 resonances at 178.7, 179.0 and 179.4 ppm can tentatively be assigned to the single glutamine residue and 2 of the 3 aspartate residues in the protein, respectively, though the aspartates are somewhat downfield from their usual position. Similarly, the resonance at 181.0 ppm must arise from the C-terminal alanine residue while the most downfield resonance at 182.5 ppm can be assigned to the single glutamic acid residue at position 23, which Kimura et al. (9) found to be a site of possible microheterogeneity. All of these tentative assignments can easily be verified by pH titration studies since aspartate, glutamate, and C-terminal residues exhibit characteristic titration shifts while glutamine does not titrate (3,10,11).

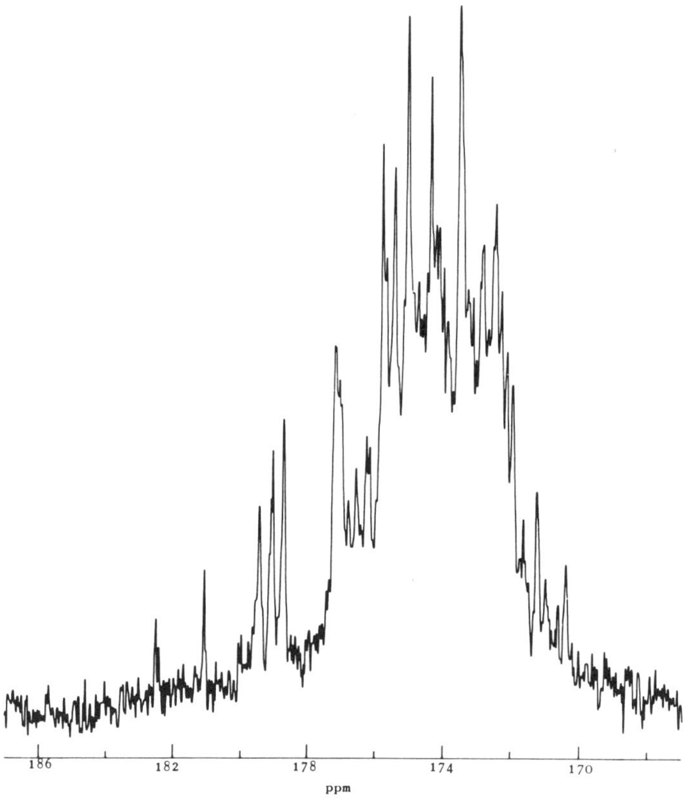

FIGURE 1: Carboxyl and carbonyl region of the natural abundance ^{1}H noise-decoupled ^{13}C NMR spectrum at 67.9 MHz of rabbit liver metallothionein-II (∿13 mM), pH 8, T=25°C. 120,000 transients were accumulated with an acquisition time of 0.58 sec. (total time = 19.5 hrs).

The aliphatic region of the spectrum is shown in Figure 2. Again, excellent resolution of individual resonances is observed. By using reported chemical shift values of amino acids in peptides and proteins in conjunction with the tentative sequence, it is possible to make many preliminary assignments of resonances to individual groups of carbon atoms. For example, the region between 44 and 62 ppm contains all 61 α carbon resonances as well as the 9 serine β carbons. Unfortunately, due to the intense resonances derived from the Tris buffer at 62.3 and 60.7 ppm, the serine β carbons are obscured. The prominent resonance at 54.0 ppm probably arises from the 20 cysteine α carbons. Of the methylene carbons which appear further upfield, the prominent peaks at 40.7, 30.5, 27.4, and 23.1 ppm most likely correspond to the ε, β, δ, and γ carbons of the 8 lysine residues, respectively. Again, these tentative assignments can be easily verified by pH titration. Since cysteine β carbons should give rise to a resonance at about 27.6 ppm, it is possible that the 27.4 ppm resonance also contains intensity from the 20 cysteine β carbons. However, Cd^{2+} and Zn^{2+} ligation to the cysteine sulfur in glutathione has been reported to lead to downfield C_β shifts of about 3 ppm, suggesting that similar shifts might be expected in metallothionein. If this is the case, there is no prominent resonance near 30 ppm which would account for 20 carbon atoms, suggesting there may be considerable restriction in mobility and/or heterogeneity in the environments of the cysteine residues. Finally, the spectral region between 14 and 21 ppm contains the methyl carbons belonging to the threonine, alanine, isoleucine, and methionine residues. Resolved resonances for each of these 13 methyl groups are observed, leading to the possibility of individual assignments such as was recently reported for BPTI (3).

All of the above tentative resonance assignments have been made solely on the basis of chemical shift comparisons with literature values. Given the large number of well-resolved resonances, however, it should not be difficult to make definitive assignments once additional ^{13}C spectral data on metallothionein is obtained as a function of pH, selective ^{1}H decoupling, metal ion content, and chromatographic species. Once these assignments have been made, it should be possible by chemical shift criteria and relaxation behavior to identify the non-cysteine metal ligands as well as to assess the function of the ubiquitous lysine, arginine, and serine residues in the protein.

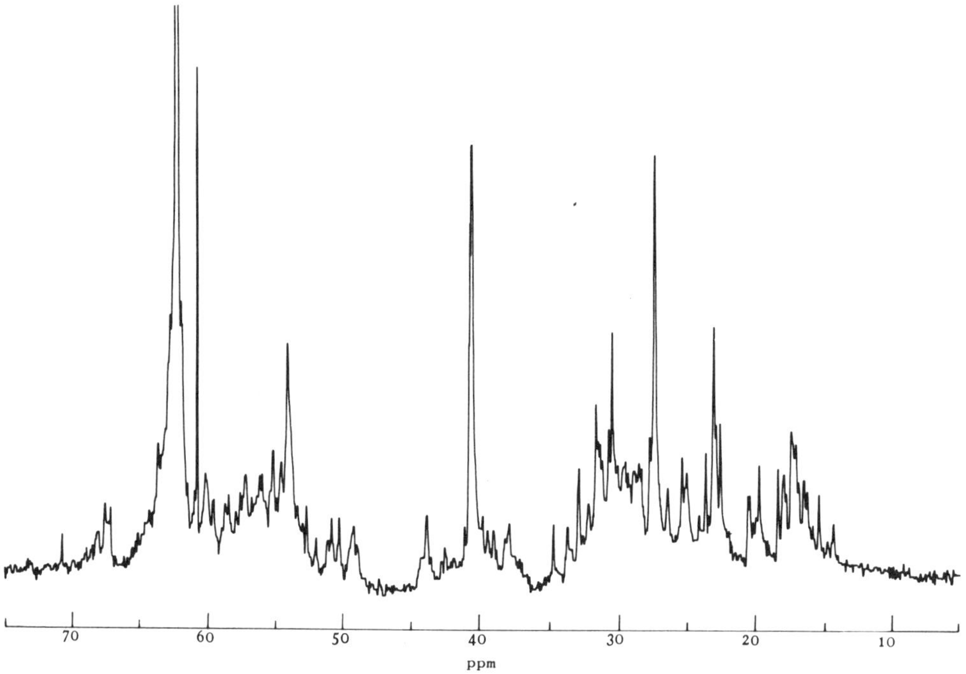

FIGURE 2: Aliphatic region of the natural abundance ^{1}H noise-decoupled ^{13}C NMR spectrum of rabbit liver metallothionein-II. Conditions were the same as in Fig. 1.

^{113}Cd NMR spectrum

The natural abundance ^{113}Cd NMR spectrum of the same metallothionein sample as was used for the ^{13}C NMR spectrum is shown in Figure 3. Despite the relatively low signal to noise ratio, 4 resonances are reproducibly observed at 614, 624, 643, and 668 ppm. From the chemical shift values of these resonances alone, several inferences may be made concerning the nature of the Cd^{2+} binding sites. First, the fact that the ^{113}Cd shifts are all located downfield of 600 ppm immediately indicates that each Cd^{2+} ion is liganded to at least 2 and most likely 3 or more sulfur ligands. A more exact assessment of the extent and geometry of sulfur ligation is difficult at this time due to the lack of cadmium chemical shift data on well-characterized Cd-thiolate model complexes. 1:2 Cd:glutathione complexes have been reported to give ^{113}Cd resonances at 318 (5) and 527 ppm (6) while a 1:4 complex at alkaline pH gave a shift of 610 ppm (6). The well-defined $[Cd_{10}(SCH_2CH_2OH)_{16}]^{4+}$ complex exhibits ^{113}Cd shifts of 380 and 520 ppm for the 2 classes of Cd atoms liganded to 3 sulfur and 4 sulfur ligands, respectively (5). However, an exact correspondence between the degree of deshielding and extent of sulfur ligation is complicated in this case by the fact that each sulfur ligand is shared by two or more cadmium ions. Based on this data, it is considered unlikely that the metal ions in metallothionein are arranged in clusters which involve extensive cysteine sulfur bridging. It is not possible, however, to distinguish separated metallothionein metal binding sites consisting of 3 nonbridging cysteine sulfur ligands from sites containing 4 sulfur ligands, one of which bridges 2 adjacent cadmium ions.

Another interesting aspect of the ^{113}Cd metallothionein spectrum is the presence of at least 4 resonances indicating that the 5 to 6 Cd^{2+} ions per protein molecule occupy at least 4 chemically distinct environments. Due to the great sensitivity of the ^{113}Cd shift to a variety of environmental perturbations, it cannot be stated at this time whether the distinct cadmium resonances arise from differences in the extent of sulfur ligation, differences in the nature of the non-sulfur ligands, heterogeneity in protein populations due to different relative Zn^{2+} and Cd^{2+} contents, or to differences in the Cd^{2+} environments due to protein tertiary structure. We hope to be able to distinguish these possibilities by further study of appropriate model compounds and metallothionein species containing different relative amounts of Zn^{2+} and Cd^{2+}. Spectral interpretation should also be greatly aided by the 8-fold increase in sensitivity which can be achieved by substituting > 96% enriched $^{113}Cd^{2+}$ for natural abundance Cd^{2+} - either _in vitro_ or _in vivo_ (by injecting the animals with $^{113}CdCl_2$).

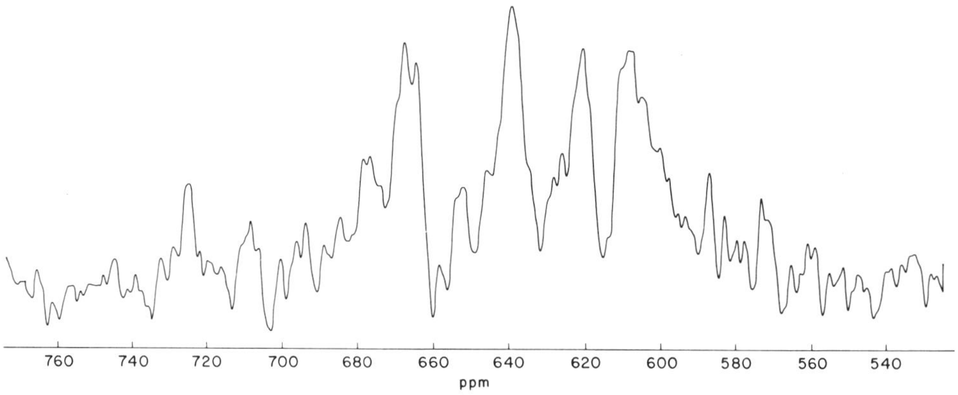

FIGURE 3: Natural abundance ^{113}Cd NMR spectrum at 19.96 MHz of rabbit liver metallothionein-II (∿13 mM), pH 8, T=25°C. 280,000 transients were accumulated with acquisition time = 0.20 sec and pulse delay = 0.30 sec (total time = 39 hrs)

Acknowledgement

This work was supported by NIH Grant ES 01674 from the National Institute of Environmental Health Sciences and NIH Grant 18778. The high field ^{13}C NMR work was supported by NIH Grant PR 00798 from the Division of Research Resources. J.D.O. wishes to thank Bruker Instruments, Inc. for travel support.

References

1. Rupp, H., Voelter, W., and Weser, V. (1974) FEBS Lett. 40, 176-179.
2. Norton, R.S., Clouse, A.O., Addleman, R., and Allerhand, A. (1977) J. Amer. Chem. Soc. 99, 79-83.
3. Richarz, R. and Wüthrich (1978) Biochemistry 17, 2263-2269.
4. Cardin, A.D., Ellis, P.D., Odom, J.D., and Howard, J.W., Jr. (1975) J. Amer. Chem. Soc. 97, 1672-1679.
5. Haberkorn, R.A., Que, L., Jr., Gillum, W.O., Holm, R.H., Liu, C.S., and Lord, R.C. (1976) Inorg. Chem. 15, 2408-2414.
6. Birgersson, B., Carter, R.E., and Drakenberg, T. (1977) J. Magn. Resonance 28, 299-302.
7. Armitage, I.M., Schoot Uiterkamp, A.J.M., Chlebowski, J.F., and Coleman, J.E. (1978) J. Magn. Resonance 29, 375-392.
8. Kägi, J.H.R., Himmelhoch, S.R., Whanger, P.D., Bethune, J.L., and Vallee, B.L. (1974) J. Biol. Chem. 249, 3537-3542.
9. Kimura, M., Otaki, N., and Imano, M. (1979)This volume, p. 163-168.
10. Shindo, H. and Cohen, J.S. (1976) Proc. Natl. Acad. Sci. USA 73, 1979-1983.
11. Howarth, O.W. and Lilley, D.M.J. (1978) Progr. NMR Spectroscopy 12, 1-40.

METABOLISM OF METALLOTHIONEIN

Metallothionein
Jeremias H. R. Kägi and Monica Nordberg, eds.
Birkhäuser Verlag, Basel/Boston/Stuttgart, 1979

IN VIVO AND *EX VIVO* INDUCTION OF RAT LIVER METALLOTHIONEIN

Frank O. Brady, Myna Panemangalore, and Frances A. Day
Division of Biochemistry, Physiology, and Pharmacology
University of South Dakota School of Medicine
Vermillion, South Dakota, USA 57069
and
A.M. Fiskin and G. Peterson
VA Hospital, Kansas City, Missouri, USA

Metallothioneins are low molecular weight, inducible cytoplasmic proteins synthesized in certain eucaryotic tissues in response to transition metal ions. Most effective as inducers of metallothionein in rat liver are cadmium and zinc, although mercury and silver (1,2) also are effective. Copper has been implicated in the synthesis of another low molecular weight protein, chelatin, in both liver and kidney (3), although there are reports of copper thionein induction in pig (4) and rat liver (5), as well as its purification from human fetal liver (6). Under normal conditions, zinc thionein has been detected in rat testis (7), in horse liver (8), and in human liver (9).

Metallothionein has been implicated as functioning in cadmium and mercury detoxification and in zinc homeostasis. As the latter function is probably the most fundamental physiologically, the efforts of our laboratory have been directed at elucidating the biochemical and physiological factors involved in the induction, biosynthesis, metabolism and function of metallothionein. In addition, we have made some parallel studies on copper and chelatin.

IN VIVO INDUCTION

In these studies relatively low, subtoxic dosages of metal, administered as a single injection intraperitoneally, were used to induce metallothionein and chelatin. Animals were killed at appropriate times after injection, their livers were excised, homogenized, and a high speed supernatant was chromatographed on a column of Sephadex G-75. Fractions were collected and analyzed and quantitated for metal content by atomic absorption spectroscopy. Results were calculated and expressed as micrograms metal present in metallothionein or chelatin per gram wet weight of liver. Metallothionein and chelatin were differentiated by their behavior when subjected to DEAE-cellulose chromatography and polyacrylamide gel electrophoresis.

Submitted to the "First International Meeting on Metallothionein and Other Low Molecular Weight Metal-binding Proteins", Zürich, July 17 - 22, 1978

Our results (10,11) are summarized in Table I for the induction of rat liver metallothionein and chelatin by cadmium, zinc, copper, and mercury salts.

Table I

Summary of induction of rat liver metallothionein and chelatin

Treatment	Dosage (mg/kg b.w.)	ZnMT peak	ZnMT $t_{1/2}$	CdMT peak	CdMT $t_{1/2}$	Cu-Chelatin peak	Cu-Chelatin $t_{1/2}$	Ref.
Zn(II)	5	18hr	10.1hr					(10)
Cu(II)	0.5	9hr	18.2hr			9hr	8.6hr	(10)
Cd(II)	0.25	24hr	4.53d	18hr	6.80d			(10)
Hg(II)*	0.5	18hr	9.4hr					(11)

*No mercury was detected in the zinc thionein obtained from these animals.

Several interesting observations were obtained in these studies. Zinc thionein was inducible in rat liver by the administration of each of the four metals, the protein having a relatively rapid turnover ($t_{1/2}$ < 20hr) in all cases except for cadmium ($t_{1/2}$ = 4.53d). Induction by cadmium resulted in the synthesis of a protein containing both cadmium and zinc, while induction by mercury resulted in the synthesis of zinc thionein in addition to the expected synthesis of copper chelatin. Zinc thionein, but not cadmium thionein or copper chelatin in all cases was super-inducible by the administration of actinomycin D (1 mg/kg b.w.) at the peak of induction, compared with induced animals not receiving the drug.

Zinc homeostasis and status must be of central importance in the induction of these proteins. If it is true that exogenously administered copper, cadmium, and mercury are disturbing the endogenous pool of zinc, then the characteristics of induction of zinc thionein observed in these animals could be explained by the need for copper, cadmium, and mercury concentrations in the liver first to become and then to remain elevated, possibly resulting in an increase in the level of intracellular free zinc. The final metal content of the induced metallothionein may only reflect the relative binding affinities for thionein and the abundance of those metals present.

EX VIVO INDUCTION

As a first step toward determining whether or not a fluctuation in intracellular zinc levels is an important mechanism in the induction of metallothionein, we have chosen to study the induction and synthesis of metallothionein *ex vivo* in the isolated perfused rat liver (12). In using this technique we have attempted to exclude the confusion that could arise from fluctuations in the endogenous metal pool which could lead to the migration of metal ions from peripheral tissues into the blood and thus to the liver. The responses obtained in the *ex vivo* system, we hope, simulates the *in vivo* responses and is thus preferable to an *in vitro* system.

The technique of isolated liver perfusion has been described in detail and reviewed elsewhere (13,14). The method we followed was that developed by Miller (15). A commercially manufactured perfusion apparatus was used in these studies (MRA Corp., Clearwater, Florida, USA). The perfusion medium, suitable for obtaining net protein synthesis in the perfused liver (14), was freshly prepared before use. It consisted of 250 mg glucose, 2 g bovine serum albumin (Fraction V), 6.3 mg heparin, 7.7 mg L-cysteine-HCl, 312.3 mg amino acid mixture (16), 5.1 units insulin, and 5 mg hydrocortisol in 64 ml of Krebs-Ringer bicarbonate buffer (pH 7.4) to which were added 35 ml of washed erythrocytes prepared from outdated human blood (15). The perfusion medium was continuously recirculated and oxygenated with humidified 95% O_2/5% CO_2.

To determine if metallothionein could be induced by Cd (II) in the isolated perfused liver, livers from normal stock diet fed rats were perfused with exogenous Cd(II) for a period of 150min following equilibration of the livers for 30min. The results of a dose response experiment are presented in Figure 1. All of the Cd(II) was absorbed by the perfused livers within the first 15min. As can be seen, the amount of Cd(II) in metallothionein increased with increasing dosage of Cd(II) added to the perfusion medium, saturation occurring at 75 µg. The Zn(II) in the metallothionein of Cd(II)-exposed livers decreased to about 60% of the level found in control livers perfused without Cd(II), and this level was not altered as the amount of Cd(II) in the perfusate was increased.

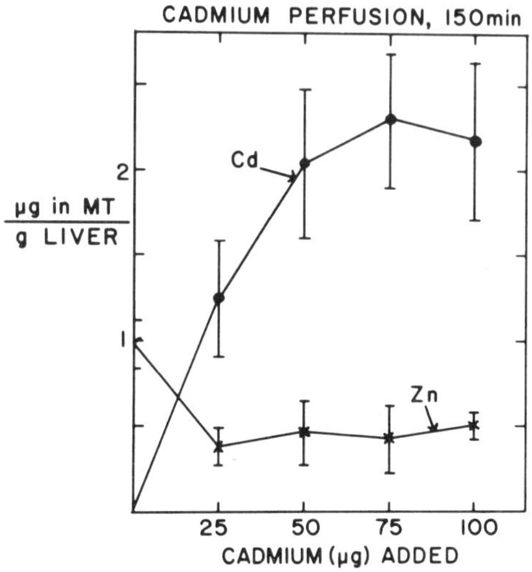

Figure 1. Induction of metallothionein, cadmium dose response curve (12). Cadmium (●) and zinc (X) contents in metallothionein (µg/g liver) from isolated perfused rat livers presented as a function of increasing dosages of Cd(II) added to the perfusion media. Data are expressed as the means ± standard error for four perfusions.

The results of a time course of induction of metallothionein by 75 µg Cd(II) in isolated perfused livers is presented in Figure 2. The amount of Cd(II) in metallothionein increased with the length of time of perfusion, and it was still increasing at 150min though somewhat more slowly than initially. After an initial decrease at 30min the Zn(II) in metallothionein of Cd(II)-exposed livers perfused for 90min and 150min increased slightly, but this increase was not statistically significant. In this experiment we also observed that the amount of Cd(II) bound to high molecular weight proteins in the liver cytosol decreased as the duration of perfusion increased, and more Cd(II) was found associated with metallothionein, reflecting the transfer of Cd(II) from high molecular weight proteins to metallothionein as the time of perfusion increased. All of the administered Cd(II) in these two experiments could be accounted for in the high speed supernatants plus the excreted bile.

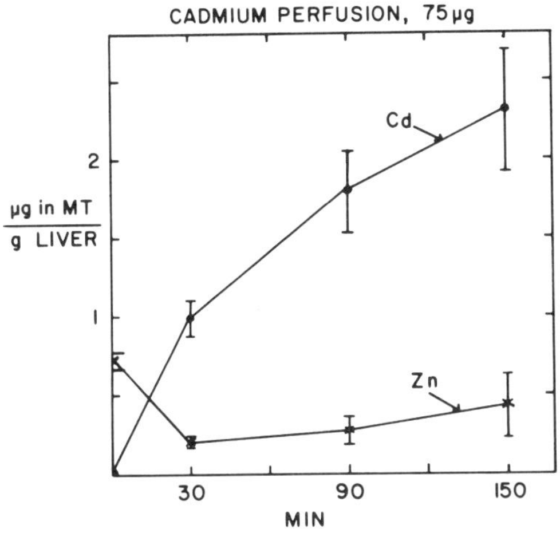

Figure 2 (12). Induction of metallothionein, time course curve. Cadmium (●) and zinc (X) contents in metallothionein (μg/g liver) from isolated perfused rat livers, perfused with 75 μg Cd(II), presented as a function of the length of time of perfusion with cadmium. Data are expressed as the means ± standard error for four perfusions.

To ascertain if *de novo* synthesis of cadmium thionein was occurring in the isolated perfused livers in response to exposure to Cd(II), the effects of transcriptional and translational inhibitors on the synthesis of cadmium thionein in perfused livers were studied. The inhibitors used were actinomycin D (1 mg/kg b.w.), cordycepin (10 mg/kg b.w.) and cycloheximide (10 mg/kg b.w.) and these were administered *in vivo* to the rats prior to removal of the livers for perfusion or *ex vivo* in the perfusion medium. The inhibitors were either administered intraperitoneally to the rats 4hr before the removal of livers for perfusion or in the perfusion medium 60min prior to the addition of Cd(II).

All these inhibitors significantly inhibited ($P<0.001$) the incorporation of cadmium and of zinc into metallothionein in the isolated perfused livers whether administered *in vivo* or *ex vivo* except for *ex vivo* administered actinomycin D, which did not significantly inhibit the incorporation of cadmium into metallothionein. Perhaps 60min *ex vivo* exposure to actinomycin D is not sufficient to produce changes similar to those obtained by a 4hr *in vivo* exposure. The synthesis of metallothionein in the isolated perfused rat liver, thus, appears to be regulated at both transcriptional and translational levels.

With the capabilities of the isolated liver perfusion technique now well in hand and routine in our laboratory, we are pursuing several avenues concerning metallothionein and chelatin. The capability of livers from zinc deficient rats to synthesize cadmium thionein is currently being assessed, with the observation that these livers are much more sensitive to a given dose of cadmium as compared with livers from zinc sufficient animals (17). We have been unsuccessful as yet in getting livers from zinc sufficient animals to take up net amounts of zinc, although $^{65}Zn(II)$ is taken up to some extent, and is excreted in the bile, and is found associated with metallothionein (17). Recently, we have been successful in obtaining net uptake of exogenously administered copper by the perfused livers with the concomitant synthesis of a low molecular weight protein containing copper (18). This protein has not yet been fully characterized by us, so we do not know whether the copper is associated with thionein or with chelatin.

Radioimmunoassay of metallothionein

A radioimmunoassay for rat liver cadmium thionein has recently been reported by Vander Mallie (19-21). Anti-sera obtained from immunized rabbits was equally capable of precipitating both species of cadmium thionein from rat liver. Radiolabeled metallothionein of high specific activity was obtained by iodinating the purified protein with the Bolton-Hunter reagent. With this assay a sensitivity in the range of 5-100 ng cadmium thionein was obtained.

We also have developed in our laboratory a radioimmunoassay for metallothionein, using a simple method for radiolabeling the protein to a high specific activity with $^{109}Cd(II)$ (22). The best results were obtained by incubating zinc thionein (1-10 μg) with 10^7-10^8 cpm of $^{109}CdCl_2$ in 0.02 M Tris·HCl, pH 8.6, containing 0.5 mM dithiothreitol (DTT), overnight (14-16hr) at 4°. Table II compares the effectiveness of the various methods we have used to try to radiolabel metallothionein.

Table II

Radiolabeled Metallothionein (22)

Label	Method	Yield (%)	Specific Activity (cpm/μg protein)
$^{109}CdCl_2$	0 DTT	7	5.9×10^4
	0.5 mM DTT	45-76	$0.64\text{-}1.85 \times 10^6$
	2.5 mM DTT	65	5.5×10^5
	5.0 mM DTT	54	4.8×10^5
$^{65}ZnNO_3$	0	0	0
	0.5 mM DTT	0	0
	5.0 mM DTT	0	0
$^{109}CdCl_2$	*In vivo*	4.1	190-460
[^{35}S]-cystine	*Ex vivo*	0.3	990-1210

Using this radiolabeled metallothionein, a double antibody radioimmunoassay has been developed which is sensitive in the range of 30-300 ng of metallothionein. Zinc thionein and cadmium (3.9), zinc(3.1)-thionein are essentially equal in their ability to displace ^{109}Cd-labeled zinc thionein in this assay. The primary antibody was obtained from rabbits immunized with DEAE-cellulose purified rat liver zinc thionein, and the secondary antibody was obtained from goats immunized with ammonium sulfate purified rabbit immunoglobulins.

DARK-FIELD ELECTRON MICROSCOPY OF CADMIUM-METALLOTHIONEIN

In collaboration with A. Max Fiskin and colleagues at the VA Hospital, Kansas City, Missouri, USA, an examination of the ultrastructure of rat liver cadmium metallothionein was conducted using conventional dark-field electron microscopy (23) with a Philips Model EM 300, as described by Ottensmeyer (24). The dark-field images were recorded at 80,000x and were enlarged in three stages on Kodak Lantern Plates (Medium) to give moderately contrast enhanced images for characterization by optical autocorrelation (25).

Cadmium metallothionein was unambiguously visualized with this technique. The "native" structure of the protein is reflected in rectangular projections of unstained protein, corresponding to a hexahedral dessicated shape, 3.6x2.5x1.6 nm. The protein was observed in side-by-side and end-to-end aggregation, as well as in relaxed, partially denatured forms. Distinct domains, proposed as corresponding to pairs of $Cd\text{-}(CyS)_3$

0.9 nm apart, were seen as strong electron scattering centers. Figure 3 presents a schematic of the types of projections seen in these studies.

Space filling models of amino acid sequences of the nature X-X-CyS-X-X-CyS-X-X-CyS were constructed and were seen to readily fold into trigonal planar coordination found in cadmium complexes with sulfhydryl compounds. In light of the numerous amino acid sequences of various metallothioneins currently being reported, it will be interesting to see how these reflect on model building based on our dark-field electron microscopy data.

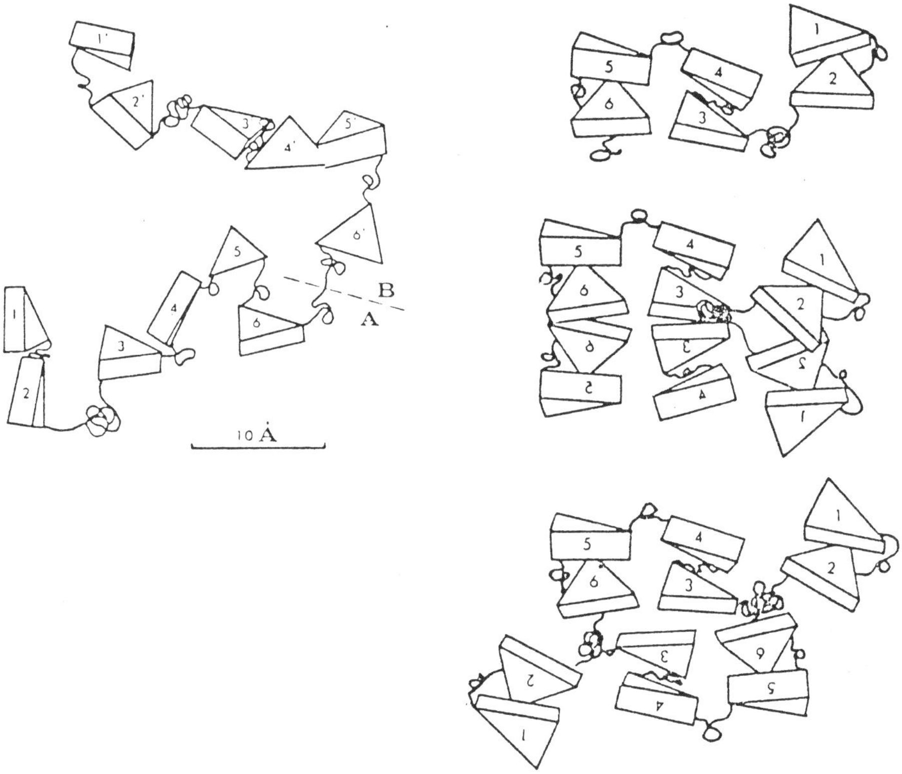

Figure 3. Schematic of the strong scattering centers in Cd-metallothionein. (Left), denatured or extended dimer. (Right, top), tightly folded monomer. (Right, middle and bottom), two tightly folded and aggregated dimers.

COPPER CHELATIN AND COPPER THIONEIN

Rajagopalan's group at Duke University has described the purification to homogeneity and properties of a copper containing protein, which they have named chelatin, which was synthesized in rat liver and in yeast following the administration of exogenous copper (3,26,27). It was characterized as having a M_r < 8000, a low content of aromatic amino acids, a cysteine content of 14.6 mol %, a decreased mobility on PAGE at pH 8.9 and different behavior on anion exchange chromatography, as compared with metallothionein.

Weser's group at Tübingen has reported the isolation of a copper thionein from yeast (28) and the conversion of chicken liver metallothionein into a copper thionein (29). Bremner's group at Aberdeen has reported the isolation of a copper, zinc thionein from pig liver (4) and rat liver (5) and its occurrence in sheep and calf liver (30). These proteins have been suggested as being similar to the fetal and neonatal copper proteins isolated by Riordan and Gower (31-32). The purification and partial sequencing of a copper thionein from human fetal liver has also been reported (6).

In our hands, copper treated rats produced a protein in liver which, when isolated, exhibited properties similar to those seen with chelatin, as reported (3,26,27). We have not been able to obtain a thionein to which was bound more than trace amounts of copper. Yet, we believe that both copper chelatin and copper thionein exist, and that arguments between laboratories about the existence of one or the other protein may be counterproductive. The zinc thionein status of an animal may be central to an understanding of this controversy. Clearly, copper can displace cadmium and zinc from metallothionein *in vitro* (33). Might this not also occur *in vivo* under conditions of high metallothionein levels <u>and</u> high tissue concentrations of copper? High metallothionein levels are known to occur in certain species at certain ages. High copper levels occur in fetal liver (34), as well as in adult liver after the administration of exogenous copper (27,31,35). Such an hypothesis is testable, and our group is currently engaged in experiments to answer these questions. Early results (18) indicate that rats with high liver zinc thionein levels take up 3-4 times as much exogenous copper in the liver as do rats with normal low levels of liver zinc thionein. The absorbed copper is associated with low M_r proteins in both cases, but we have not yet finished characterizing them, so their identity with chelatin and/or thionein remains to be determined.

ACKNOWLEDGMENTS

The work from the authors' laboratory reported and discussed in this paper was supported by a research grant from the USPHS, DHEW (ES 01288) with funds provided by an interagency grant from the EPA (EPA-IAG-DA-E772). Frank O. Brady is a recipient of a Research Career Development Award from the National Institute of Environmental Health Sciences (ES 00022).

REFERENCES

1. Winge, D.R., Premakumar, R., and Rajagopalan (1975) Arch. Biochem. Biophys. 170, 242-252.
2. Bremner, I., and Davies, N.T. (1974) Biochem. Soc. Trans. 2, 425-427.
3. Winge, D.R., Premakumar, R., Wiley, R.D., and Rajagopalan, K.V. (1975) Arch. Biochem. Biophys. 170, 253-266.
4. Bremner, I., and Young, B.W. (1976) Biochem. J. 155, 631-635.
5. Bremner, I., and Young, B.W. (1976) Biochem. J. 157, 517-520.
6. Ryden, L., and Deutsch, H.F. (1978) J. Biol. Chem. 253, 519-524.
7. Webb, M. (1975) Biochem. Soc. Trans. 3, 632-634.
8. Kägi, J.H.R., Himmelhoch, S.R., Whanger, P.D., Bethune, J.L., and Vallee, B.L. (1974) J. Biol. Chem. 249, 3537-3542.
9. Bühler, R.H.O., and Kägi, J.H.R. (1974) FEBS Letters 39, 229-234.
10. Day, F.A., Coles, B.J., and Brady, F.O. (1978) Bioinorg. Chem. 8, 93-105.
11. Martin, M., and Brady, F.O. (1977) Proc. S.D. Acad. Sci. 56, 72-76.
12. Panemangalore, M. and Brady, F.O. (1978) J. Biol. Chem. 253, 7898-7904.
13. Exton, J.H. (1975) Meth. Enzymol. 39, 25-36.
14. Bartosek, I., Guaitanim A., and Miller, L.L., editors (1973) Isolated Liver Perfusion and its Applications, Raven Press, New York.
15. Miller, L.L. (1973) in "Isolated Liver Perfusion and its Applications," Bartosek, I., Guaitani, A., and Miller, L.L., eds., Raven Press, New York, pp. 11-52.
16. John, D.W., and Miller, L.L. (1969) J. Biol. Chem. 244, 6134-6142.

17. Panemangalore, M., and Brady, F.O., unpublished results.

18. Day, F.A., and Brady, F.O., unpublished results.

19. Vander Mallie, R.J. (1977) Fed. Proc. 36(3), Abs. 4860.

20. Vander Mallie, R.J. (1978) Fed. Proc. 37(6), Abs. 2989.

21. Vander Mallie, R.J. (1978) Immunochem., in press.

22. Brady, F.O., and Kafka, R.L., submitted for publication.

23. Fiskin, A.M., Peterson, G., and Brady, F.O. (1977) Ultramicroscopy 2, 389-395.

24. Ottensmeyer, F.P., Whiting, R.F., Schmidt, E.E., and Clemens, R.S. (1975) J. Ultrastruct. Res. 52, 193-201.

25. Fiskin, A.M. (1972) Proc. Electron Microsc. Soc. Amer., 30th Mtg., Abs. 598.

26. Premakumar, R., Winge, D.R., Wiley, R.D., and Rajagopalan, K.V. (1975) Arch. Biochem. Biophys. 170, 267-277.

27. Premakumar, R., Winge, D.R., Wiley, R.D., and Rajagopalan, K.V. (1975) Arch. Biochem. Biophys. 170 , 278-288.

28. Prinz, R., and Weser, U. (1975) H.S.Z. Physiol. Chem. 356, 767-776.

29. Rupp, H., and Weser, U. (1974) FEBS Letters 44, 293-297.

30. Bremner, I., and Marshall, R.B. (1974) Br. J. Nutr. 32, 293-300.

31. Riordan, J.R., and Gower, I. (1975) Biochim. Biophys. Acta 411, 393-398.

32. Riordan, J.R., and Gower, I. (1975) Biochem. Biophys. Res. Commun. 66, 678-686.

33. Rupp, H., Voelter, W., and Weser, U. (1975) H.S.Z. Physiol. Chem. 356, 755-765.

34. Cunningham, I.J. (1931) Biochem. J. 25, 1267-1294.

35. Bloomer, L.C., and Sourkes, T.L. (1973) Biochem. Med. 9, 78-91.

Metallothionein
Jeremias H. R. Kägi and Monica Nordberg, eds.
Birkhäuser Verlag, Basel/Boston/Stuttgart, 1979

Factors Influencing the Occurrence of Copper-Thioneins in Tissues

Ian Bremner
Rowett Research Institute, Bucksburn, Aberdeen

Identification of copper-thioneins

There has been considerable controversy as to the nature of the low molecular weight copper-binding protein present in liver and other tissues. Although it has been claimed (1-4) that the copper-induced protein in rat liver is not metallothionein there is a marked lack of agreement as to the amino acid composition and charge properties of the protein. The cysteine content has ranged from 2 to 14% and the protein has been reported to bind reversibly (3), irreversibly (2) or not at all (4) to anion-exchange columns. In contrast with these findings, Bremner and Young were able to isolate and unequivocally identify copper-thioneins from the livers of copper-injected rats (4) and also from pigs (5). Copper-thioneins have also been isolated from the livers of neonatal calves (6) and the kidneys of copper-poisoned sheep (7) and growing rats (8).

In trying to understand the reasons for the diverse claims as to the nature and properties of the copper-induced protein, it should be noted that no attempt was made in many of the studies (1-4) to protect against copper-induced oxidative changes in the protein, whereas Bremner and Young (5,6) took care to maintain the copper proteins at low temperature and under anaerobic conditions throughout the isolation procedure.

The unsuitability of some of the isolation procedures for the purification of the copper-proteins has been demonstrated in studies where the livers of copper-injected rats were fractionated in parallel by the procedures of Bremner and Young (5,6) or of Winge et al.(2). In the former case only copper-thionein was obtained, in high yield, whereas in the latter case only a copper fraction with

Submitted to the "First International Meeting on Metallothionein and Other Low Molecular Weight Metal-binding Proteins", Zürich, July 17 - 22, 1978

cysteine content of 5% and leucine and isoleucine content of 6-9% was isolated. It seems likely, therefore, that the use of the latter procedure (2) led to significant modification of the copper-thionein.

It may be significant that multiple injections of copper were used in some studies (1-4), whereas only a single injection was used when copper-thionein was successfully isolated (5). The multiple injection procedure would increase the copper:zinc ratio in the induced protein and exacerbate the difficulties encountered in its isolation. In support of this, the selective loss of zinc which occurred during isolation of renal copper-thionein (8) was associated with a substantial increase in the molecular weight distribution of the protein.

Further confirmation that the non-thionein proteins may be artefacts which are formed during the isolation procedure was obtained in studies on the incorporation of [^{3}H]leucine into the copper-induced protein in rat liver (10). The absence of any significant incorporation of ^{3}H into the partially-purified protein is consistent with the presence in the copper-protein of only a very small proportion of leucine residues, as has been found in copper-thioneins (5,6). These observations cannot be reconciled with the claim that the copper-induced protein contains about 7% leucine residues (1-4). The similar antigenic properties of cadmium-thionein and of one of the non-thionein proteins (3) also indicates that the copper-protein is similar, if not identical to metallothionein (11).

Induction of Synthesis of Copper-Thionein

The formation of copper-thionein in rat liver and the incorporation of [^{35}S] cysteine into the protein are stimulated by injection or dietary supplementation with copper and are inhibited by administration of cycloheximide (10,12). This suggests that copper can induce synthesis of the protein *de novo*. An inhibitory effect of actinomycin D on the production of one of the non-thionein copper-proteins has also been reported (13). If, as suggested above, this protein is a form of copper-thionein, it suggests that induction by copper of the synthesis of metallothionein may be under transcriptional control.

The stimulatory effect of copper on the incorporation of [^{35}S]cysteine into copper-thionein has also been demonstrated in rat kidneys (10), but there is still some uncertainty about the ability of copper to induce synthesis of metallothionein in the intestinal mucosa. It was reported (14,15) that intragastric administration of copper in the rat stimulated the incorporation of copper, [^{3}H]leucine and [^{14}C]lysine into the protein. Furthermore as both the amino acid incorporation and copper absorption were inhibited by cycloheximide treatment, it appeared that the protein had a regulatory function in the control of copper absorption and that its synthesis could be induced by copper. However it is not clear how specific was this stimulatory effect on protein synthesis, as metallothionein contains very little leucine (5,6). Furthermore the rats used were starved for 18 h prior to the experiment, which would have resulted in increased concentrations of zinc-thionein in the intestinal mucosa (16). In view of the ability of copper to displace zinc from metallothionein (17), it cannot be excluded that the incorporation of copper into intestinal metallothionein within the relatively short time of 2 h arose from displacement of zinc from pre-synthesised protein. It is noteworthy that parenteral administration of copper had little effect on production of intestinal copper-thionein whereas zinc administration greatly stimulated incorporation of copper into the protein (16). Further studies are clearly required to establish the mechanisms governing production of intestinal copper-thioneins.

Effect of Zinc on the Accumulation of Hepatic Copper-Thionein

Most studies on copper-thionein have concentrated on the role of the protein in liver, where it is thought to function in the initial uptake of the metal from plasma and in its temporary storage or cellular detoxication (12). The copper-protein is formed under normal physiological conditions (18) and is even found in foetal tissue (7,19), which indicates that the genetic potential to respond to increased copper and zinc concentrations is advanced at an early stage of development. Increases in liver copper content usually result in increases in copper-thionein production. Up to 40% of the copper present in pig liver has been found to be in this form (18), but this proportion can vary both between and within species. For example, there is a decrease from 42 to 20% in the proportion of hepatic copper present as metallothionein in the developing sheep foetus between 80 and 140 days gestational age (19). In adult sheep and in calves, the proportion of hepatic copper-thionein is extremely variable (17,20) ranging from 5 to 20% of the total liver copper. This variation has been related to differences in the liver zinc content (17), the total amount of copper and zinc present as metallothionein being a function of the liver zinc concentration, and the relative copper and zinc contents of the liver determining which metals are found on the protein. When liver zinc concentrations are low, the copper in the liver cytosol may therefore be bound to proteins of high molecular weight rather than to metallothionein.

Although these results have been interpreted in terms of possible competitive displacement of zinc from metallothionein within the cell, it cannot be excluded that some redistribution of the metals could have occurred during the processing of the livers. However, there was no evidence of similar competition between copper and zinc for binding sites on metallothionein in porcine liver (18) and it would appear that sheep and calves have a limited capacity to accumulate copper-thioneins in their liver. This could be of some significance with regard to the greater susceptibility of these animals to copper toxicosis.

The close relationship between the binding of copper and zinc to metallothionein is also shown by the incorporation of substantial amounts of zinc into the protein when its synthesis has been induced by copper (5,10,12). However, zinc is apparently not essential for the induction process as synthesis of the protein was also induced by copper in zinc-deficient rats, without any appreciable binding of zinc to the protein. Nevertheless it was found that no copper-thionein accumulated in the livers of rats (12), pigs (18) or sheep (20) receiving copper supplements for several weeks if the animals were zinc-deficient. This was related to the faster disappearance of copper from the protein in zinc-deficient rats which had been injected with copper and to the decreased biological stability of copper-thionein in these animals (10,12). The reason for the shorter half-life of the protein in the zinc-deficient rats, 12 h instead of 17 h as found in zinc-supplemented animals, has not been established but it could be connected with the low zinc content of the protein and possible differences in conformation.

It is of interest that the rates of disappearance of copper and zinc from these metallothioneins were exactly the same as those for the catabolism of the proteins. This suggests that the utilisation of the metals after their incorporation into metallothionein was associated with the degradation of the protein, as has been reported for the zinc-induced protein, which has a half-life of 19-20 h (21). It is evident that in studying the binding of metals to metallothionein it is not sufficient to consider only the relative binding affinities of the metals. It is also necessary to consider the relationship between the bound metal and the biological stability of the protein. The rapid degradation of the copper-protein is of some interest as it is consistent with the temporary nature of any storage function of the protein for copper (12).

Copper-Thioneins in Kidney

Relatively little attention has been paid to the occurrence of copper-thioneins in tissues other than liver, but it has recently been found that about half of the copper in the kidneys of copper-poisoned sheep occurs in this form (8). Very large amounts of the protein accumulate suddenly at the time of the haemolytic crisis which is characteristic of this syndrome, when blood copper and liver copper-thionein concentrations are high (22) and considerable damage occurs to the renal proximal convoluted tubules. It is not known how far these events are related but it is known that extracellular cadmium-thionein is rapidly taken up by the kidneys and has nephrotoxic properties in the rat (23). Furthermore, the metabolism of parenterally administered copper-thionein (9) is similar to that of the cadmium-protein, with most of the copper-protein either being excreted in intact form in the urine or being reabsorbed by the kidneys where it is rapidly degraded. Although it seemed possible that a similar process could be taking place in the copper-poisoned sheep, no traces of copper-thionein could be found in the blood of these animals (8).

Copper-thioneins also accumulate in the kidneys of rats under normal physiological conditions. Indeed concentrations of the protein in rat kidneys are generally greater than those in the liver. The accumulation of copper-thionein was found to be dependent on both the age and sex of the animals (9) with concentrations of copper present as metallothionein increasing steadily from about 1 μg/g kidney at weaning to 4.9 and 13.8 μg/g in male and female rats respectively at 3 months of age. In contrast, there was no consistent change in the concentrations of zinc present in this form and these remained at about 3 μg/g throughout. Copper-thionein generally accounted for about 65% of the increased kidney copper content of all samples.

These results indicated that the renal accumulation of copper-thionein might be under some form of hormonal control. This was supported by the prevention of the increase in kidney copper-thionein concentrations when the female rats were ovariectomized at 7 weeks of age (9). Furthermore, administration of medroxyprogesterone to the ovariectomized rats, and also to the intact controls, had a stimulatory effect on copper-thionein production in the kidney, but not in the liver (I. Bremner & R. B. Williams, unpublished observations). At this stage it is not known whether the changes in renal copper-thionein content represent direct effects on the synthesis of the protein or whether they are secondary to other changes in copper metabolism. There was no obvious correlation between the sex-linked changes in plasma copper concentrations and in renal copper-thionein content.

These findings suggest some degree of tissue specificity in the accumulation of different metallothioneins, with copper having a greater tendency to bind to metallothionein in kidneys. Unfortunately the metabolic role of renal copper-thionein is not yet established, although it has been speculated (9) that it could be involved in tubular reabsorption processes and the control of urinary copper excretion.

Effect of Cadmium on the Binding of Copper to Metallothionein

Rats receiving long term dietary supplements of cadmium also incorporated large amounts of copper with cadmium into renal metallothionein (24). Relatively small amounts of zinc occurred in the renal protein and no copper was present in the hepatic protein. From data on the total kidney concentrations of these metals (25) it would appear that renal concentrations of cadmium- and copper-thioneins were probably greater in female than in male rats, which is of some interest in view of the sex-linked differences in the renal accumulation of the copper-protein (9).

It has been suggested that the inhibition of copper absorption by cadmium and zinc is a consequence of displacement of copper from intestinal metallothionein. However, the decreased copper-64 absorption in rats receiving low dietary concentrations of cadmium was later found to be associated with increased binding of copper-64 to a low molecular weight protein in the intestinal mucosa (26). Dietary supplementation with zinc also decreases copper absorption but increases the binding of copper to intestinal metallothionein (16). These observations are consistent with the regulatory role proposed for this protein in copper absorption (15) but indicate that copper displaces cadmium or zinc from the pre-synthesized protein in the mucosa and not vice versa.

REFERENCES

1. Evans, G. W., Wolenetz, M. L., and Grace, C. I. (1975) Nutr. Rep. Int. 12, 261-269.
2. Winge, D. R., Premakumar, R., Wiley, R. D. and Rajagopalan, K. V. (1975) Arch. Biochem. Biophys. 170, 253-266.
3. Riordan, J. R. and Gower, I. (1975) Biochim. Biophys. Acta 411, 393-398.
4. Irons, R. D. and Smith, J. C. (1977) Chem.-Biol. Interact. 18, 83-89.
5. Bremner, I. and Young, B. W. (1976) Biochem. J. 157, 517-520.
6. Bremner, I. and Young, B. W. (1976) Biochem. J. 155, 631-635.
7. Hartmann, H. J. and Weser, U. (1977) Biochim. Biophys. Acta 491, 211-222.
8. Bremner, I. and Young, B. W. (1977) Chem.-Biol. Interact. 19, 13-23.

9. Bremner, I., Hoekstra, W. G., Davies, N. T. and Williams, R.B. (1978) In: "Trace Element Metabolism in Man and Animals-3" M. Kirchgessner ed., Arbeitskreis für Tierernährungsforschung Weihenstephan, p. 44-51.
10. Bremner, I., Hoekstra, W. G., Davies, N. T. and Young, B. W. (1978) Biochem. J. (in press).
11. Madapallimatam, G. and Riordan, J. R. (1977) Biochem. Biophys. Res. Comm. 77, 1286-1293.
12. Bremner, I. and Davies, N. T. (1976) Brit. J. Nutr. 36, 101-112.
13. Premakumar, R., Winge, D. R., Wiley, R. D. and Rajagopalan, K. V. (1975) Archs. Biochem. Biophys. 170, 267-277.
14. Evans, G. W. and LeBlanc, F. N. (1976) Nutr. Rep. Int. 14, 281-288.
15. Evans, G. W. (1978) In: "Trace Element Metabolism in Man and Animals-3" M. Kirchgessner ed., Arbeitskreis für Tierernährungsforschung Weihenstephan, p. 98-105.
16. Hall, A. C., Young, B. W. and Bremner, I. unpublished observations.
17. Bremner, I. and Marshall, R. B. (1974) Brit. J. Nutr. 32, 293-300.
18. Bremner, I. (1976) Brit. J. Nutr. 35, 245-252.
19. Bremner, I., Williams, R. B., and Young, B. W. (1977) Brit. J. Nutr. 38, 87-92.
20. Bremner, I. and Marshall, R. B. (1974) Brit. J. Nutr. 32, 283-291.
21. Feldman, S. L. and Cousins, R. J. (1976) Biochem. J. 160, 583-588.
22. Bremner, I., Young, B. W. and Mills, C. F. (1976) Brit. J. Nutr. 36, 551-561.
23. Cherian, M. G., Goyer, R. A. and Delaquerriere-Richardson, L. (1976) Toxicol. Appl. Pharmac. 38, 399-408.
24. Stonard, M. D. and Webb, M. (1976) Chem.-Biol. Interact. 15, 349-363.
25. Murthy, L., Rice, D. P. and Petering, H. G. (1978) In: "Trace Element Metabolism in Man and Animals-3" M. Kirchgessner ed., Arbeitskreis für Tierernährungsforschung Weihenstephan, p. 557-560.
26. Davies, N. T. and Campbell, J. K. (1977) Life Sci. 20, 955-960.

Metallothionein
Jeremias H. R. Kägi and Monica Nordberg, eds.
Birkhäuser Verlag, Basel/Boston/Stuttgart, 1979

NUTRITIONAL AND ENVIRONMENTAL FACTORS AFFECTING METALLOTHIONEIN LEVELS.

P.D. Whanger and S.H. Oh

Department of Agricultural Chemistry
Oregon State University
Corvallis, Oregon 97331, USA

The first evidence for the presence of that most unusual protein called metallothionein (MT) was presented in 1957 from Dr. Vallee's laboratory (1). A more detailed study in this laboratory on the physical and chemical properties of MT include low molecular weight, high sulfhydryl and metal content, and the absence of aromatic amino acids. Although once thought to be present only in the kidney, it is now known to be a ubiquitous protein. This protein from animals not intentionally exposed to metals contains predominantly zinc when isolated from the liver but contains more cadmium when isolated from the kidney (4). Because of the chemical properties of MT, many researchers have speculated that it is involved in heavy metal detoxification (5,6,7,8,9), but evidence is beginning to accumulate for its involvement in zinc metabolism (10,11,12). The present report covers some of the dietary factors, including metals, sulfur levels and stress influencing the MT levels in tissues.

EFFECT OF METALS ON MT LEVELS.

One of the problems of studying the effects of various factors on MT levels is that a good analytical method is not available for its measurement. A radiochemical method based on estimation of MT with ^{203}Hg radiotracer has been described (13) and an antibody method has been investigated (14), but these methods have their limitations. We routinely obtain an estimation of the amount of MT present by either

Submitted to the "First International Meeting on Metallothionein and Other Low Molecular Weight Metal-binding Proteins", Zürich, July 17 - 22, 1978

determining the amount of radioactivity incorporated into MT after ^{14}C-cystine injection or by determining the amount of zinc present in MT per gram tissue. Under most circumstances, these two methods of determining the relative concentration of MT are highly correlated.

Since the amount of cadmium in MT increased with exposure to this element, it was assumed to stimulate its synthesis. By use of radioactive cystine and cadmium, evidence was found to indicate that cadmium does indeed stimulate MT synthesis (15,16,17,18). Inhibitors of protein synthesis were shown to block the incorporation of radioactive cystine into MT upon cadmium exposure (16), providing further evidence for its enhanced synthesis by this element. Our work shows that when cadmium is incorporated into MT, it remains constant in this fraction for extended periods of time and does not disappear as the MT protein moiety is degraded (18). The accumulation of cadmium in MT also causes zinc to accumulate in this fraction (19). In contrast to cadmium, however, zinc disappears from the MT fraction at a parallel rate with the degradation of the MT protein moiety (18). We interpreted this to indicate that MT is more importantly involved with zinc metabolism than cadmium.

Subsequent work revealed that zinc accumulates only in the MT fraction when the zinc content of the diet is elevated (10,20). When the soluble fraction from tissues is eluted through gel filtration columns (G-75 Sephadex), essentially 3 zinc containing peaks are obtained. The first peak elutes with the void volume of the column, the second one at molecular weight (MW) of about 30,000 and the third one at the position where MT elutes. Since this is the peak we routinely take for further purification of MT, we are sure that the zinc in this peak is associated with MT. With ^{14}C-cystine studies, we have shown that more than 90% of the zinc and ^{14}C in this fraction from this gel filtration step is associated with MT. In some preliminary work, we found that zinc accumulated only in the MT fraction in tissue cytosols, and thus proposed MT serves as a zinc storage protein analogous to ferritin and iron (20).

In a more detailed study, we confirmed that zinc accumulates only with MT in the liver and kidney cytosols when excess zinc is present in the diet (10). The accumulated zinc in this fraction, however, was depleted to pre-supplementation levels after only 3 to 4 days of feeding a zinc deficient diet (Fig. 1). Similar patterns were found for the liver and kidneys in chicks, but this element disappeared

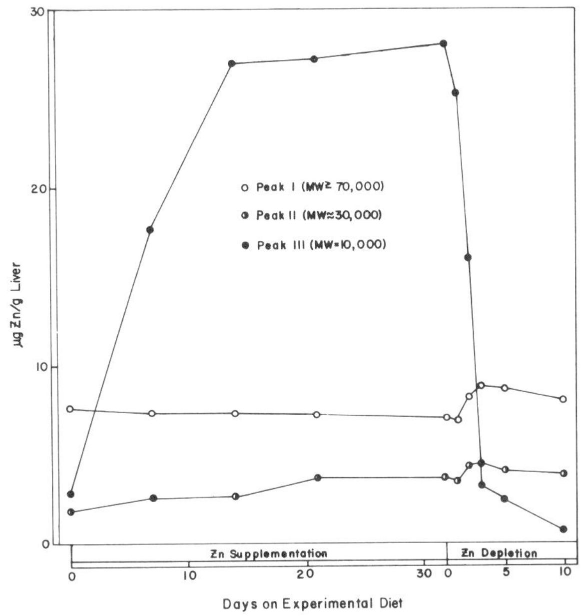

Figure 1. Accumulation and depletion of zinc with time in liver cytosol proteins of rats fed a diet containing 2,000 ppm zinc followed by feeding a zinc deficient diet. Rats were fed a diet containing 2,000 ppm zinc for 32 days, and were killed at various times for the determination of the amount of zinc associated with the liver cytosol proteins. Additional rats were then fed the zinc deficient diet and killed at various times for determining zinc in cytosol proteins. Similar patterns were found for the kidneys. These data are taken for reference 10.

much more slowly from the pancreas during this depletion period. This raises some interesting possibilities, since many of the enzymes secreted by this organ require zinc for activity. During this depletion phase in the rat study, significant increased excretion of the metal occurred in the feces and urine. No evidence was found to indicate that zinc was excreted in the bile in association with MT. We concluded from this study that the involvement of MT as a storage protein for zinc may be a tenuous hypothesis. However, subsequent work with pregnant rats indicate that extremely high levels of zinc is associated with liver MT in newborn rats with even 20 ppm zinc in the diet. With weanling rats fed diets for 8 weeks, zinc did not start to accumulate in tissue MT until levels above 500 ppm were used. Thus, the amount of zinc stored with MT as well as the degree of inducement by zinc appears to be age dependent. Consequently, MT could be serving as a storage protein for zinc in the newborn animals.

With the use of radioactive cystine, zinc has been shown to stimulate the synthesis of MT (18,21), which is consistent with work showing inhibitors of protein synthesis will block this stimulatory effect of zinc (22). In other work, increases of liver zinc content were found to be synchronous with and associated almost completely with increased zinc-binding to MT (23). Thus, there is little doubt concerning the ability of zinc to stimulate MT synthesis.

Other elements which have been shown to stimulate MT synthesis include copper, mercury, silver and possibly tin (18,24,25). Although these metals will stimulate MT synthesis the method of administration determines whether they are stored with MT in the liver. Mercury is bound to a low MT protein, presumably MT, when it is injected (26,27), but not upon repeated exposures to this element (28) or when included in the diet (29). It does not appear to make a difference in the kidney, however, since mercury is associated with a low MT protein, presumably MT, in the kidney regardless of whether it is injected or given orally. Similar patterns were found for copper in our laboratory. Copper was associated with a protein of about 30,000 MW in the liver cytosol when this element was included in the diet at elevated levels, but was associated with a protein eluting at the position of MT from the gel filtration column when copper was injected into rats. Whether copper is bound to MT in the liver when it is injected is presently a controversial issue (30,31,32). In contrast, copper is associated with a low MW protein, presumably MT, in the kidney regardless of whether it is injected or given orally. Therefore, the route of administration of some elements must be taken into consideration before valid conclusions can be drawn.

Some evidence is available to indicate that selenium may alter the influence of metals on MT levels (33). This is not unexpected since selenium has been shown to divert the binding of some metals like mercury (27) from MT to higher MW proteins. Mercury administration resulted in almost a three-fold increase of MT levels in kidneys, but the simultaneous administration of selenium with this element abolished its stimulative effect on MT levels (33).

The biological half-life is also influenced by the kind of metal. In our initial work, an injection of cadmium resulted in a half-life of about 4.2 days (18). However, another research group reported the half-life of MT to be 2.8 days in the liver when cadmium was injected into the animal (8). These workers, however, found the half-life of MT in kidneys to be about 5.0 days. We reinvestigated the influence of metals on the half-life of MT, and fed rats diets with high zinc levels (2,000 ppm), this high zinc diet then a basal diet, or a diet with cadmium (100 ppm). The half-lives of liver MT were respectively 2.7, 1.3 and 3.3 days and those of the kidney MT were respectively 2.2, 1.4 and 3.6 days (34). Thus, the MT induced by cadmium has a longer half-life than that induced by zinc in both liver and kidney. The major individual species of MT, which were further separated by DEAE cellulose chromatography, had similar half-lives to the preparations isolated by gel filtration. The half-life of liver MT was found to be about 18 hours when zinc was injected into rats (21). Zinc disappeared from the MT fraction at a parallel rate with the protein moiety, indicating the release of zinc from zinc-thionein probably occurs simultaneously with the degradation of the protein moiety. We have also shown this for the liver MT. However, we have found that this is not the case with kidney MT (34). Thus, the half-life of zinc in the MT fraction cannot be used to determine the half-life of MT under all circumstances. The differences in these results are probably reflective of the various conditions under which these experiments were conducted.

EFFECT OF SULFUR CONTENT OF DIET.

Since MT has a very high cystine content, sulfur deficiency would be expected to limit its synthesis. An Experiment was conducted with rats fed diets containing various levels of methionine with 1,000 ppm zinc. Figure 2 indicates that sulfur deficiency does limit the synthesis of MT in liver, as indicated by the amount of zinc in this protein. After 8 days on the experiment, the amount of zinc in MT reached a plateau in those rats fed the 2 lowest levels of sulfur. The zinc continued to accumulate in MT of those rats fed the diet containing 1.0% methionine, but the highest level of methionine prevented the accumulation of zinc with MT, as compared to the 1.0% methionine diet. Similar patterns were noted for the kidneys. The greatest amount of MT in rat kidneys was again in those fed the diet containing 1.0% methionine, but the higher level of this sulfur amino acid was inhibitory to this accumulation. This effect of dietary sulfur upon MT levels was confirmed by ^{14}C-cystine studies.

In order to obtain further information on the influence on dietary sulfur, the half-lives of MT were determined in sulfur supplemented and deficient rats. The half-life of MT in liver of the respective groups was 1.6 and 1.3 days, and the respective half-life values in kidneys were 2.7 and 4.3 days. Thus sulfur deficiency did not affect the degradation of MT in liver, but markedly affected it in the kidneys. This data further indicate that the conditions under which the experiment is conducted have a marked influence on the values obtained for MT half-lives.

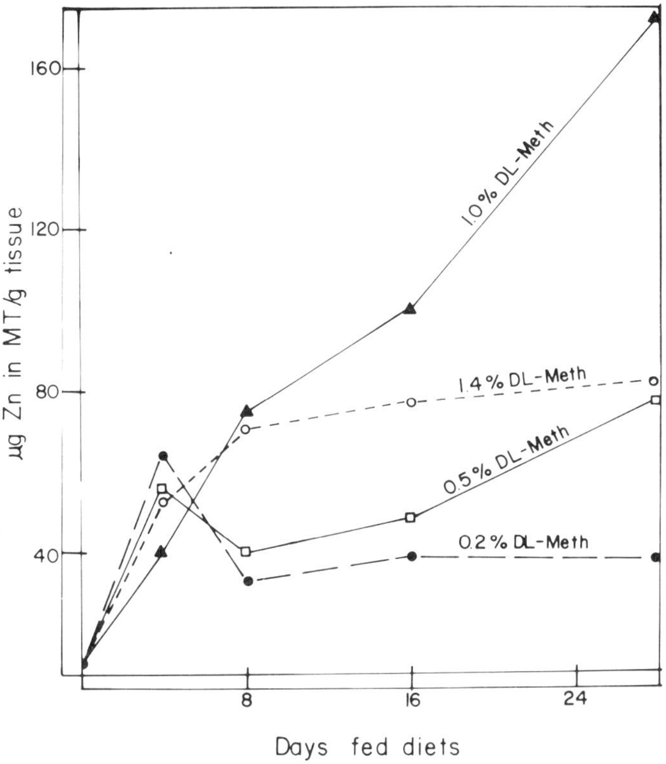

Figure 2. Effects of various levels of dietary sulfur upon the MT concentrations in liver as judged by the amount of zinc with this protein. Rats were fed diets containing 1,000 ppm zinc with four levels of methionine, and killed at various times during the 28-day experiment for determination of the level of liver MT. Crystalline amino acid mixture was used as the source of protein. Similar patterns were also noted for the kidneys.

INFLUENCE OF STRESS AND INFECTION.

Since other workers (11) indicated that starvation would increase MT levels, we studied the influence of a number of stresses on MT levels in rats. The stresses studied were cold environment, hot environment, heat burn, streuous exercise, and CCl_4 injection (35). As markers of MT synthesis, ^{14}C-cystine and ^{65}Zn were injected parenterally into rats, and MT in liver and kidney cytosols separated by Sephadex G-75 column chromatography. Among the stresses tested, the greatest stimulation of MT synthesis in liver was observed under cold environment, strenuous exercise, and CCl_4 intoxication, as evidenced by the ^{65}Zn incorporation and amount of stable zinc in MT (Table 1). In contrast to liver only small responses were observed in MT synthesis in kidneys. These same patterns were confirmed by ^{14}C-cystine incorporation in MT (35).

Table 1. Effect of various stresses on metallothionein levels in liver and kidneys of rats.

	Liver		Kidney	
Stress	Zn in MT*	^{65}Zn in MT**	Zn in MT*	^{65}Zn in MT**
Control	16.9 (1.3)	5.4	7.2 (2.8)	13.2
Cold	71.3 (5.9)	17.0	13.0 (4.4)	17.4
Heat	26.3 (2.4)	9.5	6.3 (2.5)	10.4
Exercise	53.8 (5.2)	16.3	6.0 (2.5)	11.2
Burn	19.7 (1.5)	5.5	5.6 (2.3)	9.5
CCl_4	115.0 (9.1)	29.3	5.8 (2.1)	10.0

* µg Zn in MT per whole tissue. Values in parentheses are µg Zn in MT per g tissue.
** Percentage of total tissue ^{65}Zn in MT.

Bacterial endotoxin and a substance derived from glycogen-stimulated rabbit peritoneal exudate cell preparation have been shown to enhance MT synthesis (36). An inhibitor of protein synthesis was shown to prevent this stimulatory effect of these substances on MT synthesis, providing evidence for increased de novo synthesis due to these agents. Some evidence has been obtained to indicated that infection results in different half-life values of MT in these animals.

Therefore, the results in this communication indicate that a number of conditions will influence MT levels in tissues, and this may be the reason different results have been obtained from various laboratories.

SUMMARY

A number of elements including zinc, cadmium, mercury and copper will enhance the synthesis of MT, but only zinc and cadmium are deposited with this protein at higher levels than with other proteins in the liver when these are given orally. However, mercury and copper are deposited with MT in the kidney regardless of whether given orally or by injection. As the level of zinc is increased in the diet a corresponding increase of zinc concentration occurs only in the MT fraction of tissue cytosols. Zinc disappears from the MT fraction to the supplementation levels within 3 to 4 days when rats which have been fed a high zinc diet are changed to a zinc deficient diet. The evidence indicates that MT is involved with zinc metabolism. The half-life of MT is longer when cadmium is used to induce its synthesis than when zinc is used. Dietary sulfur deficiency limits the synthesis of MT. Various stresses, in which strenuous exercise, cold environment, and CCl_4 injection were most effective, and infections were found to enhance MT synthesis, suggesting a role for it in the defense mechanism.

References

1. Margoshes, M. and Vallee, B.L. (1957) J. Amer. Chem. Soc. 79, 4813-4814.
2. Kagi, J.H.R. and Vallee, B.L. (1960) J. Biol. Chem. 235, 3460-3465.
3. Kagi, J.H.R. and Vallee, B.L. (1961) J. Biol. Chem. 236, 2435-2442.
4. Kagi, J.H.R., Himmelhoch, S.R., Whanger, P.D., Bethune, J.L. and Vallee, B.L. (1974) J. Biol. Chem. 249, 3537-3542.

5. Nordberg, G.F., Nordberg, M., Piscator, M. and Versterberg, O. (1972) Biochem. J. 126, 491-498.
6. Winge, D.R. and Rajagopalan, K.V. (1972) Arch. Biochem. Biophys. 153, 755-762.
7. Weser, V., Rupp, H., Donay, F., Linnemann, F., Voelter, W., Voetsch, W., Jung, G. (1973) Eur. J. Biochem. 39, 127-140.
8. Shaikh, Z.A. and Smith, J.C. (1976) Chem. Biol. Inter. 15, 327-336.
9. Kimura, M., Otaki, N., Yoshiki, S., Suzuki, M., Horiuchi, N. and Suda, T. (1974) Arch. Biochem. Biophys. 165, 340-348.
10. Chen, R.W., Vasey, E.J. and Whanger, P.D. (1977) J. Nutr. 107, 805-813.
11. Bremner, I. and Davies, N.T. (1975) Biochem. J. 149, 733-738.
12. Richards, M.P. and Cousins, R.J. (1976) J. Nutr. 106, 1591-1599.
13. Zelazowski, A. and Piotrowski, J.K. (1977) Acta Biochim. Polon. 24, 97-103.
14. Chen, R.W., Vasey, E.J. and Whanger, P.D. (1975) Pacific Slope Biochem. Conf. 17, 42.
15. Shaikh, Z.A. and Smith, J.C. (1976) Chem. Biol. Intern. 15, 327-336.
16. Squibb, K.S. and R.J. Cousins (1974) Environ. Physiol. Biochem. 4, 24-30.
17. Weser, V., Donay, F. and Rupp, H. (1973) FEBS Lett. 32, 171-174.
18. Chen, R.W. Whanger, P.D. and Weswig, P.H. (1975) Biochem Med. 12, 95-105.
19. Deagen, J.T., Oh, S.H. and Whanger, P.D. (1978) Chem. Biol. Intern. (In preparation).
20. Chen, R.W., Eakin, K.J. and Whanger, P.D. (1974) Nutr. Rep. Int. 10, 195-200.
21. Feldman, S.L. and Cousins, R.J. (1976) Biochem. J. 160, 583-588.
22. Richards, M.P. and Cousins, R.J. (1975) Bioinorg. Chem. 4, 215-224.
23. Bremner, I. and Davies, N.T. (1975) Environ. Physiol. Biochem. 149, 733-738.
24. Sabbioni, E. and Marafante, E. (1975) Environ. Physiol. Biochem. 5, 132-141.
25. Winge, D.R., Premakumar, R. and Rajagopalan, K.V. (1975) Arch. Biochem. Biophys. 170, 242-252.

26. Wisniewska-Knypl, J.M., Trojanowska, B.B., Piotrowski, J.K. and Jablonska, J.K. (1972) Acta Biochem. Vol 19, 11-18.
27. Chen, R.W., Whanger, P.D. and Fang, S.C. (1974) Pharmac. Res. Comm. 6, 571-579.
28. Piotrowski, J.K., Trojanowska, B. and Sapota, A. (1974) Arch. Toxicol. 32, 351-360.
29. Chem, R.W., Whanger, P.D. and Weswig, P.H. (1975) Fed. Proc. 34, 927.
30. Bremner, I. and Young, B.W. (1976) Biochem. J. 157, 517-520.
31. Evans, G.W., Wolenetz, M.L. and Grace, C.I. (1975) Nutr. Rep. Int. 12, 261-269.
32. Winge, D.R., Premakumar, R., Wiley, R.D. and Rajagopalan, K.V. (1975) Arch. Biochem. Biophys. 170, 253-266.
33. Piotrowski, J.K., Bem, E.M. and Werner, A. (1977) Biochem. Pharm. 26, 2191-2192.
34. Oh, S.H., Deagen, J.T., Whanger, P.D. and Weswig, P.H. (1978) Bioinorgan. Chem. (In press).
35. Oh, S.H., Deagen, J.T., Whanger, P.D. and Weswig, P.H. Amer. J. Physiol. (In press). (1978) 234, E282-285.
36. Sobocinski, P.Z., Canterbury, W.J., and Mapes, C.A. (1977) Fed. Proc. 36, 1100.

Metallothionein
Jeremias H. R. Kägi and Monica Nordberg, eds.
Birkhäuser Verlag, Basel/Boston/Stuttgart, 1979

SYNTHESIS AND DEGRADATION OF LIVER METALLOTHIONEIN

Robert J. Cousins
Department of Nutrition
Rutgers - The State University of New Jersey
New Brunswick, New Jersey 08903

The function of metallothionein has been debated for years. However, now it is clear that at least one of its roles is in hepatic zinc metabolism. In the liver, metallothionein appears to be related to the mechanism of zinc uptake into hepatocytes and may function as a binding ligand to help regulate the intracellular concentration of free zinc ions (1). Therefore, since the dietary intake of zinc is variable and zinc metabolism at the cellular level is dynamic, it is reasonable to assume that metallothionein is subject to changes in synthesis and degradation in response to zinc status.

METALLOTHIONEIN BIOSYNTHESIS

It was clear from the early work of Shaikh and Lucis that rat liver metallothionein could be labeled *in vivo* with C^{14}-cystine following administration of cadmium (2). This suggested that perhaps metallothionein was an inducible protein. Experiments from my laboratory (3,4) as well as those of Bremner (5) and of Webb (6) have supported that concept. Specifically, it was shown that injections of either zinc or cadmium brought about the incorporation of labeled amino acids and zinc or cadmium into liver metallothionein. These experiments also showed that both the A and B form of metallothionein are "induced" (5,7). We were able to show that actinomycin D, cordycepin and cycloheximide, when administer-

Submitted to the "First International Meeting on Metallothionein and Other Low Molecular Weight Metal-binding Proteins", Zürich, July 17 - 22, 1978

ed to animals prior to an injection of either zinc or cadmium, would block the apparent induction of metallothionein (3,4,8-10). In contrast, when these inhibitors of protein synthesis were administered after the injection of either metal, the induction process could not be inhibited. These data strongly suggested that both metals were capable of stimulating the increase in metallothionein synthesis via a mechanism that required the transcription of new messenger RNA.

In order to more closely define the mechanism of the apparent induction process, we conducted experiments to measure the rate of metallothionein synthesis following administration of zinc (10). We found that the rate of synthesis was not increased in linear fashion, but instead it was rapidly stimulated following administration of the metal. A maximum rate was reached by 6 hr post-injection and it subsequently returned to a near basal level by 15 hr after a single injection of metal had been given. In contrast, the amount of total metallothionein-bound zinc accumulated within liver cells reached a plateau about 18 hr after administration of the metal. Since the rate of synthesis was increased, we proposed that the amount of metallothionein mRNA synthesized following an injection of zinc was increased. Thus, one would expect to find more metallothionein mRNA in the polysomes of those animals that are synthesizing metallothionein at an elevated rate. Using a cell-free protein synthesizing system, in which polysomes were utilized as the source of metallothionein mRNA, we were able to show that polysomes from zinc-injected animals did, in fact, direct the synthesis of more metallothionein-like polypeptides than did polysomes from control animals (11). These data provided the first cell-free evidence to support the concept that indeed zinc is able to induce the synthesis of metallothionein via changes in the translatable pool of metallothionein mRNA.

In order to more firmly establish that metallothionein synthesis is regulated via a flow of information starting at transcription it is necessary to demonstrate that more metallothionein mRNA is synthesized following administration of zinc. Metallothionein synthesis could be inhibited *in vitro* by cordycepin, indicating that metallothionein mRNA contains a poly(A) sequence (10). This characteristic has allowed us to isolate metallothionein mRNA with other poly(A+) mRNA by oligo(dT)-cellulose chromatography. Once the poly(A+) RNA's have been isolated, they can be translated in a cell-free protein synthesizing system employing wheat germ extract (12,13). A ^{3}H-labeled species with chromatographic properties that are identical with metallothionein synthesis *in vivo* can be generated in this system. Identity of metallothionein was established by Activated Thiol Sepharose 4B (11) and gel filtration chromatography (13). Recently, we have identified nascent chains of ^{3}H-labeled metallothionein in the cell-free product by immunoprecipitation and electrophoresis (12). In order to accomplish this rabbit anti-metallothionein serum was mixed with the cell-free products. The immunoprecipitated, ^{3}H-labeled protein (metallothionein) was isolated by polyacrylamide slab gel electrophoresis.

Using a combination of these characterization techniques, we have found that the amount of translatable metallothionein mRNA in liver following an injection of zinc is not uniform (Fig. 1), but increases to a maximum about 6 hr after administration of the metal (12,13).

Therefore, this measurable increase in the amount of polysomal metallothionein mRNA supports our previous estimate of the maximal rate of metallothionein synthesis using *in vivo* techniques (10). Administration of actinomycin D, prior to isolation of the polysome blocked the enrichment of polysomal metallothionein mRNA following a zinc injection (13). These data provide strong evidence that metallothionein biosynthesis is induced via a flow of information that requires the *de novo* transcription of metallothionein mRNA (Fig. 2).

REGULATION OF ZINC ACCUMULATION IN HEPATOCYTES BY GLUCOCORTICOIDS

On numerous occasions we have shown that in intact animals, there is a direct correlation between the uptake of zinc into the liver from the plasma and the binding of newly deposited zinc to metallothionein (4,8,14,15). When the synthesis of metallothionein is inhibited, via administration of actinomycin D or cordycepin, accumulation does not occur and zinc remains circulating in the plasma. In order to investigate this phenomenon in more detail, we have studied liver cells in isolation, in an attempt to establish the parameters involved in the uptake and storage process. Collagenase perfusion and selective attachment techniques were used to obtain primary cultures of rat liver parenchymal cells in high yield and high viability. These experiments have demonstrated that the accumulation of hepatic zinc is an energy dependent, temperature dependent and concentration dependent phenomena (16). Moreover, actinomycin D blocked the accumulation of zinc when added in vitro to isolated hepatocytes (17).

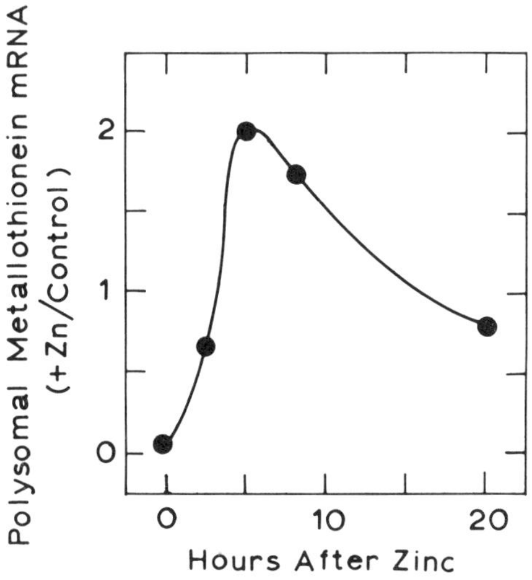

Figure 1. Polysomal metallothionein mRNA content following an injection of zinc (2.0 mg). Rat liver polysomes were isolated and the poly(A+) mRNA was translated in a cell-free protein synthesizing system (wheat germ) containing ^{3}H-glycine, lysine and serine. ^{3}H-labeled metallothionein was isolated by activated thiol Sepharose chromatography.

It was feasible, via the cell culture technique, to screen a wide variety of hormones for their influence on zinc accumulation by liver cells (16,17). We found that glucocorticoids are specific stimulators of zinc accumulation. The amount of zinc accumulated was directly proportional to the potency of the glucocorticoid. Other steroids, including both male and female sex hormones, aldosterone as well as a wide variety of peptide hormones, have no influence on the accumulation of zinc in isolated parenchymal cells. The glucocorticoid-stimulated increase in zinc accumulation appears to influence influx and/or retention processes, but does not influence the exit of zinc from these cells. Once stimulated the glucocorticoid-treated cells can accumulate additional zinc in absence of the hormone (17).

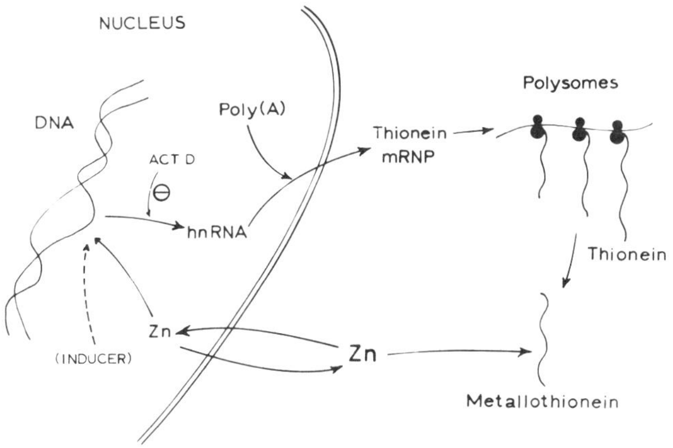

Figure 2. Proposed mechanism for hepatic metallothionein induction.

It is of paramount importance to point out that the additional zinc accumulated in hepatocytes, as a result of glucocorticoid treatment, is bound to metallothionein. These in vitro experiments clearly suggest that metallothionein has a definite role in the intracellular compartmentalization of zinc. We should point out that they were conducted with concentrations of zinc that are comparable to those found in normal rat plasma. If the concentration of zinc in the cell culture medium was elevated to approximately 10 times the normal plasma concentration, the amount of zinc incorporated into the cells (as metallothionein) is about the same as that following treatment of the cells with glucocorticoids. This latter evidence lends support to the concept that zinc is the direct inducer of metallothionein synthesis.

The observation that glucocorticoids are able to stimulate the accumulation of hepatic zinc (as metallothionein) has important implications in a wide variety of stress situations. It has been demonstrated that traumas such as serious burns, bacterial infections and environmental stresses lead to a measurable decrease in the plasma zinc concentration (18,19). Therefore, considering the important role of glucocorticoids in regulating certain aspects of liver function it is not surprising that they have a key role in hepatic zinc metabolism.

METALLOTHIONEIN DEGRADATION

We have investigated the degradation rate of zinc-metallothionein, labeled with both ^{65}Zn and ^{35}S-cystine (7), and cadmium-metallothionein, labeled with both ^{115m}Cd and ^{3}H-cystine (20), in intact animals. As shown in Fig. 3, the half life ($t\frac{1}{2}$) of induced zinc-metallothionein is between 18 and 20 hr. Moreover, it appears that zinc is removed from the polypeptide simultaneously with degradation of the thionein polypeptide chain. In contrast, the $t\frac{1}{2}$ of the thionein polypeptide of induced cadmium-metallothionein appears to be about 3.5 days, but the cadmium content of the protein remains virtually constant. In an attempt to explain this difference in $t\frac{1}{2}$, we propose that the polypeptide chain of cadmium-thionein is indeed degraded, but that cadmium is bound to nascent thionein polypeptide chains and is thus retained as metallothionein-associated metal. This hypothesis is tenable since it has been demonstrated that Cd^{2+} is able to replace Zn^{2+} in metallothionein (21). It appears from these data that we cannot explain the long term accumulation of cadmium in tissues on the basis of a lack of degradation of cadmium-metallothionein. Instead, it is likely that the marked ability of cadmium to bind to newly synthesized thionein polypeptides is the phenomenon that accounts for cadmium accumulation in some tissues.

We have also investigated the degradation of various metallothioneins using in vitro techniques. Specifically, we have tested the susceptibility of cadmium-metallothionein, zinc-metallothionein and thionein to digestion by trypsin, pronase and lysosomal proteases (22). Extracts of purified liver lysosomes were able to degrade all three forms of metallothionein, but the rate of degradation was markedly influenced by metal binding: thionein > zinc-metallothionein > cadmium-metallothionein. In contrast, neutral proteases were very effective in degrading thionein, but had little effect on metallothioneins. Thus _in situ_ it is likely that thionein is degraded in the cytoplasm, which would account for its low concentration in that fraction, whereas metallothioneins are degraded in the lysosomes.

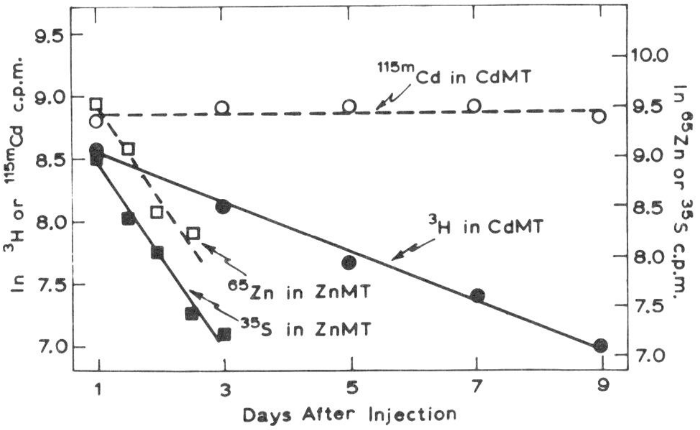

Figure 3. Degradation of liver metallothionein following induction by cadmium or zinc. The t½ of ^{65}Zn and ^{35}S-cystine labeled zinc-metallothionein was 20 and 18 hours for ^{65}Zn and ^{35}S, respectively. The t½ of ^{115m}Cd and ^{3}H-cystine labeled cadmium-metallothionein was $\gg$ 9 days and 3.6 days for ^{115m}Cd and ^{3}H, respectively.

Metallothionein appears to have an important role in hepatic zinc metabolism. It is clear that the synthesis of this protein is stimulated in response to zinc, via a mechanism that increases the amount of translatable poly(A)-containing metallothionein messenger RNA. Glucocorticoids, which specifically enhance the accumulation of zinc within hepatocytes, appear to indirectly stimulate the synthesis of this protein. Thionein seems to be the species of metallothionein with the shortest $t\frac{1}{2}$. Binding of cadmium and/or zinc markedly increases the $t\frac{1}{2}$ and concomitantly the metallothionein content of hepatocytes. This report has not focused on the important function of intestinal metallothionein in regulating zinc absorption (8,14,15). In assessing the potential role of metallothionein in biochemical processes, we cannot neglect its key role in zinc metabolism.

ACKNOWLEDGMENTS

The work from the author's laboratory discussed in this paper was supported by NIH Grants ES 00777 from the National Institute of Environmental Health Sciences; AM 18555 and AM 20485 from the National Institute of Arthritis, Metabolism and Digestive Diseases as well as by the New Jersey Agricultural Experiment Station.

REFERENCES

1. Cousins, R. J. (1978) In: "Trace Element Metabolism in Animals - III", M. Kirchgessner ed., Arbeitskreis fur Tierernahrungsforschung, Weihenstephan, W. Germany, p. 57-63.
2. Shaikh, Z. A. and Lucis, O. J. (1970) Fed. Proc. 29, 301 abs.
3. Squibb, K. S. and Cousins, R. J. (1974) Environ. Physiol. Biochem. 4, 24-30.
4. Richards, M. P. and Cousins, R. J. (1975) Bioinorg. Chem. 4, 215-224.
5. Bremner, I. and Davies, N. T. (1975) Biochem. J. 149, 733-738.
6. Webb, M. (1972) Biochem. Pharmacol. 21, 2751-2765.
7. Feldman, S. L. and Cousins, R. J. (1976) Biochem. J. 160, 583-588.
8. Richards, M. P. and Cousins, R. J. (1975) Biochem. Biophys. Res. Comm. 64, 1215-1223.
9. Richards, M. P. and Cousins, R. J. (1977) Proc. Soc. Exp. Biol. Med. 156, 505-508.
10. Squibb, K. S., Cousins, R. J., and Feldman, S. L. (1977) Biochem. J. 164, 223-228.

11. Squibb, K. S. and Cousins, R. J. (1977) Biochem. Biophys. Res. Comm. 75, 806-812.
12. Shapiro, S. G., Markowitz, L. A., Squibb, K. S., and Cousins, R. J. (1978) Fed. Proc. 37, 214 abs.
13. Shapiro, S. G., Squibb, K. S., Markowitz, L. A. and Cousins, R. J. (1978) Biochem. J. (in press). 175, 833.
14. Richards, M. P. and Cousins, R. J. (1976) J. Nutrition 106, 1591-1599.
15. Richards, M. P. and Cousins, R. J. (1976) Proc. Soc. Exp. Biol. Med. 153, 52-56.
16. Failla, M. L. and Cousins, R. J. (1978) Biochim. Biophys. Acta 538, 435-444.
17. Failla, M. L. and Cousins, R. J. (1978) Biochim. Biophys. Acta (in press). 543, 293-304.
18. Flynn, A., Pories, W. J., Strain, W. H. and Hill, D. A. (1973) Science 173, 1035-1036.
19. Beisel, W. R., Pekarek, R. S., and Wannemacher, R. W., Jr. (1976) In: "Trace Elements in Human Health and Disease" Vol. 1, A. S. Prasad ed., Academic Press, New York, p. 87-106.
20. Feldman, S. L., Squibb, K. S., and Cousins, R. J. (1978) J. Toxicol. Environ. Health (in press). 4, 805.
21. Kagi, J. H. R. and Vallee, B. L. (1960) J. Biol. Chem. 236, 2435-2442.
22. Feldman, S. L., Failla, M. L., and Cousins, R. J. (1978) Biochim. Biophys. Acta (in press). 544, 638-646.

Metallothionein
Jeremias H. R. Kägi and Monica Nordberg, eds.
Birkhäuser Verlag, Basel/Boston/Stuttgart, 1979

PARTIAL PURIFICATION AND TRANSLATION IN VITRO OF METALLOTHIONEIN MESSENGER RNA

Robert D. Andersen and Ulrich Weser

Physiologisch-chemisches Institut
Universität Tübingen
D7400 Tübingen, German Federal Republic

The rate of biosynthesis of metallothioneins has been found to increase after exposure to heavy metals (1-5), after restriction of food intake (6), during hypozincemia (7), and in response to environmental stress (8). This effect has been attributed either to translational control of protein biosynthesis (9) or to transcriptional control of mRNA biosynthesis (10,11). A homologous cell-free system (12) and a heterologous cell-free system using wheat germ (13) have been used for translation *in vitro* of metallothionein mRNA from rat liver. We report here the results of a preliminary characterization of the poly(A)$^+$ (polyadenylated) mRNA for metallothionein from rat liver, the translation of this mRNA *in vitro* in the wheat germ system, and the effects of the heavy-metal ions Cd^{2+} and Zn^{2+} on changes in the cytoplasmic concentration of this mRNA.

MATERIALS AND METHODS

Polyribosomes were isolated from rat liver homogenates, either after injection of the animals 5h previously with 10 µmol $CdCl_2$ or $ZnCl_2$ per kg body wt., or from controls, and purified by sedimentation through a 0.5M-2.0M-sucrose step gradient. Both free and membrane-bound polyribosomes were collected as a result of detergent treatment. Polyribosome fractions were prepared by sedimentation through a 0.5M-1.5M-sucrose linear gradient. Poly(A)$^+$ mRNA was isolated from sodium dodecyl sulphate-dissociated polyribosomes or polyribosome fractions by chromatography through oligo(dT)-cellulose (14). Poly(A)$^+$ mRNA was used in the wheat germ translation system either without further purification or after fractionation by sedimentation through a 5-30% sucrose linear gradient. Samples to be translated were ethanol precipitated, washed in 70, 90 and 100% ethanol and redissolved in sterile water.

Messenger RNA samples were translated in a wheat germ cell-free system (15, 16) and the translation products were carboxymethylated (17). Analysis of translation products and of purified and carboxymethylated rat liver metallothioneins 1 and 2 (5) was carried out by sodium dodecyl sulphate/20% poly-

Submitted to the "First International Meeting on Metallothionein and Other Low Molecular Weight Metal-binding Proteins", Zürich, July 17 - 22, 1978

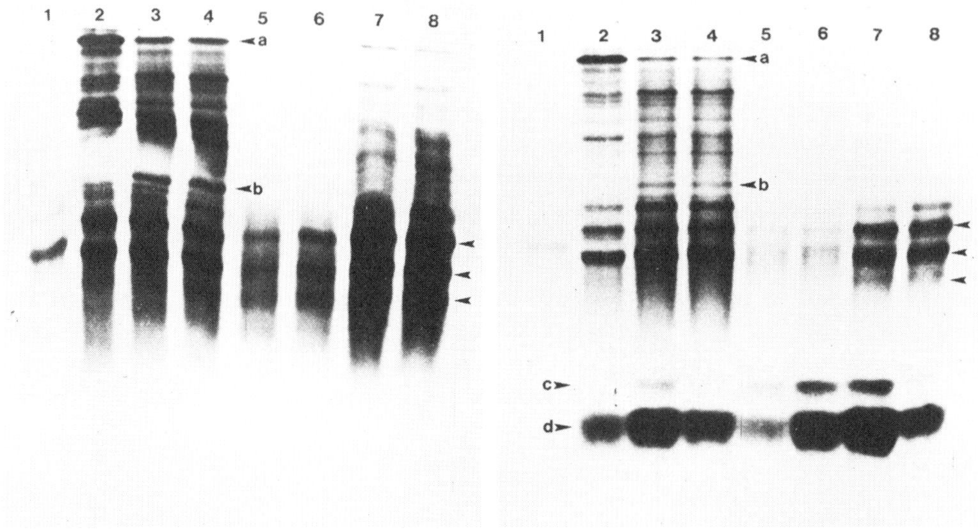

Plate 1. Electrophoresis of [^{3}H]leucine (A, left) and [^{3}H]serine (B, right) translation products produced in vitro by poly(A)$^+$ mRNA species from rat liver polyribosomes and n=2-5 polyribosome fractions.

Poly(A)$^+$ mRNA additions to each translation assay were: slot 1, none; slot 2, mRNA from controls; slot 3, mRNA from $CdCl_2$-treated animals; slot 4, mRNA from $ZnCl_2$-treated animals; slots 5, 6, 7 and 8 (n=2, 3, 4 and 5 ribosomes/polyribosome), mRNA species from $CdCl_2$-treated animals.

acrylamide-gel electrophoresis (18). Slab gels were analyzed by fluorography (19), and quantitated by digital densitometry. Further details of the above procedures have been reported (20).

RESULTS AND DISCUSSION

Metallothioneins were identified among the products of biosynthesis *in vitro*, directed by rat liver poly(A)$^+$ mRNA, on the basis of the electrophoretic mobility of their carboxymethylated derivatives in comparison with authentic metallothioneins isolated from rat liver (Plate 1B, band d). We attribute the high mobility of carboxymethylmetallothioneins in the sodium dodecyl sulphate/20% polyacrylamide-gel system to the high negative charge density on these proteins. This is due both to the large number of carboxymethylcysteine residues in these derivatives, in comparison with other rat liver proteins, and to the small size of the metallothioneins. The absence of any translation products with this high electrophoretic mobility incorporating [^{3}H]leucine (Plate 1A) and the very high specific activity of [^{3}H]serine in the metallothioneins (Plate 1B) is in agreement with the amino acid composition of these proteins from rat liver (5). In addition, a double-label experiment in which [^{3}H]leucine and [^{35}S]cysteine were both incorporated into rat liver proteins using the wheat germ system showed all ^{3}H radioactivity to be incorporated into proteins migrating more slowly than the metallothioneins and over half of the ^{35}S radioactivity to be incorporated into metallothioneins (20).

Total [^{35}S]cysteine radioactivity incorporated into metallothioneins directed by mRNA species from control, Cd^{2+}-treated and Zn^{2+}-treated rats was determined relative to total [^{3}H]leucine incorporation. A 3.7-fold increase in metallothionein biosynthesis was found using mRNA from Cd^{2+}-treated animals and a 1.7-fold increase was found with Zn^{2+}-treatment. Similar results are seen with [^{3}H]serine incorporation (Plate 1A and 1B, slots 2-4). Quantification of mRNA activity in a cell-free system is often complicated by the nonlinear response frequently encountered with increasing concentrations of nucleic acid (21). Nevertheless, the observed increase in yield of metallothionein translation products in the wheat germ cell-free system, relative to total products synthesized other than metallothioneins, is qualitative evidence for an increase in the cytoplasmic concentration of this mRNA after metal treatment. The accurate quantification of cytoplasmic metallothionein mRNA concentrations will require the use of a complementary DNA probe in a hybridization assay. On the basis of translation in the wheat germ system, we estimate that metallothionein mRNA is approx. 1% of total rat liver mRNA after induction by Cd^{2+}.

The size of the metallothionein mRNA was determined by two methods. In the first, rat liver polyribosomes were fractionated on a linear sucrose gradient and poly(A)$^+$ mRNA isolated from each of the polyribosome fractions. The translation products directed by these mRNA species can be seen in Plate 1B, slots 5-8, corresponding to mRNA species from two, three, four and five ribosomes/polyribosome respectively. The [^{3}H]serine fluorogram reveals that the metallothionein mRNA binds three or four ribosomes. The presence of consi-

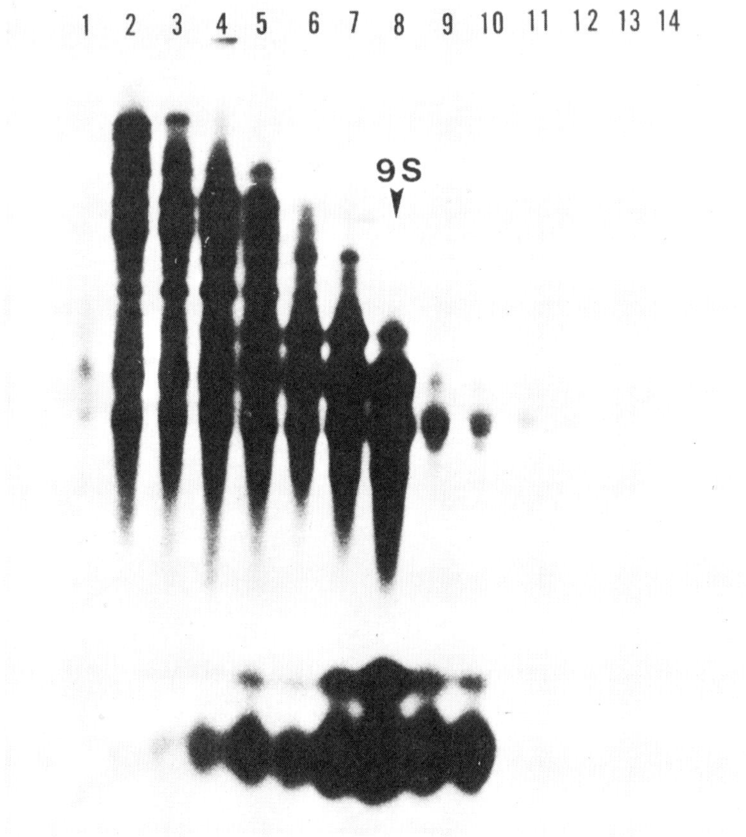

Plate 2. Analysis of sucrose gradient ultracentrifugation of poly $(A)^+$ mRNA from rat liver polyribosomes isolated from $CdCl_2$-treated rats by electrophoresis of the translation products produced in vitro.

Poly $(A)^+$ mRNA additions to each translation assay were: slot 1, none; slots 2-14, mRNA species in decreasing size from part of a 5-30% sucrose gradient fractionation of mRNA.

derable [^{3}H]leucine radioactivity (Plate 1A, slots 5-8) in higher mol. wt. proteins suggests that the poly(A)$^+$ mRNA species from these fractions contain both metallothionein and other mRNA species. We estimate a 14-fold purification of metallothionein mRNA from rat liver tetraribosomes. A second method for sizing the metallothionein mRNA was by sucrose gradient ultracentrifugation of rat liver poly(A)$^+$ mRNA isolated from $CdCl_2$-treated animals. Messenger fractions were found in which metallothionein biosynthesis in vitro was directed by 9S mRNA (Plate 2). From the distribution of translation products seen in Plate 2 we conclude that the metallothionein mRNA is one of the smallest messengers in rat liver.

The kinetics of metallothionein biosynthesis after $CdCl_2$-treatment have been found to be transient, with a maximum biosynthetic response occurring 5h after a single subcutaneous injection of this metal ion (5). The time-dependence of stimulation of metallothionein mRNA concentration in rat liver cytoplasm was found to parallel the in vivo stimulation of biosynthesis of this protein (Figure 1). In both cases there was a return to control values by 17h. This suggests that the metallothionein mRNA may have a half-life of only 6h or less. We have found a half-life for the zinc metallothioneins in rat liver of 4-5h (5). The similarity in these values suggests that the cellular controls over biosynthesis and degradation of metallothioneins are tightly coupled and that these proteins provide a rapidly expandable and contractible pool for temporary storage of zinc. This could provide the organism with a means of maintaining tight control over zinc homeostasis during periods of physiological stress resulting from exposure to toxic agents, injury or infection (7,8).

At least one other mRNA species increases upon exposure to Cd^{2+} or Zn^{2+} (Plate 1A and 1B, band b). This is a translation product with apparent mol. wt. of 22000 and contains both leucine and serine. Unlike metallothionein mRNA, the mRNA for this product remains elevated in concentrations for at least 17h after metal treatment. Another mRNA species, coding for a 65000-70000-dalton protein, is markedly decreased in concentration after metal-treatment (Plate 1A and 1B, band a). The size and amount of this translation product suggest that it may be preproalbumin, whose mRNA constitutes 11% of total rat liver mRNA (22).

Species of mRNA which sediment at 9S (eg. rabbit globin [23], human β-globin [24], and uteroglobin [25]) are found to contain approx. 600-650 nucleotides. Metallothioneins whose amino acid sequences have been determined are 61 amino acid residues long (26-28). This is considerably less than the length of globin chains (141 residues in the α-chain, 146 residues in the β-chain) or the preuteroglobin chain (80-90 residues), suggesting that metallothionein mRNA may have correspondingly larger untranslated regions, about 400 nucleotides. The identity in electrophoretic mobility between metallothioneins translated in vitro and native metallothioneins, and the absence of post-translational processing in the wheat germ system, suggest that a premetallothionein is not synthesized.

Alternatively, the metallothionein mRNA is large enough to accomodate two independent cistrons of approx. 180 nucleotides each. The existence of metallothionein tetraribosomes (Plate 1B, slot 7) suggests that at least 300 nucleotides are covered by bound ribosomes. There is no known precedent for

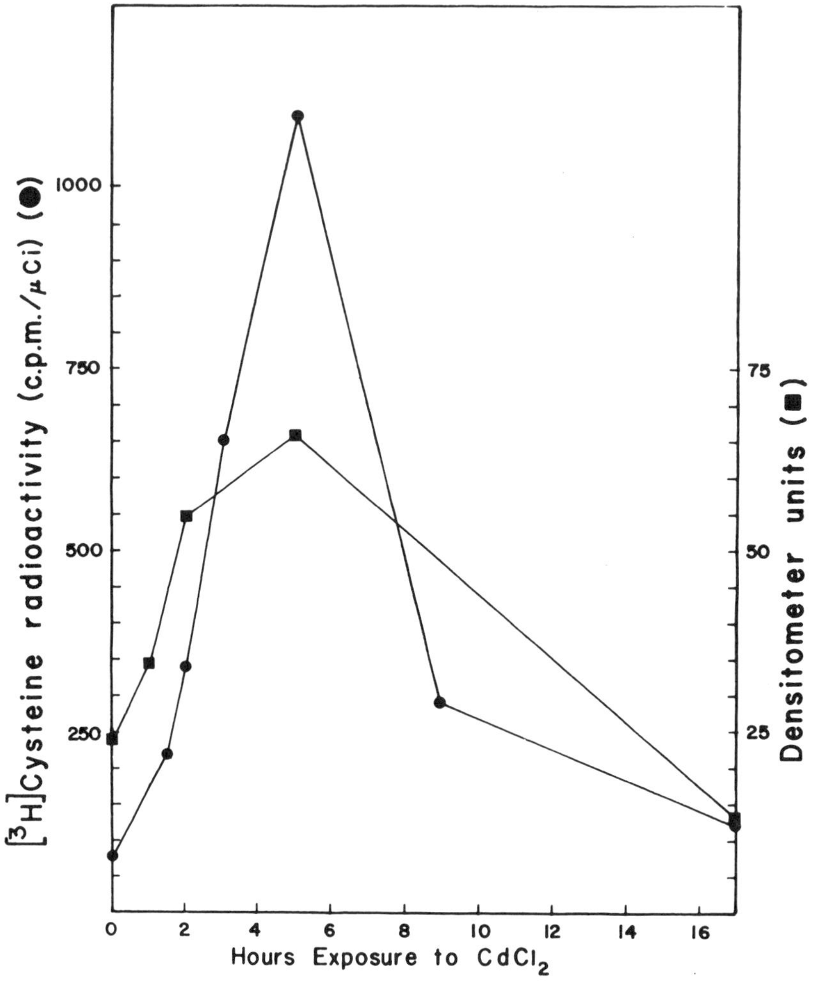

Figure 1. Time-dependence of metallothionein biosynthesis and mRNA formation in rat liver after a single injection of $CdCl_2$.

Rats were injected subcutaneously with 10μmol of $CdCl_2$/kg body wt. at the times indicated. Biosynthesis of metallothioneins in vivo was determined by a 1h labelling period with [^{3}H]cysteine before determining total radioactivity incorporated into metallothioneins. Poly $(A)^+$ mRNA was isolated in a separate experiment and translated in the wheat germ system. The radioactivity incorporated into metallothioneins was determined by densitometry of the fluorogram.

Plates 1 and 2 and Figure 1 are reproduced with the permission of Biochemical Journal.

a eukaryotic mRNA with two initiation sites. It is not known whether the presence of two initiation sites would be required for the synthesis of the two isometallothioneins found in liver. The existence of a polycistronic messenger in the case of the metallothioneins may be required to facilitate post-transcriptional processing of the mRNA and to ensure the stability of the polyribosome during translation of these low mol. wt. proteins.

The translation product identified as band c in Plate 1B, and which does not contain leucine (Plate 1A), may result from read-through of the metallothionein mRNA into a second cistron leading to the synthesis of a double-length molecule. Further analysis of these translation products will be necessary to identify the nature of the material in band c and to substantiate the claim for a polycistronic metallothionein mRNA. We are presently engaged in cloning the metallothionein mRNA sequence in a chimeric bacterial plasmid (29). This should provide us with more conclusive evidence on the origin of the two major isoproteins of metallothionein, and allow further investigations to be carried out into the mechanism of regulation of metallothionein gene activation.

ACKNOWLEDGEMENT

Parts of this study were aided by a grant from the Deutsche Forschungsgemeinschaft (We 401/14). We thank Dr. Ulrich Weidle, Professor Hans Probst, Dr. Stephan Jenke and Arno Fraterman of the Physiologisch-chemisches Institut for technical assistance and many stimulating discussions during the course of this work. We gratefully acknowledge the help of Dr. Klaus Bartel of the Max Planck Institut, Martinsried, Germany, and Richard Fisher of the Molecular Biology Institut, University of California (Los Angeles), for making available the digital densitometry equipment and associated computing facilities for quantification of [^{3}H]serine fluorograms.

REFERENCES

1. Shaikh, Z. A. and Lucis, C. J. (1970) Fed. Proc. (Abstracts) 29, 301.
2. Squibb, K. S. and Cousins, R. J. (1974) Environ. Physiol. Biochem. 4, 24-30.
3. Chen, R. W., Whanger, P. D. and Weswig, P. H. (1975) Biochem. Med. 12, 95-105.
4. Shaikh, Z. A. and Smith, J. C. (1975) Fed. Proc. (Abstracts) 34, 271.
5. Andersen, R. D., Winter, W. P., Maher, J. J. and Bernstein, I. A. (1978) Biochem. J. 174, 327-338.
6. Bremner, I. and Davies, N. T. (1975) Biochem. J. 149, 733-738.
7. Sobocinski, P. Z., Canterbury, W. J., Jr., Mapes, C. A. and Dinterman, R. E. (1978) Am. J. Physiol. 234, E399-E406.
8. Oh, S. H., Deagen, J. T., Whanger. P. D. and Weswig, P. H. (1978) Am. J. Physiol. 234, E282-E285.
9. Webb, M. (1972) Biochem. Pharmacol. 21, 2767-2771.

10. Richards, M. P. and Cousins, R. J. (1975) Biochem. Biophys. Res. Commun. 64, 1215-1223.
11. Squibb, K. S., Cousins, R. J. and Feldman, S. L. (1977) Biochem. J. 164, 223-228.
12. Squibb, K. S. and Cousins, R. J. (1977) Biochem. Biophys. Res. Commun. 75, 806-812.
13. Shapiro, S. G., Markowitz, L. A., Squibb, K. S. and Cousins, R. J. (1978) Fed. Proc. (Abstracts) 37, 253.
14. Kabat, D. (1975) J. Biol. Chem. 250, 6085-6092.
15. Marcu, K. and Dudock, B. (1974) Nucleic Acids Res. 1, 1385-1397.
16. Roberts, B. E. and Paterson, B. M. (1973) Proc. Natl. Acad. Sci. (USA) 70, 2330-2334.
17. Schechter, I. and Burstein, Y. (1976) Proc. Natl. Acad. Sci. (USA) 73, 3273-3277.
18. Laemmli, U. K. (1970) Nature 227, 680-685.
19. Laskey, R. A. and Mills, A. D. (1975) Eur. J. Biochem. 56, 335-341.
20. Andersen, R. D. and Weser, U. (1978) Biochem. J. (in press). 175, 841-852.
21. Rhoads, R. E., McKnight, G. S. and Schimke, R. R. (1973) J. Biol. Chem. 248, 2031-2039.
22. Tse, T. P. H. and Taylor, J. M. (1977) J. Biol. Chem. 252, 1272-1278.
23. Chantrenne, H., Burny, A. and Marbaix, G. (1967) Prog. Nucl. Acid Res. and Mol. Biol. 7, 173-194.
24. Marotta, C. A., Forget, B. C., Cohen-Solal, M. and Weissman, S. M. (1976) Prog. Nucl. Acid Res. and Mol. Biol. 19, 165-175.
25. Atger, M. and Milgrom, E. (1977) J. Biol. Chem. 252, 5412-5418.
26. Kojima, Y., Berger, C., Vallee, J. H. R. (1977) Proc. Natl. Acad. Sci. (USA) 73, 3413-3417.
27. Kissling, M. M. and Kägi, J. H. R. (1977) FEBS Letters 82, 247-250.
28. Huang, I-Yih and Yoshida, Akira (1977) J. Biol. Chem. 252, 8217-8221.
29. Andersen, R. D. and Herschman, H. R. (work in progress).

ROLE OF METALLOTHIONEIN
IN METAL METABOLISM AND TOXICITY

Metallothionein
Jeremias H. R. Kägi and Monica Nordberg, eds.
Birkhäuser Verlag, Basel/Boston/Stuttgart, 1979

FUNCTIONS OF HEPATIC AND RENAL METALLOTHIONEINS IN THE CONTROL OF THE METABOLISM OF CADMIUM AND CERTAIN OTHER BIVALENT CATIONS

M. Webb

(Toxicology Unit, M.R.C. Laboratories, Woodmansterne Road, Carshalton, Surrey, SM5 4EF, England).

As the first metallothionein to be isolated (1) contained a high content of Cd^{2+}, it was a logical assumption that one function of thionein might be to provide a defence mechanism against this toxic cation. This possibility, amongst others, was considered by Kägi and Vallee (2,3) and was developed by Piscator (4) as an hypothesis to explain the progressive accumulation of biologically inert Cd^{2+} in the livers and kidneys of rabbits, exposed to repeated small doses of $CdCl_2$. The detoxification function of thionein against low level exposure to either parenterally, or orally, administered Cd^{2+}, was confirmed in a number of later investigations (5-10, see also ref.11). These investigations established clearly that much of the high body burden of Cd^{2+} accumulated by the experimental animal under these conditions was present as the metallothionein in the livers and kidneys, and thus was unavailable for interaction with processes with important biological functions. Other studies suggested that the inducible synthesis of thionein in these organs is not limited to Cd^{2+}, but probably functions under conditions of either chronic exposure to Hg^{2+} (12), or increased tissue concentrations of the essential, but potentially toxic, Zn^{2+} and Cu^{2+} cations (e.g. refs. 13 & 14).

Pretreatment of experimental animals with a low dose of Cd^{2+}, or higher dose of Zn^{2+}, is known to protect them against a subsequent, normally lethal, dose of Cd^{2+} (15) and also to prevent the Cd^{2+}-induced testicular damage (16,17), placental haemorrhage (18) and foetal malformations (19-21). As both pretreatment cations stimulate the synthesis of the corresponding metallothioneins, protection against such acute effects of Cd^{2+} also has been attributed to these metalloproteins (6,8,22-25). Since Zn^{2+} is a common component of the metallothioneins that are induced by both Cd^{2+} and Zn^{2+}, it is possible that such protection could result from the replacement of this cation by Cd^{2+}. Alternatively, pretreatment might eliminate the lag in thionein synthesis (26), such that further production of the protein is an immediate response to subsequent challenge with Cd^{2+}. There is some experimental support for both of these possibilities. Thus Cd^{2+} has been shown to displace Zn^{2+} from zinc-thionein (7,8) and from (cadmium, zinc)-thionein (22,24), in the livers of animals pretreated with Zn^{2+} and Cd^{2+} respectively. Yoshikawa (27) and Suzuki and Yoshikawa (22) consider that this

Submitted to the "First International Meeting on Metallothionein and Other Low Molecular Weight Metal-binding Proteins", Zürich, July 17 - 22, 1978

replacement of Zn^{2+} in presynthesized (cadmium, zinc)-thionein leads to a more rapid accumulation and immobilization of Cd^{2+} in the liver of the pretreated animal and, as a result of this, uptake of the cation by other organs is decreased. According to Leber and Miya (24) the hepatic concentration of (cadmium, zinc)-thionein in mice, and thus of replaceable Zn^{2+}, increases with the pretreatment dose, as does the tolerance to Cd^{2+} on subsequent exposure. A positive correlation between the hepatic content of this metallothionein and the LD_{50} of Cd^{2+} in mice after pretreatment with different doses of the cation also has been reported by Probst et al. (25). Treatment of rats with an oral dose of Cd^{2+} (20 mg/kg) was found by Squibb et al. (23) to protect against a second (100 mg Cd^{2+}/kg), given after 24h. and to increase the uptake of Cd^{2+} not only in the liver, but also in the kidneys and testes (cf. Yoshikawa (27), Suzuki and Yoshikawa (28)). In non-pretreated (control) animals, hepatic thionein synthesis exhibited the usual lag and thus, during the first few hours after the oral administration of Cd^{2+} (100 mg/kg), most of the cation that was taken up by the liver was associated with the high molecular weight proteins of the cytosol. In the pretreated animals, however, the lag phase was eliminated and, from the earliest times after the administration of the second dose, all of the Cd^{2+} in the soluble fraction of the liver was bound to thionein (23).

Whilst these investigations seem to support the concept of a protective function of pre-induced (cadmium, zinc)-thionein against the acute toxicity of Cd^{2+} there are other observations which are difficult to reconcile with this view. For example, protection can be obtained with, and is effective against, other cations that do not induce thionein synthesis. Thus In^{3+} and Mn^{2+} protect against acute doses of Cd^{2+} (28), and pretreatment with Pb^{2+} confers protection against toxic doses of Pb^{2+} (29). Also, protection against Cd^{2+} in rats has been shown to be maximal 1-3 days after pretreatment with a low dose of Cd^{2+} and then to decrease with time (30). Both the content of the pre-induced metallothionein, as well as the capacity for the immediate synthesis of this metalloprotein, however, were retained in the liver of the pretreated animal for a much longer period. In agreement with the observations of Yoshikawa (27), Suzuki and Yoshikawa (22), Squibb et al. (23), and Cherian and Vostal (31), hepatic uptake of Cd^{2+} was found to be much greater in the pretreated animals than in the non-pretreated controls. Uptake of the cation by the kidney, spleen, pancreas, brain and heart, however, was unaltered, whilst faecal excretion was decreased. The increased retention in the liver, therefore, probably was due to decreased biliary excretion of Cd^{2+} that is known to result from pretreatment (31).

The synthesis of hepatic zinc-thionein is stimulated by restriction of food intake (32), but Webb and Verschoyle (31) have shown that the intravenous LD_{50} of Cd^{2+} in starved rats is the same as that in normally fed animals. Also, even though weanling rats contain very high concentrations of zinc-thionein in their livers, the LD_{50} of intraperitoneally administered Cd^{2+} is not significantly different from that in adult females (G.P. Samarawickrama and M. Webb, unpublished observations).

The above discussion suggests, therefore, that the acute toxicity of Cd^{2+} is not determined by the hepatic concentration of presynthesized zinc-, or (cadmium, zinc)-thionein. Webb and Magos (33) also conclude that the presence of the latter metallothionein in the kidneys of Cd^{2+}-pretreated rats cannot explain the resistance of these animals to normally nephrotoxic doses of Hg^{2+}. In this work, for example, Cd^{2+}-pretreatment was found to increase not only the renal content of thionein-bound-Hg^{2+}, as observed previously by Suzuki (34) and Shaikh et al. (35), but also the contents of Hg^{2+} in other cellular components. As, at normally nephrotoxic doses, increased Hg^{2+}-incorporation into these components was greater than that into the metallothionein, it seems that there is no obvious reason to attribute the protective effect of Cd^{2+} to the induction of thionein synthesis, and other mechanisms (cf. e.g. 36) seem more probable.

Whilst, therefore, a role for thionein in the detoxification of certain heavy metal ions seems to be established clearly only at continuous low level exposure, this is unlikely to be the normal physiological function of the protein. It has been suggested that either thionein itself (37), or its copper derivative (38), may be the biologically active molecule with functions in the maintenance of redox potentials (37), ion transport (37), metabolic pools of cysteine residues (39), and in bioenergetic systems, particularly when the contents of cytochrome c oxidase are low (38). There is also much evidence that the primary biological function of this inducible protein is related to the metabolism of the essential Zn^{2+} cation (7,8,32,40-44). According to Chen et al. (43,45), zinc-thionein, which accumulates in the livers of rats that are maintained on a diet with high levels of Zn^{2+}, is eliminated within 3 days when the animals are transferred to a Zn^{2+}-deficient diet, the loss being associated with increased urinary and faecal excretion of low molecular weight Zn^{2+} complexes. They conclude, therefore, that thionein has a fundamental role in the accumulation of excess Zn^{2+}, rather than in the storage of the cation for subsequent utilization. A conservation function, however, is indicated by the formation of hepatic zinc-thionein in the rat during post-surgical trauma (46) and in response to starvation (32,44).

Regulation of Zn^{2+} absorption in rats, attributed by Cotzias et al. (47) to negative-feedback control, was shown by Evans et al. (48) to be mediated, at least in part, by the Zn^{2+}-content of the intestinal mucosa which, in turn, is regulated by the Zn^{2+}-concentration of the plasma. Previously a function of thionein in the regulation of the absorption and/or transport of both Zn^{2+} and Cu^{2+} seemed to have been established by the isolation of metalloproteins, tentatively identified as zinc- and copper-thioneins from the intestinal mucosa of the chick (49), rat (50) and bovine (51). Later work by Evans and collaborators (48,52-55), however, established that Zn^{2+}-absorption is determined not by the formation of zinc-thionein, which is absent from the intestinal mucosa of the rat when the rate of absorption is high, but by a low molecular weight peptide (55), that acts as a ligand for Zn^{2+} in the mucosa, and the degree of saturation of the cation-binding sites of the carrier protein, albumin (53,56), or transferrin (57), in serum. This work was confirmed and extended by Richards and Cousins (44,58,59) who produced evidence that absorption of Zn^{2+} in the rat was related directly to the presence in the intestinal mucosa of the Zn^{2+}-chelate of the low molecular weight ligand and inversely to the synthesis of zinc-thionein. These authors suggest that, in cells of both the liver and intestinal mucosa, the contents of this metallothionein are correlated with the serum Zn^{2+}-concentration; in the former, synthesis of zinc-thionein is considered to control uptake and storage of Zn^{2+} and, in the latter, to form an alternative cation-binding species, which acts competitively with the Zn^{2+}-carrier peptide to regulate the transfer of the cation to the blood.

At present, there are difficulties in the application of this regulatory hypothesis to the pregnant animal, in which transfer of Zn^{2+} from the mother to the foetus must be related to the Zn^{2+}-concentration in the maternal blood. As was observed initially by Kägi (60), metallothioneins seem to be present in large amounts in certain foetal tissues. High concentrations of zinc-thionein, for example, occur in the livers of foetal rats (61, G.P. Samarawickrama and M. Webb, unpublished observations) rabbits and human-beings (A. Bakka and M. Webb, unpublished observations), and of (copper,zinc)-thioneins in the livers of murine (A. Bakka and M. Webb, unpublished observations), bovine (62) and probably porcine (63) foetuses. In the liver of the foetal rat, the concentration of thionein-bound-Zn^{2+} may exceed 70 µg/g wet wt. tissue and yet be at or near the limit of detection by conventional methods of analysis in the maternal liver (G.P. Samarawickrama and M. Webb, unpublished observations). Although the variability in cation contents of different foetal metallothioneins might be considered to be indicative of a storage or protective role, there is some evidence to suggest that hepatic zinc-thionein of the rat foetus, at least, may be of functional significance (A. Bakka, G.P. Samarawickrama and M. Webb, unpublished observations). Thus, in these foetal livers, the content of thionein-bound-Zn^{2+} increases rapidly from the 16th day of gestation and, at birth, may be between 70 and 100 μg/g wet wt. tissue. After birth, the hepatic concentration of the metallothionein at first increases, but at a slower rate, to a maximum at about 7 days. It then falls almost to zero at 18 days. Intravenous administration of Cd^{2+} (1.25 mg/kg) to the pregnant rat on the 16th day of gestation prevents subsequent accumulation of zinc-thionein in the foetal livers. This seems to be due to the inhibition of Zn^{2+}-transport by Cd^{2+} in the maternal placenta, and not to the small amount of the latter cation that enters the foetus at this dose level, and which is bound by the metallothionein of the foetal liver. Inhibition of Zn^{2+}-transport persists for the remainder of gestation, although it decreases slowly (e.g. from 84% at 4h. after Cd^{2+}-administration to 66% at 48h.) with time.

If accumulation of zinc-thionein provides a defence mechanism against excess Zn^{2+} during normal foetal development, the content of the metalloprotein might be expected to remain low in the livers of the newborn pups of the Cd^{2+}-treated mothers. In these litters, however, the hepatic concentration of zinc-thionein increases to reach approximately the same maximum (60-80 μg thionein-bound-Zn^{2+}/g wet wt. tissue) at the same time as that in normal newborn animals. Fractionation of the liver cytosol of the weanling animal shows that, once the content of hepatic zinc-thionein reaches its maximum and begins to decrease, the contents of both Zn^{2+} and protein in one of the high molecular weight fractions increase. The gain in Zn^{2+} by this heterogeneous fraction which, as shown originally by Bremner and Marshall (64) and Bremner and Davies (46), contains superoxide dismutase and carbonic anhydrase, is not stoichiometric with the loss from the metallothionein; an observation that possibly is to be expected from the findings of Chen _et al_. (43,45) on the elimination of thionein-bound-Zn^{2+} from the liver of the Zn^{2+}-loaded adult rat. These results, therefore, indicate but, at present, provide no firm evidence to support, the possibility that accumulation of hepatic zinc-thionein during late foetal and early post-natal life in the rat may be related to subsequent requirements for the cation in the synthesis of other metalloproteins at later stages of development. Such a function has been considered previously for neonatal mitochondrocuprein (65), which is regarded as a polymeric form of partially loaded copper-thionein (65,66), and may act as a reservoir of Cu^{2+} for the subsequent formation of cytochrome-c oxidase, the content of which increases rapidly during the neonatal period.

It is possible that, if copper-thionein undergoes polymerization in vivo to mitochondrocuprein, the presence of zinc-thionein in some foetal livers, and of (copper, zinc)-thionein in others, may be correlated with the content of this insoluble copper-protein. This would imply that copper- and zinc-thioneins occur as separate species, and thus would be contrary to current concepts that both cations are present, though in different ratios, in all forms of (copper, zinc)-thioneins (67,68). Whilst this may be true when the Cu:Zn ratio is high, it has been shown that crude preparations of (copper, zinc)-thionein (atomic ratio Cu:Zn = 1:1) from 1-8 day old pig liver, yield zinc-thionein as a single molecular species on preparative electrophoresis (63). No evidence has been obtained, however, that the hepatic content of mitochondrocuprein in newborn pigs is related to the tissue content of Zn^{2+} (63). Such a relationship might be expected if Zn^{2+} either prevents the polymerization of copper-thionein to mitochondrocuprein, or plays a role in the metabolism and elimination of this insoluble protein from the liver of the newborn.

REFERENCES

1. Margoshes, M. and Vallee, B.L. (1957) J. Amer. Chem. Soc. 79, 4813-4814.
2. Kägi, J.H.R. and Vallee, B.L. (1960) J. Biol. Chem. 235,3460-3465.
3. Kägi, J.H.R. and Vallee, B.L. (1961) J. Biol. Chem. 236,2435-2442.
4. Piscator, M. (1964) Nord. Hyg. Tidskr. 45, 76-82.
5. Nordberg, G.F., Piscator, M. and Lind, M. (1971) Acta Pharmacol. Toxicol. 29, 456-470.
6. Nordberg, G.F. (1971) Environ. Physiol. 1, 171-187.
7. Webb, M. (1972) Biochem. Pharmacol. 21, 2751-2765.
8. Webb, M. (1972) Biochem. Pharmacol. 21, 2767-2771.
9. Squibb, K.S. and Cousins, R.J. (1974) Environ. Physiol. Biochem. 4, 24-30.
10. Stonard, M. and Webb, M. (1976) Chem.-Biol. Interact. 15, 349-363.
11. Friberg, L., Piscator, M., Nordberg, G.F. and Kjellström, T. (1974) Cadmium in the Environment, 2nd Ed. Cleveland, CRC Press.
12. Piotrowski, J.K., Trojanowska, B. and Sapota, A. (1974) Arch. Toxicol. 32, 351-360.
13. Bremner, I and Davies, N.T. (1973) Rep. Rowett Inst. 29, 126-135.
14. Webb, M. (1977) In "Clinical Chemistry and Chemical Toxicology of Metals", S. Brown ed., Amsterdam. Elsevier p.51-64.
15. Terhaar, C.J., Vis, E., Roudabush, R.L. and Fassett, D.W. (1965) Toxicol. Appl. Pharmacol. 7, 500.
16. Pařizek, J. (1960) J. Reprod. Fertil. 1, 294-309.

17. Gunn, S.A., Gould, T.C. and Anderson, W.A.D. (1961) Arch. Pathol. 71, 274-281.
18. Pařizek, J. (1964) J. Reproduct. Fertil. 7, 263-265.
19. Ferm, V.H. and Carpenter, S.J. (1967) Nature (Lond) 216, 1123.
20. Ferm, V.H. and Carpenter, S.J. (1968) Lab. Invest. 18, 429-432.
21. Semba, R., Ohta, K. and Yanamura, H. (1974) Teratol. 10, 96.
22. Suzuki, Y. and Yoshikawa, H. (1974) Ind. Health 12, 141-151.
23. Squibb, K.S., Cousins, R.J., Silbon, B.L. and Levin, S.(1976) Exptl. Mol. Pathol. 25, 163-171.
24. Leber, A.P. and Miya, T.S. (1976) Toxicol. Appl. Pharmacol. 37, 403-414.
25. Probst, G.S., Bosquet, W.F. and Miya, T.S. (1977) Toxicol. Appl. Pharmacol. 39, 61-69.
26. Cempel, M. and Webb, M. (1976) Biochem. Pharmacol. 25, 2067-2071.
27. Yoshikawa, H. (1973) Ind. Health 11, 113-119.
28. Yoshikawa, H. (1970) Ind. Health 8, 184-191.
29. Sanai, G., Hasegawa, T. and Yoshikawa, H. (1972) J. Occup. Med. 14, 301-305.
20. Webb, M. and Verschoyle, R.D. (1976) Biochem. Pharmacol. 28, 673-679.
31. Cherian, M.G. and Vostal, J.T. (1974) Toxicol. Appl. Pharmacol. 29, 141.
32. Bremner, I., Davies, N.T. and Mills, C.F. (1973) Biochem. Soc. Trans. 1, 982-985.
33. Webb, M. and Magos, L. (1976) Chem. Biol. Interact. 14, 357-369.
34. Suzuki, Y. (1972) Ind. Health 10, 56-58.
35. Shaikh, Z.A., Lucis, O.J. and Coleman, B.L. (1973) Proc. 7th Conf. Trace Substances in Environmental Health. Columbia Univ. Missouri, Mo. p 313-321.
36. Kempson, S.A., Ellis, B.G. and Price, R.G. (1977) Chem.-Biol Interact. 18, 217-234.
37. Kägi, J.H.R., Himmelhoch, S.R., Whanger, P.O., Bethune, J.L. and Vallee, B.L. (1974) J. Biol. Chem. 249, 3537-3542.
38. Prinz, R. and Weser, U. (1975) Hoppe-Seyler's Zeit. Physiol. Chem. 356, 767-776.
39. Weser, U., Rupp, H., Donay, F., Linnemann, F., Voelter, W., Voetsch, W. and Jung, G. (1973) Eur. J. Biochem. 39, 127-141.
40. Bühler, R.H.O. and Kägi, J.H.R. (1974) FEBS lett. 39, 229-233.
41. Richards, M.P., Bluestone, J.A., Newkirk, M.D. and Cousins, R.J. (1974) Fed. Proc. 33, 699.
42. Chen, R.W., Eakin, D.J. and Whanger, P.D. (1974) Nutr. Rep. Internat. 10, 195-200.
43. Chen, R.W., Whanger, P.D. and Weswig, P.H. (1975) Fed. Proc. 34, 927.
44. Richards, M.P. and Cousins, R.J. (1976) J. Nutr. 106, 1591-1599.

45. Chen, R.W., Vasey, E.J. and Whanger, P.D. (1977) J. Nutr. 107, 805-813.
46. Bremner, I. and Davies, N.T. (1975) Biochem. J. 149, 733-738.
47. Cotzias, G.C., Borg, D.C. and Selleck, B. (1962) Amer. J. Physiol. 202, 359-363.
48. Evans, G.W., Grace, C.I. and Hahn, C. (1973) Proc. Soc. Exptl. Biol. Med. 143, 723-725.
49. Starcher, B.C. (1969) J. Nutr. 97, 321-326.
50 Evans, G.W. and Cornatzer, W.E. (1970) Fed. Proc. 29, 695.
51. Evans, G.W., Majors, P.F. and Cornatzer, W.E. (1970) Biochem. Biophys. Res. Commun. 40, 1142-1148.
52. Hahn, C.J. and Evans, G.W. (1973) Proc. Soc. Exptl. Biol. Med. 144, 793-795.
53. Evans, G.W., Grace, C.I. and Hahn, C.J. (1974) Bioinorganic Chem. 3, 115-120.
54. Evans, G.W., Wolenetz, M.L. and Grace, C.I. (1975) Nutr. Rep. Internat. 12, 261-269.
55 Evans, G.W. and Hahn, C.J. (1974) Adv. Exptl. Med. Biol. 48, 285-297.
56. Suso, F.A. and Edwards, H.M. (1971) Proc. Soc. Exptl. Biol. Med. 137, 305-309.
57. Evans, G.W. and Winter, T.W. (1975) Biochem. Biophys. Res. Commun. 65, 1218-1224.
58. Richards, M.P. and Cousins, R.J. (1975) Biochem. Biophys. Res. Commun. 64, 1215-1223.
59. Richards, M.P. and Cousins, R.J. (1975) Bioinorganic Chem. 4, 215-224.
60. Kägi, J.H.R. (1970) Proc. 5th Internat. Cong. Biochem. p.130-131.
61. Kaszpar, B.W., Piotrowski, J.K., Marciniak, W. and Sielczynska, M. (1976) Bromat. Chem. Toksykol. 9, 315-326.
62. Hartmann, H.J. and Weser, U. (1977) Hoppe Seyler's Zeit. Physiol. Chem. 491, 211-232.
63. Webb, M., Plastow, S. and Magos, L. (1978) in preparation.
64. Bremner, I. and Marshall, R.B. (1974) Br. J. Nutr. 32, 283-291.
65. Porter, H. (1974) Biochem. Biophys. Res. Commun. 56, 661-668.
66. Rupp, H. and Weser, U. (1974) FEBS Lett. 44, 293-297.
67. Bremner, I. and Young, B.W. (1976) Biochem. J. 155, 631-635.
68. Bremner, I. and Young, B.W. (1976) Biochem. J. 157, 517-520.

Metallothionein
Jeremias H. R. Kägi and Monica Nordberg, eds.
Birkhäuser Verlag, Basel/Boston/Stuttgart, 1979

METALLOTHIONEIN IN INTESTINAL COPPER METABOLISM

Gary W. Evans

United States Department of Agriculture
Human Nutrition Laboratory
Grand Forks, North Dakota, U.S.A.

Approximately a decade after Margoshes and Vallee (1) first identified metallothionein, Evans et al. (2) suggested that this sulfhydryl-rich protein may be involved in the metabolism of copper in both the liver and the intestine. Recently, Bremner and Young (3) demonstrated that copper is bound to metallothionein in mammalian liver. Subsequently, after obtaining some erroneous results (4), this author finally succeeded in demonstrating that copper is bound to metallothionein in rat intestine (5).

Since copper is bound to intestinal metallothionein, we have attempted to delineate the function of this protein in intestinal copper metabolism. The following report describes the relationship between metallothionein and copper in rats fed varying levels of copper and in rats given varying levels of oral copper.

MATERIALS AND METHODS

Isolation and Identification of Intestinal Copper-Thionein

To determine the effect of varying dietary copper levels on intestinal copper content and intestinal copper-

Submitted to the "First International Meeting on Metallothionein and Other Low Molecular Weight Metal-binding Proteins", Zürich, July 17 - 22, 1978

binding proteins, 30-day old male Sprague-Dawley rats were fed a copper-deficient diet (6) that contained 0.57 μg Cu/g diet. After four weeks, the animals were supplemented with copper by the addition of 2.5 μg Cu/ml, 25 μg Cu/ml and 50 μg Cu/ml in the form of copper acetate to the drinking water. Ten days after the addition of copper to the water, the rats were decapitated and a 25 cm segment of the small intestine, beginning at the pylorus, was removed from each animal. The intestinal segments were rinsed with 100 ml 0.9% NaCl after which the segments were prepared for copper analysis or gel-filtration chromatography.

To determine the copper concentration of the intestine from rats fed various levels of copper, the intestines from four rats in each group were freeze-dried, weighed dry and subsequently wet-ashed with a mixture of nitric acid, hydrogen peroxide and sulfuric acid. Blanks were prepared in an identical manner. The copper content of the ashed samples and blanks was determined with a Varian[1] atomic absorption spectrometer set at 324.7 nm. Results are expressed as μg Cu/g dry wt.

Copper-binding proteins in the intestinal cytosol were detected and subsequently purified by gel-filtration chromatography. Rinsed intestines from four rats in each of the three groups were pooled and homogenized in 1.5 volumes 10 mM Tris-acetate buffer, pH 7.4, that contained 100 mg NaN_3/l. The buffer solution was saturated with N_2. After homogenization, the intestinal preparations were centrifuged at 105,000 g for 60 min. Following centrifugation, the supernatant solutions were removed and applied to a 2.5 x 90 cm column packed with Sephadex G-75 (Pharmacia Fine Chemicals, Piscataway, NJ) that had been equilibrated with the N_2-saturated Tris-acetate buffer described above. The samples were eluted from the column with the same buffer and 6.0 ml fractions were collected. The copper content of the

---

[1]Mention of a trademark or proprietary product does not constitute a guarantee or warranty of the product by the U.S. Department of Agriculture, and does not imply its approval to the exclusion of other products that may also be suitable.

eluted fractions was assayed directly by atomic-absorption spectrophotometry.

Since the copper in the intestinal preparations from each of the three groups of rats was eluted from Sephadex G-75 in fractions that corresponded to a single, identical peak, the copper-containing fractions from the three groups were pooled and concentrated in an Amicon Ultrafiltration Cell equipped with a Diaflo Ultrafiltration Membrane, Type UM2 (Amicon Corporation, Lexington, MA). Ultrafiltration was carried out under N_2 at 60 psi. Following ultrafiltration, the intestinal copper-binding protein was purified by ion-exchange chromatography on DEAE-Sephadex (Pharmacia Fine Chemicals) and gel-filtration chromatography on Bio-Gel P-10 exactly as described for hepatic copper-binding proteins by Bremner and Young (3).

Following purification, the protein sample was oxidized with performic acid and subsequently hydrolyzed with 6N-HCl at 110°C for 21 hr. Amino acid analysis was carried out with a Beckman Model 119 Amino Acid Analyzer.

Incorporation of Labeled Amino Acid Into Intestinal Metallothioneins

Adult male Sprague-Dawley rats (250-300 g) that had been fed a diet of Purina Lab Chow were fasted for 18 hr. Thereafter, 4 rats each were stomach-tubed with a 1.0 ml solution of 0.9% NaCl, 0.9% NaCl that contained 1 μg Cu^{2+} ($CuSO_4 \cdot 5H_2O$), or 0.9% NaCl that contained 20 μg Cu^{2+} ($CuSO_4 \cdot 5H_2O$). Thirty minutes after administration of the saline-copper solutions, each rat was stomach-tubed with 0.5 ml of 0.9% NaCl that contained 5 μCi ^{14}C-lysine (New England Nuclear, Boston, MA). Two hours after administration of the labeled lysine, the rats were decapitated and a 25 cm segment of small intestine, distal to the pylorus, was removed. The intestinal contents were rinsed out with 100 ml 0.9% NaCl, the segments were homogenized in N_2-saturated Tris-acetate buffer and the copper-thionein fraction was isolated from the cytosol by chromatography on Sephadex G-75.

The quantity of protein in the copper-thionein fraction from the intestinal segments was estimated by the method described by Lowry et al. (7). The quantity of labeled lysine incorporated into copper-thionein was determined by liquid scintillation counting (Nuclear Chicago).

Metabolism of ^{64}Cu in Rat Intestine

Sixteen adult male Sprague-Dawley rats (250-300 g) that had been fed a diet of Purina Lab Chow were fasted for 18 hr and subsequently given an oral dose of 1.0 μg ^{64}Cu (International Chemical and Nuclear Corp., Irvine, CA) by gastric intubation. The ^{64}Cu had been adjusted to pH 6.0 with NaOH and diluted to 1.0 μg ^{64}Cu/ml with 0.9% NaCl. Following administration of the labeled copper, four rats each were killed at 15 min, 30 min, 60 min, and 120 min. A 25-cm segment of small intestine and the liver were removed from each of the rats. The intestinal contents were rinsed out with 100 ml of 0.9% NaCl after which the radioactivity in both the intestinal segment and the liver was determined in a gamma well scintillation counter.

After the radioactivity in the intact intestinal segments had been determined, the segments were homogenized in the N_2-saturated Tris buffer described above after which the homogenate was centrifuged at 105,000 g for 60 min. Following centrifugation, the radioactivity in the intestinal supernatant and particulate fractions was determined.

RESULTS AND DISCUSSION

As shown in Table 1, when the copper content of the drinking water was increased 10-fold, from 2.5 ppm to 25 ppm, the copper concentration of the intestine was increased slightly but significantly ($P < .05$). When the copper content of the diet was increased from 2.5 ppm to 50 ppm, a marked increase in the copper concentration of the intestine was observed. When the soluble fraction of the intestine from the rats in each of the three groups was chromatographed on Sephadex G-75, 85-90% of the copper

TABLE 1. Copper content of the small intestine from rats fed varying levels of copper in the drinking water.

Copper Supplement µg/ml	Copper Concentration µg/g dry wt
2.5	6.34 ± 0.77
25	8.12 ± 0.29
50	16.6 ± 3.0

TABLE 2. Amino acid content of the copper-binding protein from rat intestine.

Amino Acid Residue	%[a]
Lys	13.0
Arg	9.1
Thr	4.0
Ser	13.1
Glu	6.2
Gly	8.4
Ala	8.4
Cys[b]	31.0
Val	2.0
Met	2.0
Ile	1.7
Leu	0.8
Phe	0.3

[a] Average of duplicate determinations

[b] Determined as cysteic acid

recovered was eluted in a fraction corresponding with the elution volume of proteins having a molecular weight of approximately 10,000 daltons. As shown in Table 2, the amino acid content of the purified intestinal copper-binding protein is similar to that of metallothioneins isolated from other tissues and species (3,8-11). These results demonstrate that metallothionein is involved in the metabolism of copper in the intestinal mucosa.

The incorporation of ^{14}C-lysine into intestinal metallothionein from rats given an oral dose of 1.0 μg Cu was slightly less than that from rats given only saline (Table 3). However, the incorporation of ^{14}C lysine into intestinal metallothionein from rats given an oral dose of 20 μg Cu was markedly higher than that from rats given either 1.0 μg Cu or saline (Table 3). These observations suggest that ingestion of a high level of copper results in a stimulation of metallothionein synthesis in the intestinal mucosa.

TABLE 3. Incorporation of ^{14}C-lysine into intestinal metallothionein in rats given varying oral doses of copper.

Oral Cu Dose	^{14}C-Lysine dpm/mg
0	1987 ± 135
1.0	1368 ± 238
20	3692 ± 350

The mechanism whereby copper stimulates intestinal metallothionein synthesis has not yet been delineated but this phenomenon is certainly not limited to copper.

Richards and Cousins (12,13) have demonstrated that zinc stimulates synthesis of intestinal metallothionein. The stimulation of metallothionein synthesis by copper, zinc and possibly cadmium probably represents an attempt by the intestinal cell to provide adequate ligand binding sites to sequester and detoxify excess metal.

Although metallothionein probably has a vital function in protecting the cell from the toxic effects of excess metals, the protein may also function as a ligand in the metabolism of physiological levels of copper. Following oral administration of 1.0 μg ^{64}Cu, 85-90% of the isotope recovered from the intestinal cytosol after chromatography on Sephadex G-75 was contained within the metallothionein fraction. As shown in Table 4, the quantity of ^{64}Cu in the cytosol was highest 15 min after oral administration of the isotope. Thereafter, the quantity of isotope in the cytosol decreased over the 2 hr observation period and the disappearance of ^{64}Cu from the cytosol coincided with the appearance of ^{64}Cu in the liver. In addition, approximately 60% of the total ^{64}Cu in the intestinal cell was in the cytosol, bound to metallothionein, throughout the observation period (Table 4). Thus, the copper that enters

TABLE 4. Distribution of ^{64}Cu in the intestine and ^{64}Cu content of the whole liver at varying time intervals following oral administration of 1.0 μg ^{64}Cu to rats.

	Intestine			
Time min.	Cytosol cpm	% Total	Particulate Fraction cpm	Whole Liver cpm
15	2937 ± 75	58.8	2056 ± 104	495 ± 95
30	1890 ± 58	61.6	1178 ± 76	846 ± 100
60	640 ± 50	58.8	448 ± 42	1111 ± 130
120	407 ± 100	55.8	322 ± 51	2280 ± 200

the intestinal cytosol binds reversibly to metallothionein and the copper bound to metallothionein is in equilibrium with the copper on binding sites in the particulate fraction.

The experimental results described above suggest that metallothionein is an important ligand in the metabolism of copper in the intestine. When copper enters the cytosol of the intestinal cell the metal binds to metallothionein. If the quantity of copper that enters the intestinal cell does not exceed the capacity of the mucosal transport mechanism, the copper probably occupies binding sites on preexisting metallothionein and the metal is released from metallothionein as the copper is transported across the basement membrane into the lamina propria. If, however, the quantity of incoming copper exceeds the capacity of the intestinal transport mechanism, the preexisting binding sites become saturated and this leads to a stimulation of metallothionein synthesis. By binding ionic copper, metallothionein protects the cell from the damaging effects of the metal and, in part, regulates the passage of copper into the body.

REFERENCES

1. Margoshes, M. and Vallee, B.L. (1957) J. Amer. Chem. Soc. 79, 4813-4814.
2. Evans, G.W., Majors, P.F., and Cornatzer, W.E. (1970) Biochem. Biophys. Res. Commun. 40, 1142-1148.
3. Bremner, I. and Young, B.W. (1976) Biochem. J. 157, 517-520.
4. Evans, G.W. and LeBlanc, F.L. (1976) Nutr. Rep. Intern. 14, 281-288.
5. Evans, G.W. and Johnson, P.E. (1978) Trace Element Metabolism in Animals-3 (in press). pp. 98-105.
6. Allen, K.G.D. and Klevay, L.M. (1978) Atherosclerosis 29, 81-93, 1978.

7. Lowry, O.H., Rosebrough, N.J., Farr, A.L., and Randall, R.J. (1951) J. Biol. Chem. 193, 265-275.

8. Buhler, R.H.O. and Kagi, J.H.R. (1974) FEBS Letters 39, 229-234.

9. Kagi, J.H.R., Himmelhoch, S.R., Whanger, P.D., Bethune, J.L., and Vallee, B.L. (1973) J. Biol. Chem. 249, 3537-3542.

10. Hartmann, H.J. and Weser, U. (1977) Biochim. Biophys. Acta 491, 211-222.

11. Porter, H. (1974) Biochem. Biophys. Res. Commun. 56, 661-668.

12. Richards, M.P. and Cousins, R.J. (1975) Biochem. Biophys. Res. Commun. 64, 1215-1223.

13. Richards, M.P. and Cousins, R.J. (1976) J. Nutr. 106, 1591-1599.

Metallothionein
Jeremias H. R. Kägi and Monica Nordberg, eds.
Birkhäuser Verlag, Basel/Boston/Stuttgart, 1979

THE LOW MOLECULAR WEIGHT CADMIUM-, MERCURY- AND ZINC-BINDING PROTEINS (METALLOTHIONEINS): BIOSYNTHESIS, METABOLISM, AND POSSIBLE ROLE IN METAL TOXICITY

Zahir A. Shaikh
Department of Pharmacology and Toxicology
University of Rochester School of Medicine and Dentistry
Rochester, New York, U.S.A.

Metallothionein, since its discovery in 1957, has generated considerable interest among the protein chemists, toxicologists and nutritionists, especially during the last decade. The biological function of the protein, however, remains uncertain. This paper summarizes the work on metallothionein that the author has been directly involved with since 1967. No attempt to review the literature is made here. It is anticipated that the final report of the meeting will fulfill the task of critically reviewing the entire literature.

Biosynthesis of Cadmium-Metallothionein

Metallothionein is produced in male and female rats after exposure to subtoxic amounts of Cd. This phenomenon has been observed after oral as well as parenteral administration of the element (1-3). The protein is produced in various soft tissues, and its concentration in each tissue is related to the concentration of Cd. Among the soft tissues, liver and kidney synthesize metallothionein in larger quantities than any other tissue (1-5). The presence of the protein has been detected in pancreas, spleen (3), placenta (6), erythrocytes (7), plasma (7,8), heart, testes, and duodenal mucosa of rats exposed to Cd (unpublished results). By ion exchange chromatography, the rat hepatic metallothionein is separable into two components (9). Their molecular weight, as measured by gel filtration chromatography is very similar and ranges between 10,000 and 12,000 daltons. The amino acid composition of the two components (3) is described in Table 1.

Submitted to the "First International Meeting on Metallothionein and Other Low Molecular Weight Metal-binding Proteins", Zürich, July 17 - 22, 1978

TABLE 1

AMINO ACID COMPOSITION OF RAT HEPATIC CADMIUM-METALLOTHIONEIN

Amino Acid	Number of Residues/100 Residues*	
	Cd-MTN-1	Cd-MTN-2
Lysine	13	14
Aspartic acid	8	7
Threonine	6	4
Serine	12	12
Glutamic acid	3	7
Proline	4	5
Glycine	9	7
Alanine	5	9
Cysteine	33	30
Valine	3	2
Methionine	2	1
Isoleucine	1	1
Leucine	1	1

* Nearest whole number.

The mechanism of metallothionein biosynthesis in the liver and kidney of rats injected with cadmium has been studied in detail by the use of ^{14}C and ^{35}S labeled cysteine, and it has been shown that the protein is synthesized de novo (1-5, 10). In rats injected s.c. with 30 μmoles $CdCl_2$/kg the protein is synthesized more rapidly in the kidney than in the liver. The maximum rate of incorporation of the labeled amino acid into the renal metallothionein is between 4 and 5 hr and into the hepatic protein is between 8 and 12 hr (4, 10). Using the inhibitors of protein synthesis it has been demonstrated that the biosynthesis of renal metallothionein takes place by translation of the existing mRNA, while that of the hepatic metallothionein requires the production of new mRNA (4,10). The biological half-life of the protein in the kidney is 5 days and in the liver 2.8 days (4,5).

Biosynthesis of Mercury-Metallothionein

Male and female rats exposed to $HgCl_2$ (11) or to Hg vapor (12) synthesize metallothionein in kidney but not in liver. However, Hg binds to the protein already existing in the liver of control rats or of rats previously injected with Cd (13).

In rats injected s.c. with 2 μmoles $HgCl_2$/kg the maximum rate of incorporation of ^{35}S-cysteine is achieved at 8 hr.

The incorporation of the labeled amino acid is inhibited by actinomycin D, suggesting that Hg induces the biosynthesis of the protein at transcriptional level. The half-life of the protein is determined to be 4.4 days (11). A comparison between the biosynthesis and degradation of Cd- and Hg-metallothioneins is presented in Table 2.

TABLE 2

ENHANCEMENT OF METALLOTHIONEIN BIOSYNTHESIS BY Cd AND Hg

Metal	Tissue	Maximum Incorporation (hr)	Half-life (d)	Control of Biosynthesis
Cd	Liver	8-12	2.8	Transcriptional
	Kidney	4-5	5.0	Translational
Hg	Kidney	8	4.4	Transcriptional

The renal Hg-metallothionein has been separated into three components by ion-exchange chromatography (11) and by disc gel electrophoresis (unpublished results). Only the least negatively charged component migrates similar to the renal Cd-metallothionein 2 on the polyacrylamide gel.

Metabolism of Rat Hepatic Cadmium-Metallothionein

The protein was induced in rat liver by repeated injections of $CdCl_2$. It was labeled in vivo with either ^{109}Cd or ^{35}S-cysteine. Preliminary experiments were carried out with the crude protein isolated by gel filtration chromatography (14). These showed that the i.v. injected protein was not only taken up by the kidney but was also excreted in urine. Further studies were performed using the two components of hepatic metallothionein isolated by ion exchange chromatography. The proteins were administered at dose levels of 0.01 to 2 mg/250 g rat to determine the fate of the ^{109}Cd and the protein moiety (15). The investigation reveals that the tissue distribution and urinary excretion of both components is very similar, and furthermore, that the renal accumulation and urinary excretion of the proteins is dose dependent. At 0.01 mg protein dose, for example, 80% of the Cd is retained by the kidneys. The urinary excretion accounts for only 5% of the administered dose. As the dose of the protein injected into blood is increased, less Cd is retained by the kidney and more is excreted in the urine. The Cd in the urine is excreted as intact Cd-metallothionein, as determined by ^{109}Cd and ^{35}S labeled proteins. Rapid degradation of the protein is noted in the kidney, with a half-life of only a few hours as compared to 5 days for the protein produced within the kidney (5).

The above findings and those summarized earlier do not support the dogma that metallothionein found in the kidney is synthesized in liver. The protein is released into circulation only under chronic Cd intoxication (as will be described later). Depending upon its concentration in plasma, it is either completely taken up by the kidneys or partially retained and the remaining excreted in the urine. In the kidney the protein is picked up by the lysosomes and rapidly degraded. The free Cd thus released saturates the existing renal metallothionein and also directs the biosynthesis of new protein.

Metabolism of Rat Renal Mercury-Metallothionein

The protein was obtained from kidneys of rats repeatedly injected with $HgCl_2$. It was labeled in vivo with either ^{203}Hg or ^{35}S-cysteine. The protein was isolated by gel filtration chromatography. Further separation by ion exchange chromatography was not carried out. Intravenous injections of the Hg-metallothionein were made into 250 g rats at dose levels ranging from 0.01 to 5 mg protein (16). The animals were sacrificed 3 hr later. Nearly half of the injected ^{203}Hg was found in kidneys at dose levels up to 1 mg protein. At a 5 mg dose, the retention by the kidneys was 20% lower. The excretion of ^{203}Hg in urine was proportional to the dose of injected protein. While only 2.5% of the ^{203}Hg appeared in the urine during the 3 hr period in rats given 0.01 mg protein, those receiving 5 mg protein excreted 45% of the dose. The protein was excreted intact in the urine, as observed by injection of the ^{35}S labeled protein.

The kidneys of rats given 0.5 mg ^{203}Hg-metallothionein were fractionated by homogenization and centrifugation. The cytosol was found to contain about half of the tissue ^{203}Hg 1 hr after the injection. Of this, two-thirds was bound to metallothionein. The distribution of ^{35}S was very similar 1 hr after the injection of ^{35}S-labeled metallothionein. After 48 hr, however, when 27% of the injected ^{203}Hg was bound to the renal metallothionein, only 3% of the injected ^{35}S was recovered in this protein. This indicated a rapid turnover of the exogenous metallothionein. The endogenous metallothionein in Hg-injected rats is catabolized with a half-life of 4.4 days (11). Thus, the metabolism of the crude renal Hg metallothionein was found to be very similar to the crude (14) and purified hepatic Cd-metallothioneins (15).

Cadmium-Metallothionein and the Renal Dysfunction

Male rats were given 5 μmoles $CdCl_2$/kg s.c. once a day,

5 days a week, for 14 weeks (7,8). The animals showed the presence of Cd-metallothionein in plasma as early as 4 weeks after the beginning of the treatment. The appearance of the protein in urine was delayed, however, for another 6 weeks. At this time, the animals also developed proteinuria and glucosuria. After the onset of renal damage, the liver and kidney Cd-metallothionein depleted markedly (7,8). It was not possible to determine whether the protein being excreted in urine was of hepatic or renal origin. Monitoring of Cd-metallothionein levels in the plasma under chronic Cd exposure seems to offer a potentially useful index of exposure which could be used in predicting the possibility of renal damage long before its occurrence. Whether the renal tubular damage is caused directly by the Cd-metallothionein or by the ionic Cd released by lysosomal degradation of the protein remains to be elucidated.

Cadmium-, Zinc-Metallothioneins in Cultured Cells and Other Mammalian Tissues

The Cd, Zn-metallothionein has been detected in blue-green algae, *Anacystis nidulans*, grown under laboratory conditions in the presence of $^{109}CdCl$ and $^{65}ZnCl_2$ (17). The Cd-metallothionein has also been found in human fibroblasts cultured in medium containing $^{109}CdCl_2$ (18). In mice injected with tracer doses of $^{109}CdCl_2$ and $^{65}ZnCl$, both isotopes were bound to metallothionein in liver and kidney (19). The Cd-metallothionein has been produced in rabbit tissues after injection with $^{109}CdCl_2$ (unpublished results). The proteins from rabbit liver and kidney appear to be similar to the rat proteins upon disc gel electrophoresis and upon comparison of the amino acid composition. The presence of metallothionein has been demonstrated in adult rhesus monkey liver and human liver and kidney (3,19).

Development of a Method to Measure Metallothionein

Antibodies have been produced in rabbits against rat hepatic Cd-metallothionein 2 (more negatively charged component). The antibodies have shown cross-reactivity against rat hepatic Cd-metallothionein 1, crude rat renal Cd-metallothionein, and crude human renal metallothionein (unpublished results). A radioimmunoassay procedure is currently being developed for measuring the protein quantitatively.

Acknowledgment

Supported in part by U.S. Public Health Grants ES 01247 and ES 01448.

REFERENCES

1. Shaikh, Z.A. and Lucis, O. J. (1969) Proc. Canad. Fed. Biol. Soc. 12, 101. (Abstract).
2. Shaikh, Z.A. and Lucis, O.J. (1970) Fed. Proc. 29, 298. (Abstract).
3. Shaikh, Z.A. (1972) Doctoral Dissertation, Dalhousie University, Canada.
4. Shaikh, Z.A. and Smith, J.C. (1975) Fed. Proc. 35, 266. (Abstract).
5. Shaikh, Z.A. and Smith, J.C. (1976) Chem.-Biol. Interact. 15, 327-336.
6. Lucis, O.J., Lucis, R., and Shaikh, Z.A. (1972) Arch. Environ. Health 25, 14-22.
7. Hirayama, K. and Shaikh, Z.A. (1978) Toxicol. Appl. Pharmacol. (in press). (Abstract). 45, 347.
8. Shaikh, Z.A. and Hirayama, K. (1978) Environ. Health Perspect. (in press). 28, 267-271.
9. Shaikh, Z.A. and Lucis, O.J. (1971) Experientia 27, 1024-1025.
10. Shaikh, Z.A. and Smith, J.C. (1977) Chem.-Biol. Interact. 19, 161-171.
11. Shaikh, Z.A. and Smith, J.C. (1975) Proc. Int. Conf. on Heavy Metals in the Environment, Toronto, p. 108-109. (Abstract).
12. Khayat, A.I. (1977) Master's Thesis, University of Rochester, Rochester, New York.
13. Shaikh, Z.A., Lucis, O.J., and Coleman, R.L. (1973) In: "Trace Substances in Environmental Health," vol. 7, D.D. Hemphill, ed., University of Missouri, Columbia, Missouri, p. 313-321.
14. Cherian, M.G. and Shaikh, Z.A. (1975) Biochem. Biophys. Res. Commun. 65, 863-869.
15. Shaikh, Z.A. and Hirayama, K. (1978) In: "Proceedings of the Cadmium Symposium," M. Anke and H.G. Schneider, eds. (in press). pp. 95-101.
16. Hirayama, K. and Shaikh, Z.A. (1978) Toxicol. Appl. Pharmacol. (in press). (Abstract). 45, 347.
17. MacLean, E.I., Lucis, O.J., Shaikh, Z.A., and Jansz, E.R. (1972) Fed. Proc. 31, 699. (Abstract).
18. Lucis, O.J., Shaikh, Z.A., and Embil, J.A., Jr. (1970) Experientia 26, 1109-1110.
19. Shaikh, Z.A. and Lucis, O.J. (1972) Arch. Environ. Health 24, 419-425.

Metallothionein
Jeremias H. R. Kägi and Monica Nordberg, eds.
Birkhäuser Verlag, Basel/Boston/Stuttgart, 1979

METABOLISM AND POTENTIAL TOXIC EFFECTS OF METALLOTHIONEIN

M. George Cherian, Department of Pathology
Health Science Center, University of Western Ontario
London, Ontario, Canada

The metabolism and toxicity of cadmium are closely related to a low molecular weight cytoplasmic protein, metallothionein, which is synthesized on exposure to certain divalent metals (1-5). However, the direct role of cadmium-protein bio-complexes especially that of metallothionein in the transport, metabolism and toxicity of cadmium is not completely understood. Metallothionein seems to have a central role in both the ability of cells to detoxify (6-8) and accumulate cadmium and possibly other metals without any injury (9). On the other hand, paradoxically, the extracellular metallothionein may be an important factor in producing the cell injury in chronic cadmium poisoning (10).

In this brief review the differences in metabolism and potential toxicity of cadmium-metallothionein (Cd-Mt) and cadmium chloride administered both parenterally and orally are summarized. Some of the comparative studies on zinc chloride and zinc-metallothionein (Zn-Mt) are also presented.

Metabolism and toxicity of parenterally administered metallothionein

Cadmium ion is accumulated in liver, kidney and intestine preferentially bound to metallothionein after parenteral and oral administration of cadmium salts. The induced synthesis of metallothionein by metals is particularly interesting in terms of environmental and occupational health and metal toxicology because it appears to modify the pathological effects of metals. A number of experimental studies (6-8) have suggested a protective role for Cd-Mt in cadmium toxicity. Pretreatment of experimental animals with small doses of cadmium, presumably by inducing Cd-Mt synthesis, will prevent some of the toxic effects of large doses of cadmium (6-8).

Earlier studies (7,11) demonstrated that the intravenous injection of cadmium bound to metallothionein protected mice from acute toxic effects of cadmium namely, the testicular necrosis. Cherian and Shaikh (12); and Tanaka et al. (13) have shown that the distribution of cadmium from Cd-Mt in various organs was different from that after cadmium salts. Cadmium-thionein was preferentially accumulated in the kidney with substantial urinary excretion of cadmium, while most of the cadmium salts were deposited in liver.

Table 1 compares the distribution of cadmium and zinc in various organs after the injection of equal doses of $CdCl_2$, Cd-Mt, $ZnCl_2$ and Zn-Mt.

The distribution of zinc in different organs from the two forms of zinc was similar. When administered in the form of Zn-Mt, there was a small increase in the accumulation of zinc in the kidney and the urinary excretion. However, these results were unlike the different organ distribution of radioactive ^{109}Cd from $CdCl_2$ and Cd-Mt (12). The similar tissue distribution of zinc from zinc salts and Zn-Mt and also the transfer of

Submitted to the "First International Meeting on Metallothionein and Other Low Molecular Weight Metal-binding Proteins", Zürich, July 17 - 22, 1978

radioactive zinc from injected Zn-Mt to other protein fractions in tissues suggested that zinc bound to thionein might be metabolically as active as inorganic zinc salts (14).

Table 1

Tissue Distribution of Cadmium or Zinc after Injection of $CdCl_2$, $ZnCl_2$ or as Metallothionein

	$CdCl_2$	Cd-Mt	$ZnCl_2$	Zn-Mt
	Radioactivity (^{109}Cd or ^{65}Zn) as % of total dose (Mean ± SEM)			
Liver	64.4 ±10.9	11.2 ± 2.3	23.4 ± 5.1	26.8 ± 5.4
Kidney	1.8 ± 0.1	36.7 ± 1.2	2.8 ± 0.4	5.9 ± 1.1
Spleen	0.5 ± 0.1	0.4 ± 0.1	0.6 ± 0.2	1.3 ± 0.3
Pancreas	4.4 ± 2.9	0.2 ± 0.0	2.5 ± 0.2	2.7 ± 0.5
Testes	0.3 ± 0.1	0.2 ± 0.0	0.3 ± 0.1	1.3 ± 0.7
Urine	0.01 ± 0.0	19.3 ± 2.9	0.2 ± 0.0	1.7 ± 0.7

Male rats were injected i.v. with 1.2 mg Cd/kg as $^{109}CdCl_2$ or ^{109}Cd-Mt or with 1.2 mg Zn/kg as $^{65}ZnCl$ or ^{65}Zn-Mt and·sacrificed 4 hours later. (From Cherian, M.G. et al., Toxicol. Appl. Pharmacol. 38, 399, 1976 and Cherian, M.G., J. Nutr. 107, 965, 1977).

In contrast to the Zn-Mt studies, when rat liver Cd-Mt was injected to rats, a major fraction of cadmium was deposited in the kidney. Radioactive labelling experiments revealed that the protein-metal complex can be transported intact to kidney and can be recovered from the kidney supernatant fraction (15). Metallothionein was filtered through the glomeruli and reabsorbed by the renal tubules without any degradation. However, at doses that produced tubular necrosis,Cd-Mt was excreted into urine intact suggesting a threshold for renal handling of this protein-metal complex. This type of dose response in renal reabsorption was special for metallothionein and was not observed for cadmium salts which are not filtered (16). However, this phenomenon cannot be interpreted as indicative of a transport function for metallothionein. The intracellular non-diffusible nature of this protein and its rapid clearance from the blood are not usually associated with a transport protein.

In cadmium exposed experimental animals, metallothionein was found intracellularly and was not normally detected in biological fluids such as plasma, bile and urine. It has been recently observed (17) that the synthesis of metallothionein can in fact decrease the diffusible form of cadmium in the cell thereby decreasing the biliary transport of cadmium (Table 2).

Table 2

Biliary Excretion and Tissue Distribution of Cadmium in Control and Cadmium-Pretreated Rats[a]

	Control		Pretreated	
	μg/g	% of dose	μg/g	% of dose
Biliary Excretion		5.00 ± 0.64[b]		1.03 ± 0.28[b]
Liver	15.00 ± 0.71	53.60 ± 2.93[b]	18.50 ± 1.28	64.40 ± 2.28[b]
Kidney	2.65 ± 0.13	2.08 ± 0.18	2.18 ± 0.17	1.51 ± 0.13

[a]Rats were anesthetized with pentobarbital (35 mg/kg) and injected i.v. with 1 mg Cd/kg labelled with radioactive $^{109}CdCl_2$ of known specific activity. Bile samples were collected for 3 hr. The pretreated group was injected i.p. with 0.25 mg Cd/kg 24 hr before injection of radioactivity. At the end of the experimental period, rats were sacrificed and organs and bile samples were counted. Cumulative biliary excretion of cadmium for 3 hr is expressed as percentage of the administered dose of newly introduced radioactive cadmium. The results are mean values (±SEM) from eight different experiments.

[b]The difference between the means of these groups is statistically significant, $p<0.05$.

(From Cherian, M.G., J. Toxicol. Environ. Hlth., 2: 955, 1977)

In the cadmium pretreated groups, most of the cadmium in the tissues was bound to metallothionein and the control group had little metallothionein. The accumulation of radioactive cadmium in the liver increased and the biliary excretion of cadmium decreased significantly in the pretreated group as compared to the control group. A large percentage of cadmium in the bile was partially identified as cadmium-glutathione complex (18). Thus cadmium chelation by small molecular weight sulfur containing amino acids and peptides may be an important mode of cadmium transport from one organ to another. Other possibilities like weak binding of cadmium to plasma proteins such as albumin and macroglobulins cannot be excluded in this process. One of the major arguments against a cadmium transport role for Cd-Mt was its potential nephrotoxic effect (15). Even trace amounts of injected Cd-Mt was shown to cause renal tubular damage within a short time. The toxic symptoms of prolonged cadmium exposure, especially the nephrotoxic effects appeared only after a long lag period, both in cadmium workers and in experimental animals (19). If the major transport form of cadmium to the kidney was metallothionein, the renal damage in cadmium toxicity would have appeared much earlier because of the potent nephrotoxic effects of this protein-metal complex. Thus the experimental evidence from our laboratory does not support the notion that transport of metals is a major function of thioneins.

The acute renal toxic effects of parenterally administered Cd-Mt was first reported by Nordberg et al. (11) in mice. Further studies (15) with rat liver metallothionein revealed that necrosis of nearly all renal proximal tubular lining cells occurred within 24 hours after metallothionein administration. The earliest morphological changes were the blebbing of the luminal membrane border with loss of microvillous pattern and a massive increase in subcellular vacuoles (Figure 1). These changes later progressed to necrosis of the tubular lining cells with mitochondrial damage. Deposition of dense granules and swelling of mitochondria were observed. (Figure 2).

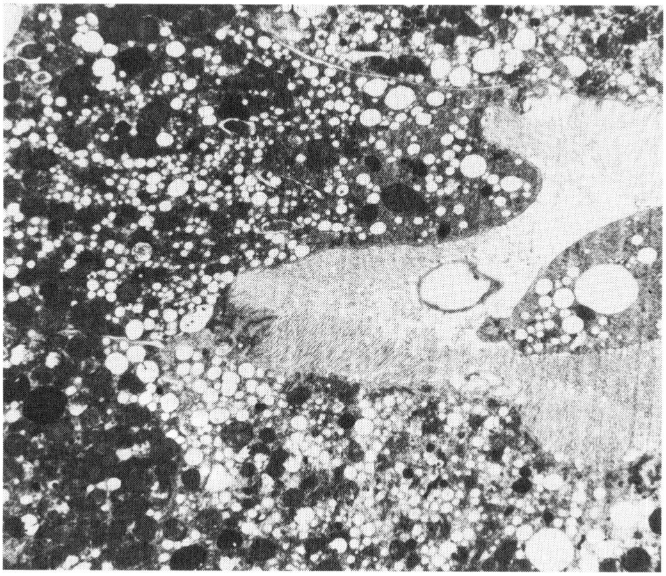

Fig. 1. Proximal renal tubule 8 hr after i.p. Cd-Mt injection (1.2 Cd/kg) Endocytotic vacuoles are increased in number and size, and extend to the basilar portion of the tubular lining cells. Uranyl and lead acetate staining, x 5200. (From Cherian et al., Toxicol. Appl. Pharmacol. 38, 399, 1976).

Recent investigations (20) on the role of cadmium-thionein in the renal toxicity of cadmium in a repeated dosing study indicated that there were two phases in chronic cadmium poisoning. The first phase was an initial tolerant phase where Cd-Mt synthesis took place and it stayed intracellularly protecting the cell from the toxic effects of cadmium. During this phase, there was neither any detectable metallothionein in plasma nor renal cellular damage. Phase two was a toxic phase characterized by proteinuria and glycosuria and this appeared later in the chronic effects of cadmium. Renal tubular damage was observed morphologically during this period and ten thousand molecular weight cadmium binding proteins were isolated from the plasma. Thus the renal damage in experimentally induced cadmium poisoning can be directly correlated to the extracellular presence of low molecular weight cadmium binding proteins.

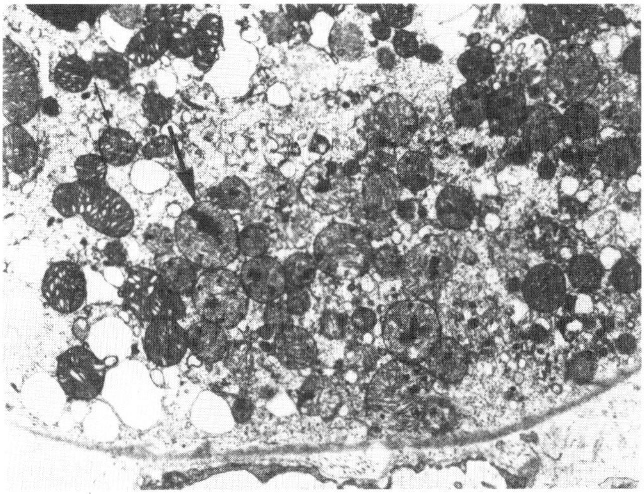

Fig. 2. Proximal renal tubular cell, 12 hr after i.p. injection of Cd-Mt. Basilar infoldings are not visible, mitochondria have lost their elongated appearance, and there are numerous vacuoles in the cytoplasm. Some mitochondria contain dark flocculent densities in the matrix (thick arrows), while others have normal cristae and matrical density (thin arrows) x 9200. (From Cherian et al., Toxicol. Appl. Pharmacol. 38, 399, 1976).

The mechanism of cadmium-thionein induced nephropathy is not understood conclusively. However, two major views are that cadmium ions released from Cd-Mt by the lysosomal systems are the toxic agent (21,22) and secondly that Cd-Mt itself is the nephrotoxin (15,20,23).

The role of the lysosomal system in the pathogenesis of Cd-Mt induced nephropathy has been proposed by Fowler and Nordberg (21). They observed increased numbers of secondary lysosomes and increased lysosomal enzyme activity in rabbit kidney after injection of Cd-Mt. According to Trump and Arstila (24), similar lysosomal changes are general phenomena in cell death and are considered to be the result, not the cause, of cell injury. The 'suicide-bag hypothesis' as the cause for the Cd-Mt induced cell injury is very controversial and its involvement in any type of cell injury is only speculative. Thus these lysomal changes are most likely a secondary response to renal cell injury caused by cadmium-protein complex rather than a primary toxic effect of Cd-Mt per se.

Even though there was a gradual degradation of injected metallothionein in the kidney, more than 95% of the cytoplasmic cadmium in the renal cell was always bound to metallothionein. The recent study (25) also showed that new thionein synthesis occurred in the kidney within a few hours after injection of metallothionein and there was no significant uptake of metallothionein by renal lysosomes. It can be concluded from the above studies (10,14,15,20,25) that the injected Cd-Mt was taken up by renal tubular lining cells actively by pinocytosis and might rupture cell membranes during this process. The metal-protein complex which had entered the cell then caused irreversible damage to mitochondria with a direct effect on energy metabolism. The rupture of cell membranes and the injury to the mitochondria can be considered as the major reasons for cell death in Cd-Mt induced nephropathy.

Distribution and toxicity of oral cadmium-metallothionein (Cd-Mt) and $CdCl_2$

The differences in absorption and metabolism of cadmium from its various biocomplexes in diet may have an important role in the toxicity of cadmium. Though the nutritional factors like iron deficiency may have a significant influence on the gastrointestinal absorption of cadmium, there is little information on the absorption of cadmium from the different cadmium containing compounds in the diet and its subsequent metabolism and toxicity. The largest source of cadmium in the diet is cereal or meat, especially in liver and kidney in the form of metallothionein. Since metallothionein is heat stable (26), this metal-protein complex is probably not destroyed by cooking and this may be the main dietary form of cadmium.

Recent perfusion studies (27) on mice intestinal loop *in situ* with identical concentration of cadmium in the form of $CdCl_2$ or Cd-Mt demonstrated that cadmium from both these forms was taken up by the intestine. However, Cd-Mt produced severe necrosis of the absorptive cells whereas cadmium chloride perfusion caused only minor abnormalities. These results (27) showed direct toxic effects of infusion of identical doses of $CdCl_2$ or Cd-Mt on intestinal mucosal cells. The disruption of microvilli, swollen mitochondria without any cristae and dilated endoplasmic reticulum were the ultrastructural evidences for severe mucosal damage with Cd-Mt. These toxic effects of Cd-Mt on the intestinal mucosa resembled its effects on the renal tubular cells after parenteral administration (15).

It is not known whether dietary Cd-Mt can reach the small intestine intact or it may be digested in the stomach. Therefore a study was recently undertaken where identical doses of cadmium as $CdCl_2$ or Cd-Mt was administered orally to mice through a stomach tube and the absorption and tissue distribution of cadmium were compared. The recovery of intact Cd-Mt from the duodenum and the identical body retention of cadmium from the two different forms suggested a similar mechanism of gastrointestinal absorption for these cadmium compounds. However, a significant increase in the renal deposition of cadmium from the oral Cd-Mt as compared to oral $CdCl_2$

was observed in these experiments suggesting the role of different forms of cadmium in the body distribution and toxicity of this metal. The results of the tissue distribution of cadmium eight days after oral administration of $CdCl_2$ or Cd-Mt in mice are shown in Table 3. These results were similar to the tissue distribution of cadmium after parenteral administration of these cadmium compounds (2,15). Though these studies suggested that Cd-Mt may be absorbed from the gastrointestinal tract intact, further studies are required to determine the exact mechanism of gastrointestinal absorption of Cd-Mt.

Table 3

Comparative Tissue Distribution of Cadmium in Mice After Oral Administration of $^{109}CdCl_2$ and ^{109}Cd-Mt

	$CdCl_2$	Cd-Mt
	% of retained cadmium (Mean ± SEM)	
Liver	66.8 ± 1.8	7.1 ± 0.7
Kidney	7.1 ± 0.6	61.1 ± 5.6
Spleen	0.8 ± 0.3	0.1 ± 0.0
Pancreas	2.9 ± 0.2	0.4 ± 0.1
Testes	0.4 ± 0.1	0.2 ± 0.0

Male mice (C57Bl/J strain) were starved overnight and 60 μg of Cd and 2 μci of ^{109}Cd were administered as $CdCl_2$ or Cd-Mt orally with a gastric tube. They were sacrificed 8 days later. The results are from 6 mice in each group.

The different factors which regulate the absorption of toxic metals from the gastrointestinal tract are not well defined. The homeostatic control mechanisms for the absorption of essential metals such as iron, zinc (28) and copper (29) were studied in different laboratories. There is no evidence for such a control mechanism in the absorption of noxious metals. The increased gastrointestinal absorption of cadmium in the newborn and young rats as compared to adult animals was reported recently (30,31). Low dietary calcium in experimental rats was also shown to increase the gastrointestinal absorption of cadmium (32). A direct correlation between the serum ferritin level and cadmium absorption was shown in human volunteers in a recent study using radioactive cadmium (33).

Since kidney is the main target organ in the chronic toxic effects of cadmium and the biological half life of renal cadmium is long, the increased cadmium deposition in kidney after oral administration of Cd-Mt may have a significant importance in deciding the maximum allowable amount of cadmium in food. It was generally assumed (19) that cadmium from the food was mainly accumulated in liver initially and gradually transported to kidney and the toxic renal effects occurred only when kidney cadmium levels increased to 200 μg/g. It is not clear whether oral Cd-Mt can cause renal disease in the general population at a faster rate than oral cadmium salts. However, if cadmium can be directly transported to kidney from oral Cd-Mt, it may have a potential danger in inducing chronic renal disease.

In summary, the renal accumulation of cadmium can be influenced by the different cadmium biocomplexes. Since both oral and parenteral Cd-Mt can selectively deposit cadmium in the kidney, the extracellular presence of metallothionein may be an important factor in the chronic toxic effects of cadmium, especially in cadmium induced nephropathy.

Acknowledgement

This study was supported by grants from the Medical Research Council of Canada and the International Lead and Zinc Research Organization.

References

1. Piscator, M. (1964). Nord. Hyg. Tidskr. 45, 76-82.
2. Webb, M. (1972). Biochem. Pharmacol. 21, 2751-2765.
3. Shaikh, Z.A. and Lucis, O.J. (1970). Fed. Proc. 29, 298.
4. Bremner, I., Davis, N.T., and Mills, C.F. (1973). Biochem. Soc. Trans. 1, 982.
5. Bremner, I. and Young, B.W. (1976). Biochem. J. 155, 631-635.
6. Terhaar, C.J., Vis, E., Roudabush, R.L., and Fasset, D. (1965). Toxicol. Appl. Pharmacol. 7, 700.
7. Nordberg, G.F. (1971). Environ. Physiol. Biochem. 1, 171-187.
8. Webb, M. and Verschoyle, R.D. (1976). Biochem. Pharmacol. 25, 673-679.
9. Goyer, R.A. and Cherian, M.G. (1977). In: 'Clinical Chemistry and Chemical Toxicology of Metals', S.S. Brown ed., Elsevier Scientific Publishing Company, p. 89-103.
10. Cherian, M.G. and Goyer, R.A. (1978). Life Sciences 23, (In Press)
11. Nordberg, G.F., Goyer, R.A., and Nordberg, M. (1975). Arch. Path., 99, 192-197.
12. Cherian, M.G. and Shaikh, Z.A. (1975) Biochem. Biophys. Res. Comm. 65, 863-869.
13. Tanaka, K., Sueda, K., Onosaka, S., and Okahara, K. (1975). Toxicol. Appl. Pharmacol. 33, 258-266.
14. Cherian, M.G. (1977). J. Nutr. 107, 965-972.
15. Cherian, M.G., Goyer, R.A., and Delaquerrier-Richardson, L. (1976). Toxicol. Appl. Pharmacol. 38, 399-408.
16. Vostal, J.J. and Cherian, M.G. (1974). Fed. Proc. 33, 519.
17. Cherian, M.G. (1977). J. Toxicol. Environ. Hlth. 2, 955-961.
18. Cherian, M.G. and Vostal, J.J. (1977). J. Toxicol. Environ. Hlth. 2, 946-954.
19. Friberg, L., Piscator, M., Nordberg, G.F., and Kjellstrom, T. (1974). Cadmium in the Environment, CRC Press, Cleveland, Ohio.
20. Cherian, M.G., Goyer, R.A. and Delaquerrier-Richardson, L. (1977). Toxicol. Appl. Pharmacol. 41, 145.
21. Fowler, B.A. and Nordberg, G.F. (1975).Proc. Intl. Conf. on 'Heavy Metals in the Environment', Toronto, Canada. p. B52-53.
22. Webb, M. and Etienne, A.T. (1977). Biochem. Pharmacol. 26, 25-30.
23. Webb, M. (1975). Biochem. Soc. Trans. 50, 632-634.
24. Trump, B.F. and Arstila, A.V. (1975). In: 'Principles of Pathology', M.F. LaVia and R.B. Hill, Jr. eds. Oxford University Press, p.9-96.

25. Cherian, M.G. (1978). Biochem. Pharmacol. 27, (In Press) 27, 1163-1166.
26. Cherian, M.G. (1974). Biochem. Biophys. Res. Comm. 61, 920-926.
27. Valberg, L.S., Haist, J., Cherian, M.G., Delaquerrier-Richardson, L. and Goyer, R.A. (1977). J. Toxicol. Environ. Hlth. 2, 963-975.
28. Richards, M.P. and Cousins, R.J. (1975). Biochem. Biophys. Res. Comm. 64, 1215-1223.
29. Evans, G.W. (1973). Physiol. Rev. 53, 535-570.
30. Sasser, L.B. and Jarboe, G.E. (1977). Toxicol. Appl. Pharmacol. 41, 423-431.
31. Kello, D. and Kostial, K. (1977). Toxicol. Appl. Pharmacol. 40, 277-282.
32. Washko, P.W. and Cousins, R.J. (1974). Fed. Proc. 33, 668.
33. Flanagan, P.R., McLellan, J.S., Haist, J., Cherian, M.G., Chamberlain, M.J. and Valberg, L.S. (1978). Gastroenterology, 74, 841-846.

Metallothionein
Jeremias H. R. Kägi and Monica Nordberg, eds.
Birkhäuser Verlag, Basel/Boston/Stuttgart, 1979

RENAL EFFECTS OF MERCURY- AND CADMIUM-METALLOTHIONEIN

Gunnar F. Nordberg
Department of Community Health and Environmental Medicine, Odense University, Denmark,
and
Department of Environmental Hygiene, Karolinska Institutet, Stockholm, Sweden.

The possibility of a role for metallothionein in inducing renal effects of cadmium and mercury has been previously discussed by several authors (1, 2, 3). Whereas injection of cadmium metallothionein (CdMT) has been shown to induce renal tubular damage in experimental animals (4, 5, 6, 7, 8), a similar effect of injected mercury-metallothionein has not previously been reported. The present studies aimed at investigating this possibility and at comparing the effects with those induced by CdMT.

MATERIALS AND METHODS

Metallothionein containing mainly cadmium (CdMT) but also small amounts of zinc was isolated from livers of cadmium-injected rabbits as previously described (9). The batch used for the present studies was identical with the one used by Fowler and Nordberg (8).

Mercury-metallothionein (HgMT) was prepared by exchange of Cd against Hg in metallothionein as described by Nordberg et al. (10) . Excess Cd was removed from the solution by ultrafiltration on an UM2 (Amicon) filter. The non-ultrafiltrable material was used for injection.

Albino mice (NMRI), 25-35 g, were injected intravenously in the tail with CdMT or HgMT. The dosage of protein was adjusted to be the same, i.e. ≈1 mg/kg in groups 2 and 8; ≈3 in 3 and 9; ≈9 in 4 and 10. The dosage of metal are given in the table.

Submitted to the "First International Meeting on Metallothionein and Other Low Molecular Weight Metal-binding Proteins", Zürich, July 17 - 22, 1978

Urine was sampled in metabolism cages during the first 24 hours (1d) or during 24-48 hours (2d) after injection. Urinary protein and creatinine were analysed and evaluated as described by Nordberg and Piscator (11). The animals were killed 1 day or 4 days after injection and samples of renal tissue were fixed in formalin, cleared, embedded and sections of 4 µg (HE) evaluated by light microscopy.

RESULTS AND DISCUSSION

Proteinuria: There was a clear difference in the renal effects of both cadmium and mercury when given bound to metallothionein, compared to the administration of the same dose of the respective divalent metal salt (Table 1). This has not previously been reported for mercury-metallothionein but is well established for cadmium. It is interesting to note that whereas zinc-thionein has been reported to be non-toxic (7) the attachment of Hg to the protein gives rise to a toxicity similar to the one of cadmium-metallothionein (Table 1). Copper-metallothionein may also induce some renal toxicity (12, 13).

Mercuric chloride is known to give rise to both glomerular and tubular renal damage and it is therefore of interest to note that the proteinuria induced by HgMT was of the tubular type. Although a qualitative or quantitative difference between the effects of CdMT and HgMT cannot be entirely ruled out by the present studies, there is no clear indication of such a difference from the present set of data.

Results from metal analyses of renal tissue from these mice are only briefly reported in the following. In animals killed 1 day after injection about 10 µg/g of Cd was found in groups 3 and 4 (CdMT injected). Animals given $CdCl_2$ in the same dose had considerably lower values i.e. about 0,7 µg/g and 3 µg/g respectively for groups 6 and 7. These values are in accord with previous observations. In animals given HgMT about 8 µg/g of Hg was found in group 9 and between 4 and 9 µg/g in group 10. Corresponding values for groups 12 and 13 ($HgCl_2$) were 4 and 7 µg/g. Like the cadmium-groups, there seemed to occur a saturation of the renal tissue in the HgMT-injected animals so that no further increase occurred in Hg-concentration in group 9. With the animals injected with $HgCl_2$ tissue concentrations increased with dosage. This course of events may be explained by the more prominent renal tubular damage that occurred at the highest dose of HgMT with subsequent loss of renal tissue into urine and a decreased ability of the renal tubules to take up mercury.

The histological evaluation demonstrated a good corre-

TABLE 1

PROTEINURIA AND ELECTROPHORETIC PATTERN

GROUP dose of Cd or Hg	time days	n	protein mg/100 ml	creatinine mg/100 ml	pattern
1 CONTROL		4	127	5,6	N
2 CdMT 0,1 mg/kg	1	4	100	5,0	N
- " -	2	3	188	6,3	N
3 CdMT 0,3 mg/kg	1	4	138	4,3	N-T
- " -	2	3	535	5,4	T
4 CdMT 0,9 mg/kg	1	4	165	5,5	T
- " -	2	3	460	4,0	T
5 $CdCl_2$ 0,1 mg/kg	1	4	74	5,0	N
- " -	2	3	131	5,5	N
6 $CdCl_2$ 0,3 mg/kg	1	4	108	5,3	N
- " -	2	3	190	7,7	N
7 $CdCl_2$ 0,9 mg/kg	1	4	86	5,0	N
- " -	2	3	174	5,5	N
8 HgMT 0,15 mg/kg	1	4	60	4,5	N
- " -	2	3	160	7,1	N
9 HgMT 0,45 mg/kg	1	4	149	5,0	N-T
- " -	2	2	109	4,3	N
10 HgMT 1,34 mg/kg	1	4	311	3,9	T
- " -	2	1	925	6,7	T
11 $HgCl_2$ 0,15 mg/kg	1	4	91	5,6	N
- " -	2	3	160	6,3	N
12 $HgCl_2$ 0,45 mg/kg	1	4	85	4,6	N
- " -	2	3	107	4,4	N
13 $HgCl_2$ 1,34 mg/kg	1	4	66	4,0	N
- " -	2	3	220	3,9	N-T*

* mixed glomerular-tubular pattern

lation with the findings by urinary analysis. Animals with clearly increased excretion of protein in urine in groups 3, 4, 9 and 10 also had evident morphological changes in their renal tissue. Various degrees of necrosis in renal tubular cells, dilation of renal tubules and formation of casts were observed.

In summary, the present studies demonstrate a renal toxicity of mercury-metallothionein similar to the one of cadmium-metallothionein. It is thus possible that metallothionein released into the blood may be involved in the elicitation of renal toxicity in the course of mercury poisoning. However, such a hypothesis still needs considerable additional experimental support before it can be regarded as plausible.

ACKNOWLEDGEMENTS

A grant from the Karolinska Institute Research Fund is gratefully acknowledged. Skilful technical assistance was provided by Mrs. Maude Pannone and Mrs. Britt Pettersson.

REFERENCES

1. Piscator, M. (1964) Nord. Hyg. Tidskr. 45, 76-82.
2. Nordberg, G. F. (1972) Environm. Physiol. Biochem. 2, 7-36.
3. Piotrowski, J., Trojanowska, B., Wisniewska-Knypl, J. M. and Bolanowska, W. (1974) Toxicol. Appl. Pharmacol. 27, 11.
4. Nordberg, G. F. (1971) Environm. Physiol. 1, 171-187.
5. Nordberg, G. F., Goyer, R. A. and Nordberg, M. (1975) Arch. Pathol. 99, 192-197.
6. Cherian, M. G., Goyer, R. A. and Delaquerriere-Richardson, L. (1976) Toxicol. Appl. Pharmacol. 38, 399-408.
7. Webb, M. and Etienne, A. T. (1977) Biochem. Pharmacol. 26, 25-30.
8. Fowler, B. A. and Nordberg, G. F. (1978) Toxicol. Appl. Pharmacol., in press. 46, 609-623.
9. Nordberg, G. F., Nordberg, M., Piscator, M. and Vesterberg, O. (1972) Biochem. J. 126, 491-498.
10. Nordberg, M., Trojanowska, B. and Nordberg, G. F. (1974) Environm. Physiol. Biochem. 4, 149-158.
11. Nordberg, G. F. and Piscator, M. (1972) Environm. Physiol. Biochem. 2, 37-48.
12. Bremner, I., Young, B. W. and Mills, C. F. (1976) Brit. J. Nutr. 36, 551- 561.
13. Bremner, I. and Young, B. W. (1977) Chem. Biol. Interactions 19, 13-23.

OTHER LOW MOLECULAR WEIGHT METAL-BINDING PROTEINS

Metallothionein
Jeremias H. R. Kägi and Monica Nordberg, eds.
Birkhäuser Verlag, Basel/Boston/Stuttgart, 1979

LOW MOLECULAR WEIGHT COPPER-BINDING PROTEINS IN CULTURED HUMAN CELLS

John R. Riordan and George Madapallimattam
Research Institute
The Hospital for Sick Children, and
Department of Clinical Biochemistry
University of Toronto
Toronto, Canada

INTRODUCTION

Not a great deal is known of the molecular mechanisms whereby heavy metals are handled intracellularly. The elucidation of these mechanisms which may be defective in certain genetic diseases or stressed in some environmental situations is essential to the understanding of both the overall homeostasis and toxicology of heavy metals. Studies of either of these aspects in humans are limited to the analysis of the readily available body fluids or pathological specimens which may be available on occasion. The use of human cells in long term culture avoids these limitations. This report describes the uptake of $^{67}Cu^{2+}$ by cultured fibroblasts and lymphoblasts and the subcellular distribution of Cu^{2+} in these cells. A significant portion of the Cu^{2+} is associated with a soluble protein fraction of similar size to that observed in tissues, particularly liver and kidney of humans as well as animals (1-17). Hence, the induction and biosynthesis of, as well as possible genetic variations among these proteins, become susceptible to investigations in human cells.

METHODS

Fibroblasts were derived from skin biopsies using the method of Goldstein and Littlefield (18) and grown as monolayers in α medium (19) containing 15% fetal calf serum. Lymphoblast lines were established by transformation of peripheral blood lymphocytes with

---

Submitted to the "First International Meeting on Metallothionein and Other Low Molecular Weight Metal-binding Proteins", Zürich, July 17 - 22, 1978

Epstein-Barr virus (20) and grown in suspension culture with the same medium used for fibroblasts.

To assess $^{67}Cu^{2+}$ uptake by fibroblasts, cells were grown on the bottoms of glass liquid scintillation vials. Prior to uptake experiments, growth medium was removed by aspiration and the monolayers washed three times with PBS (10 mM NaH_2PO_4, pH 7.4, 150 mM NaCl; 2 ml each time) and a fourth time with the same solution containing 1.5% human albumin. To begin uptake $^{67}Cu^{2+}$ (approximately 0.5 mCi/μg) prepared according to Marceau et al. (21) and an appropriate amount of $CuCl_2$ in the albumin-containing PBS was added. To terminate uptake the medium was removed and the cells rinsed three times with PBS. Cells were dissolved in 0.5 ml Protosol (NEN)and counted in a γ spectrometer as was the medium from which copper was taken up by the cells.

For subcellular fractionation, fibroblasts were grown in roller bottles, washed, harvested using a large scraper with a rubber blade and disrupted using a Stansted Cell Disruptor, Model A0612 at a pressure of 50 psi. The various subcellular fractions were obtained by sequential centrifugations as follows: nuclei, 300 g for 10 min; mitochondria, 13,000 g for 10 min; microsomes, 40,000 g for 30 min; "high speed pellet", 186,000 g for 2 h.

Lymphoblasts were grown in spinner bottles for both $^{67}Cu^{2+}$ uptake and subcellular fractionation. After incubation in $^{67}Cu^{2+}$ containing PBS (with 1.5% albumin) cells were separated from medium by centrifugation at 10,000 g for 1 min (Beckman 152 Microfuge) and washed. $^{67}Cu^{2+}$ radioactivity was determined in both cells and media. Subcellular fractionation was performed as with fibroblasts except that cells were disrupted at a pressure of 250 psi.

The "high speed supernatant" fraction from both cell types were subjected to the same procedures used for isolation of low molecular weight copper-binding proteins from liver (11) except that all solutions were saturated with N_2 and containers used throughout the fractionation kept sealed.

Copper concentrations were determined by atomic absorption spectrometry (Pye-Unicam SP 1800) and protein according to Bohlen et al (22). Polyacrylamide gel electrophoresis and amino acid analysis were performed as described previously (10).

RESULTS

Fig 1 shows the time course of uptake of $^{67}Cu^{2+}$ by both fibroblasts and lymphoblasts. Lymphoblasts appeared to take up copper at a fairly rapid rate for the first half hour, after which uptake proceeded at a considerably slower rate. Fibroblasts seemed to take up the metal at a rate similar to this slower one over the whole 3 hour time course. On the basis of the same cell numbers somewhat more copper associated with fibroblasts than with lymphoblasts. However, this difference could be accounted for to some extent by the larger size of the fibroblasts.

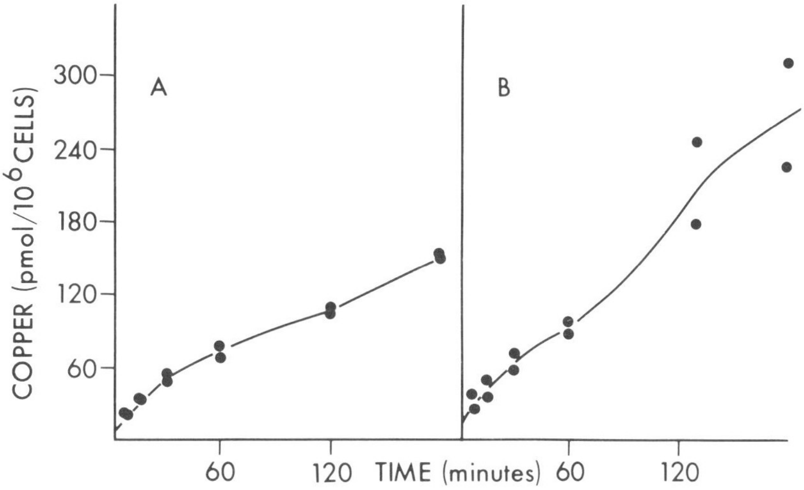

Figure 1. Time course of uptake of $^{67}Cu^{2+}$ (2.5×10^{-5}M) by lymphoblasts (A) and fibroblasts (B) from PBS containing 1.5% human albumin.

The subcellular distribution of the copper taken up by the two cell types was determined using non-radioactive copper rather than $^{67}Cu^{2+}$ both because of the longer experimental times (relative to the half life of $^{67}Cu^{2+}$) and the hazards involved in utilizing the hard γ emitting isotope throughout all the cell culture and biochemical manipulations. To provide accurately detectable copper concentrations in the subcellular fractions of a reasonable number of cells (approximately 2×10^9 lymphoblasts and 10^8 fibroblasts), 10 μg/ml Cu^{2+} was added to growth media two days prior to harvesting. This elevated the cell copper levels to about 70 μg/g cell protein in lymphoblasts and about 142 μg/g cell protein in fibroblasts without decreasing viability. Table I shows the distribution of copper on fractionation of the cellular constituents. In the case of lymphoblasts approximately 30% of the copper is retained by the nuclear pellet while nearly 70% is associated with the other cellular components (nuclear supernatant). About 15% of the copper is in the mitochondrial fraction and approximately 45% in lighter fractions, most of which remains in the post microsomal supernatant. Even when this was subjected to a higher speed centrifugation (186,000 g; 2h) little copper-associated material was sedimented. About half of the copper in the soluble fraction is greater than the equivalent of 30,000 daltons in mass (excluded by Amicon PM30). Finally the smaller (between 2,000 and 30,000 molecular weight) copper associated soluble material represents about 20% of the total cell copper.

Table I

Subcellular distribution of copper in cultured lymphoblasts and fibroblasts.

Fraction	Lymphoblasts		Fibroblasts	
	% protein	% Cu	% protein	% Cu
homogenate	100	100	100	100
nuclear pellet	20	29	18	11
nuclear sup.	69	68	85	90
mito-pellet	11	17	6	4
mito-sup	51	46	81	89
micro pellet	9	4	12	8
micro sup.	35	46	49	79
'high speed' pellet	10	6	3	5
'high speed' sup.	26	43	46	64
2-30 megadalton soluble fraction	3	20	5	26

Starting materials were 708 mg of lymphoblast protein containing 50 μg of copper and 385 mg of fibroblast protein containing 55 μg of copper.

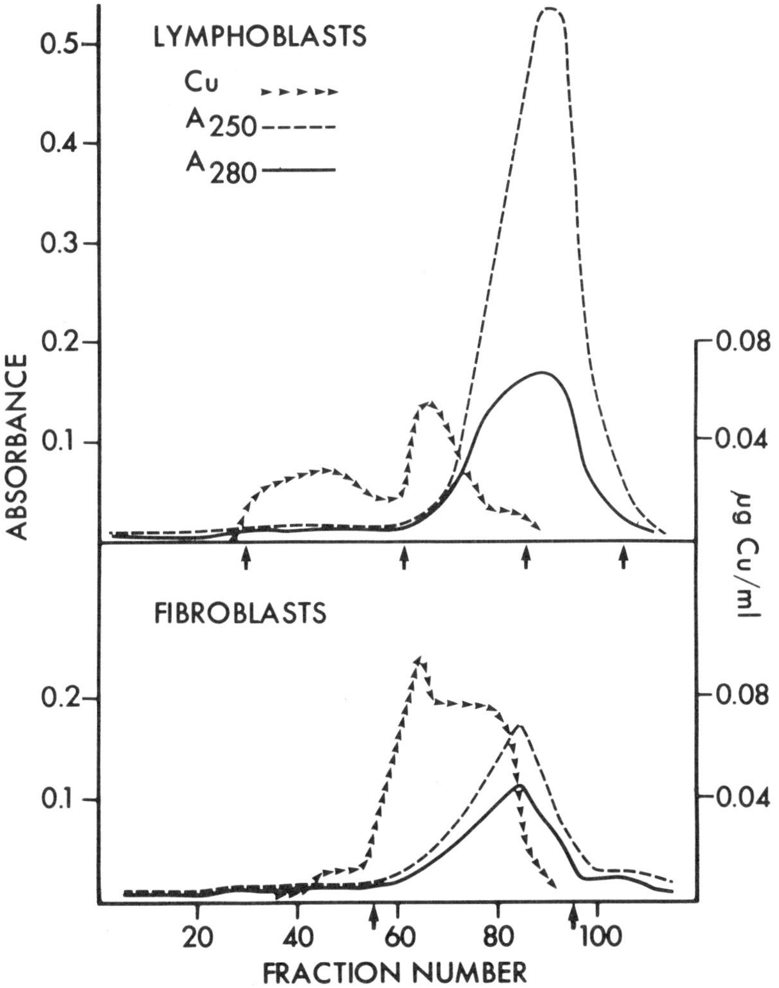

Figure 2. Elution of soluble fraction in 2,000 - 30,000 molecular weight range from a Sephadex G-75 column (2.5 x 90 cm) in 1 mM Tris-HCl, pH 8.6. Arrows indicate fractions which were pooled. A. Lymphoblasts; B. Fibroblasts.

With fibroblasts the distribution is roughly similar even though the amounts in the various fractions do not tally as well as in the experiment with lymphoblasts. Finally, however, 25% of the cell copper was recovered with the soluble fraction in the 2,000 to 30,000 molecular weight range.

The concentrated low molecular weight fractions from each type of cells was subjected to gel filtration chromatography on Sephadex G-75 (11). Fig 2 shows the profiles of absorption at 250 and 280 nm and copper concentrations. Fractions were pooled as indicated by the arrows, lyophilized and analyzed for copper and protein. Polyacrylamide gel electrophoresis in the presence of sodium dodecyl sulfate showed that in the case of lymphoblasts the major copper peak (centered at tube number 65) contained a single polypeptide with a molecular weight of approximately 7000 (Fig 3). This protein was also present in the comparable fraction from fibroblasts but some higher molecular weight material which stained for protein was also present. These fractions from both cell types contained about 1 g atom of copper per mole. Analysis of other metals have not yet been made.

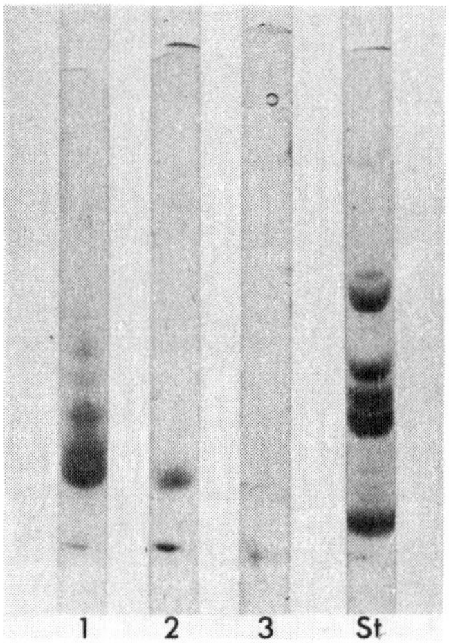

Figure 3. Protein staining bands after polyacrylamide gel electrophoresis in the presence of sodium dodecyl sulfate (10) of the first, second and third pooled fractions from Figure 2A (lymphoblasts). The fourth gel contains molecular weight markers: chymotrypsinogen A (25,000), superoxide dismutase (17,000), hemoglobin (subunit 16,750), cytochrome C (12,000) and insulin (5,734).

The amino acid compositions of these low molecular weight copper-binding proteins are shown in table II. The most obvious feature of these compositions is the similarity in the proteins from the two cell types. They are, however, distinctly different from the small copper binding proteins of human (1,17) and animal (9-16) liver. Especially evident are the very large amounts of glutamatic acid and glycine. The five half-cystine residues are similar to values for liver protein reported by some authors (9,10,13) but lower than those of others (14-17) and less than 20% of the amount which characterizes the metallothioneins (23-25). The discrepancies among cysteine contents determined by different investigators working with the same tissue are likely due to methodological variations (resulting in oxidative cleavage of cysteine in some cases; 16). However, the marked differences in nearly all residues between these cultured cell proteins and those from tissues such as liver tend to suggest that there may be genuine differences in primary structure. In view of this the similarities in size and ability to bind copper are remarkable.

Table II

Amino acid compositions of low molecular weight copper proteins.

Amino acid	Lymphoblasts (residues/100)	Fibroblasts (residues/100)
Aspartic acid	5.7	5.1
Threonine	2.2	2.2
Serine	2.2	3.6
Glutamic acid	28.3	32.3
Proline	1.8	1.4
Glycine	35.8	30.2
Alanine	5.2	6.5
Valine	2.0	2.2
Half cystine	5.4	5.2
Methionine	-	0.4
Isoleucine	1.9	1.8
Leucine	2.7	2.7
Tyrosine	-	-
Phenylalanine	1.2	1.1
Lysine	3.0	2.6
Histidine	0.9	1.2
Arginine	1.1	1.1

DISCUSSION

We have been able to demonstrate that human fibroblasts and lymphoblasts grown in culture take up approximately 100-200 pmol of copper per 10^6 cells from a buffer containing human albumin and 2.5×10^{-5} M copper during a time of one to three hours.

When the same cells were grown in the presence of 10 μg/ml Cu^{2+} prior to harvesting a major portion (>20%) of their copper content could be localized to a soluble fraction containing molecules in the 2,000 to 30,000 molecular weight range, primarily a protein of 7,000 molecular weight with about 1 g atom of bound copper per mole. The amino acid compositions of the protein from the two different cell types were remarkably similar and distinctly different from those of copper-binding proteins of similar size from tissues such as liver. However, it is important to emphasize the fact that the conditions of the current experiments in which complete subcellular fractionation was carried out were not optimized for purification of copper-thionein.

These cells should provide useful tools for the investigation of normal human copper homeostasis and abberations thereof in disease. For example, both these cell types from a patient with Menke's disease exhibited increased avidity for copper relative to cells from age and sex matched normal individuals (26). Hence, these cells seem to reflect the increased binding of the metal in this disease by all tissues including the intestinal mucosa but excluding the liver. We shall therefore make use of them to attempt to determine the molecular change responsible for the disease.

ACKNOWLEDGEMENTS

Supported by the Medical Research Council of Canada. Fibroblasts stains and lymphoblasts lines were established and grown in the laboratory of Dr. Manuel Buchwald, Dept. of Genetics. We thank D. Garthson and A. Barkas for the cell culture work.

REFERENCES

1. Morell, A.G., Shapiro, J.R. and Scheinberg, I.H. (1961) In: "Wilson's Disease: Some Current Concepts", J.M. Walsh and J.N. Cummings, eds., Blackwell, Oxford, p. 36-42.
2. Porter, H., Johnston, J., and Porter, E.M. (1962) Biochim. Biophys. Acta 65, 66-73.
3. Porter, H. (1971) Biochim. Biophys. Acta 229, 143-154.

4. Evans, G.W., Dubois, R.S. and Hambridge, K.M. (1973) Science 181, 1175-1176.
5. Bloomer, L.C. and Sourkes, F.L. (1973) Biochem. Med. 8, 78-91.
6. Evans, G.W. (1973) Physiol. Rev. 53, 535-570.
7. Porter, H. (1974) Biochem. Biophys. Res. Commun. 56, 661-668.
8. Brenner, I. and Davis, N.T. (1974) Biochem. Soc. Trans. 2, 425-427.
9. Evans, G.W. and LeBlanc, F.N. (1974) Nutr. Rep. Intl. 14, 281-288.
10. Riordan, J.R. and Gower, I. (1975a) Biochim. Biophys. Acta 411, 393-398.
11. Riordan, J.R. and Gower, I. (1975b) Biochem. Biophys. Res. Commun. 66, 678-686.
12. Premakumar, R., Winge, D.R., Wiley, R.D. and Rajagopalan, K.V. (1975) Arch. Biochem. Biophys. 170, 267-277.
13. Winge, D.R., Premakumar, R., Wiley, R.D. and Rajagopalan, K.V. (1975) Arch. Biochem. Biophys. 170, 253-266.
14. Brenner, I. and Young, B.W. (1976) Biochem. J. 157, 517-520.
15. Brenner, I. and Young, B.W. (1976) Biochem. J. 155, 631-635.
16. Hartmann, H.J. and Weser, V. (1977) Biochim. Biophys. Acta 491, 211-222.
17. Ryden, L. and Deutsch, H.F. (1978) J. Biol. Chem. 253, 519-524.
18. Goldstein, S. and Littlefield, J.W. (1969) Diabetes 18, 545.
19. Stanners, C.F., Eliceiri, G.L. and Green, H. (1971) Nature, New Biol. 230, 52.
20. Glade, P.R. and Broder, S.W. (1971) In: "In Vitro Methods in Cell Mediated Immunity", B.R. Bloom and P.R. Glade, eds. p.561.
21. Marceau, N., Kruck, T.P.A., McConnell, D.B. and Aspin, N. (1970) Int. J. Appl. Rad. Isotopes 21, 667-669.
22. Bohlen, P., Stein, S., Dairman, W. and Udenfriend, S. (1973) Archiv. Biochem. Biophys. 155, 213.
23. Kagi, J.H.R., Himmelhoch, S.R., Whanger, P.D., Bethune, J.L. and Valee, B.L. (1974) J. Biol. Chem. 249, 3537-3542.
24. Winge, D.R. and Rajagopalan, K.V. (1972) Archiv. Biochem. Biophys. 153, 755-762.
25. Madapallimattam, G. and Riordan, J.R. (1977) Biochem. Biophys. Res. Comm. 77, 1286-1293.
26. Riordan, J.R. and Madapallimattam, G. (1978) manuscript in preparation.

Metallothionein
Jeremias H. R. Kägi and Monica Nordberg, eds.
Birkhäuser Verlag, Basel/Boston/Stuttgart, 1979

RENAL METAL BINDING PROTEINS[1,2]

J.K.Piotrowski, J.A.Szymańska, E.M.Mogilnicka, A.J.Żelazowski

INTRODUCTION

The ability of metals, such as cadmium, zinc and mercury, to bind in the liver or kidneys with metallothionein or related proteins is linked with induction of biosynthesis of these proteins /1, 2/. Increased level of these proteins under the influence of metals can thus be used as indicator of essential interaction between various metals and metallothionein or related proteins.

Screening performed by our group /3, 4, 5/ using this line of evidence allows for classification presented in the Table 1.

Metal	Liver	Kidneys
Cd	++++	+++
Zn	+++	++
Cu	+	+
Co, Ag	+	-
Hg	-	+++
Bi	-	++
Au	+	+++
Rb, Cs, Be, Mg, Sr, Ba, Tl, Sn, Pb, V, As, Sb, Cr, Se, U	-	-

Table 1. Stimulating properties of various metals with regard to MTP levels in the liver and kidneys of rats.

It is striking that three of the metals, mercury, bismuth and gold, react positively mainly or exclusively in the kidneys, showing none or only slight interaction in the liver. The present report summarizes our attempts to elucidate the role of

1/Dept. of Toxicol. Chemistry, Institute of Environmental Research and Bioanalysis, Medical Academy, Narutowicza 120 a, 90-145 Łódź, Poland.

2/Supported in part by the Polish-American agreement 05-009-2 with NIOSH and grant 536/VI from the Division of Medical Sciences, Polish Academy of Sciences.

Submitted to the "First International Meeting on Metallothionein and Other Low Molecular Weight Metal-binding Proteins", Zürich, July 17 - 22, 1978

these renal proteins in binding bismuth, gold and mercury and to contribute to the understanding of their nature.

In our initial studies we assumed that these proteins are closely related to metallothionein, thus in our previous reports the term "metallothionein-like proteins" /MTP/ had been often used. In view of our more recent studies it is not certain if these renal proteins belong to the family of metallothionein. Before adequate evidence will be obtained to resolve this question, the term "renal metal binding proteins" /RMBP/ seems justified.

INDUCTION OF RMBP BY BISMUTH, GOLD AND MERCURY

Evidence for the stimulating effect of the three metals upon the RMBP has been obtained in two ways.

/a/ Multiple exposure of rats was applied /sc injections, every other day for a fortnight/ following which the level of RMBP was determined on the assumption that they belong to the family of metallothionein /Mt/. In these studies the early version of the radiochemical method /6/ was applied. Table 2 shows that all three metals enhance the level of Mt mainly in the kidneys. Only in the case of gold some involvement of the liver is apparent. For comparison, contrasting behaviour of cadmium is shown.

Metal	Dosage mg Me/kg, sc	I/N		References
		Liver	Kidneys	
Hg	7 x 0.5	0.8	5.0	1, 3
Bi	7 x 1.5	1.5	3.1	4
Au	7 x 10	3.0	5.6	8
Cd	7 x 2	39	4.1	1, 3

Table 2. The elevation of Mt in the kidneys of rats following repeated administration of mercury,bismuth and gold. I- level of Mt following exposure, N- level of Mt in non exposed rats.

/b/ Following multiple exposure of rats as under /a/, ^{35}S-cysteine was administered and the incorporation of the tracer was followed in the fraction of MW about 10.000 following filtration of the kidney extract on Sephadex G-75. Both in the case of Bi /7/ and Au /8/, the tracer was incorporated partly into the fraction of MW 10.000. In rats not exposed to metals, only negligible incorporation occured in this fraction. If the administration of ^{35}S-cysteine was preceded by cycloheximide, the incorporation of ^{35}S into the said fraction was completely abolished. Actinomycin D /inhibitor of the transcription/ was not effective in this respect following multiple exposure to metals, and in the case of single exposure was less effective than cycloheximide /7/. Evidence for induced biosynthesis of the mercury-binding protein which had been obtained earlier /9/ was confirmed now with results similar to those for bismuth and gold.

Both the above lines of evidence point to the fact, that exposure of rats to mercury, bismuth and gold elevates the level of RMBP. The process in question is conditioned upon the induction of the novo synthesis of proteins, in the same way as is induction of metallothionein by cadmium and zinc /10,11/.

A common feature, accompanying the induction of the RMBP by mercury, bismuth and gold, appeared to be a distinct elevation of the copper level in the kidneys, as shown in Fig. 1. It strongly suggests that the RMBP belong to the family of low molecular weight inducible copper-proteins.

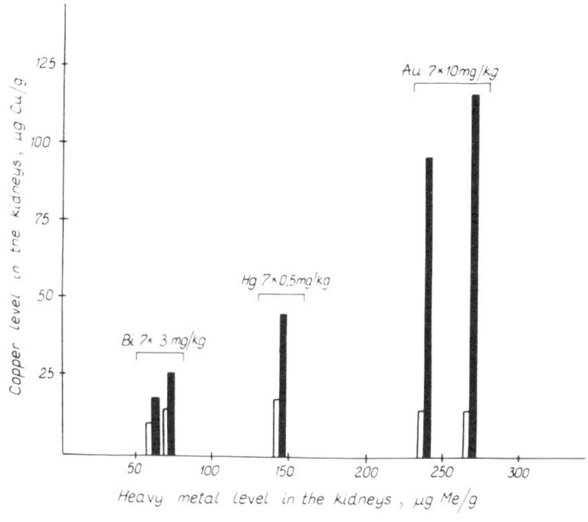

Fig. 1. The influence of bismuth, mercury and gold on the level of copper in the kidneys. Dashed columns - exposed rats, empty columns - control rats.

IN VIVO BINDING OF BISMUTH, GOLD AND MERCURY BY RMBP

In the case of all three metals the kidneys play an essential role in the systemic depositioning of the metal. Following multiple exposure the contribution of the kidneys to the total body burden amounts to approximately 80 % for mercury, 55 % for bismuth and 35 % for gold /12, 13, 14/. The complexes of mercury and bismuth with RMBP are contained mainly /or exclusively/ in the soluble fraction of the homogenate /15, 13/. The relative contribution of the metal complexes with RMBP to the total metal content of the soluble fraction has a rising tendency following multiple exposure, reaching 70 - 80 % for mercury, 60 - 70 % for bismuth, and about 50 % for gold /9, 13, 14/. Examples for bismuth and gold are shown in Fig. 2. From the above it follows that the mechanism of metal binding by RMBP is biologically essential in the case of all three metals. Its relative role decreases in the order: mercury, bismuth, gold. Although this mechanism is not unique in the process of long term retention and accumulation of these metals /15, 13/, it plays an essential role in the specific pattern of organ distribution of the metals under consideration, especially if multiple exposure is concerned.

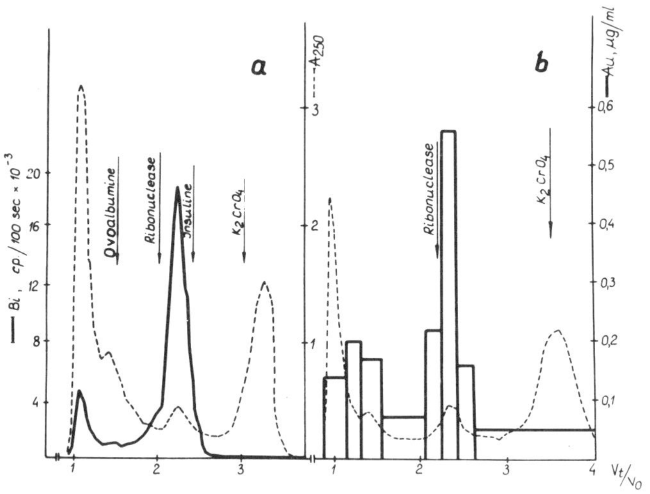

Fig. 2. Gel filtration of kidney extracts from rats, administered multiple doses of /a/ bismuth, /b/ gold.

SOME PROPERTIES OF RMBP ISOLATED FROM RATS EXPOSED TO BISMUTH, GOLD AND MERCURY

In experiments aiming at isolation of RMBP rats /in groups of 20 to 60 in various series/ were administered the metal salt /$BiCl_3$, $HgCl_2$, $HAuCl_4$ neutralized to pH 6/ sc., seven doses every other day /0.5 mg Hg/kg; 3 mg Bi/kg; 10 mg Au/kg/.Kidney homogenates were centrifuged at 10.000 g, and the supernatant was subjected to acetone fractionation. Sediments at 80 % saturation, regarded as crude RMBP, were further subjected to gel filtration on Sephadex G-75 /twice if necessary/ and separated on DEAE-cellulose in step-wise Tris-HCl gradient /Bi, Au/ or by polyacrylamide gel electrophoresis /Hg, Bi/. Determinations of mercury were performed radiochemically using ^{203}Hg, bismuth was determined radiochemically using ^{206}Bi or colorimetrically with DBDTC-Zn /16/, gold colorimetrically with rhodamine B /17/, copper colorimetrically with DBDTC-Zn /16/. Protein was determined nephelometrically with tannic acid /18/. The following observations were made:

1. Bismuth

The protein /19/, purified by filtration on Sephadex G-75 was intensely yellow. On the DEAE-cellulose it has separated into two major components, emerging at 0.1 and 1.0 M concentration. These components had a low A-250/280 ratio of only 1.5 and 0.8 and differed in spectral properties /Fig. 3/. Absorbance maxima at about 220 respectively 230 nm seem characteristic together with a second maximum /or shoulder/ at 280 respectively 270 nm.

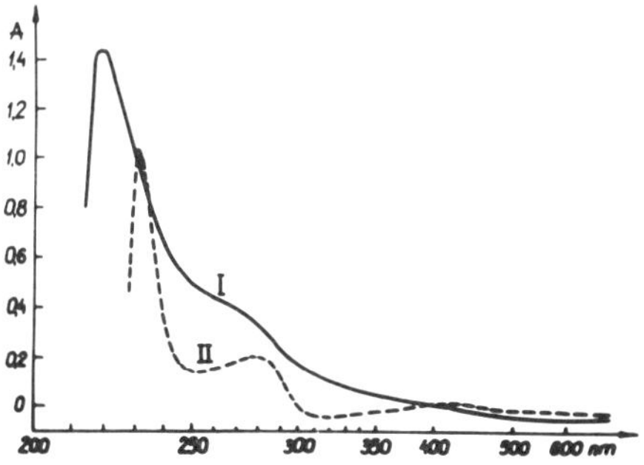

Fig. 3. Absorbance spectra of two fractions of the bismuth binding protein, obtained by DEAE-cellulose /I - fraction at 0.1 M, II - 1.0 M Tris-HCl.

Component I /yellow/ from DEAE-chromatography was further studied yielding the following characteristics: MW 7.500 /Sephadex G-75/, metal content: Bi - 4.2 %, Cu - 1.8 %, SH content determined amperometrically /20/ - 1.3 µmole/mg. Isoelectric focusing yielded three fractions of pI 3.8, 5.5, and 7.4. By aminoacid analysis in which it was directly compared with horse kidney metallothionein, it showed cysteine content lower by 1/3, presence of histidine and essential amounts of leucine, as well as of three unidentified components which were absent in metallothionein. Simultaneous comparison with metallothionein obtained from the kidneys of rats exposed to cadmium, has pointed to differences in molecular weight and spectral properties.

Alternatively, the RMBP obtained by acetone fractionation and gel filtration was subjected to electrophoresis on polyacrylamide gel. The protein yielded two major fractions /R_f 0.32 and 0.57/ and two minor fractions /R_f 0.13 and 0.80/ containing Bi, of which all contained copper.

It is suspected that the renal bismuth binding protein is not identical with metallothionein. For the major fraction a tentative name "renal chromochelatin" has been proposed /19/.

2. Gold

On gel filtration the protein was intensely yellow. The UV-spectrum, resembling those of the Bi-binding proteins, showed maximum at about 225 nm and a slight shoulder between 250 and 280 nm /Fig.4/. The crude fraction contained 1.15 % gold and 2.6 % copper. Following chromatography on DEAE-cellulose the protein yielded four fractions, emerging at 0.005, 0.10, 0.15, and 0.25 M Tris-HCl. Fractions I and IV displayed MW 8.400 and 11.700, respectively. In the course of purification the proteins tended to lose part of gold and to aggregate. All fractions of the protein contained gold in amounts of 1.1 - 3.2 %, and copper above 1.8 %. All fractions, independent of the technique of seperation, were characterized by a low A-250/280 ratio, 1,4 - 2.2.

The above data point to the fact, that the renal gold-binding proteins belong to the family of low molecular weight inducible copper-proteins.

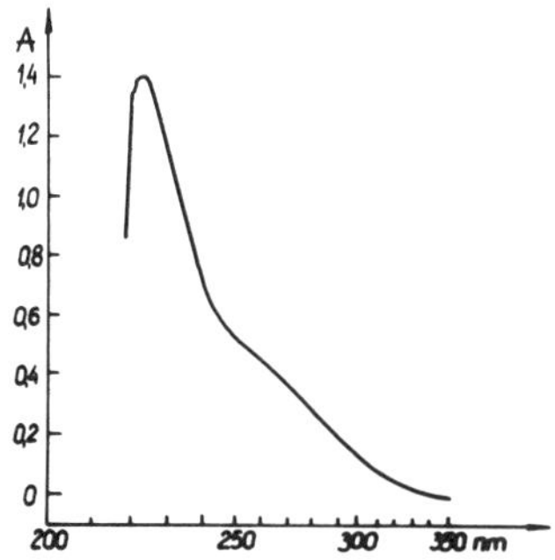

Fig. 4. UV-absorbance spectrum of the gold binding protein, following chromatography on Sephadex G-75

3. Mercury

The protein obtained through filtration on Sephadex G-75 was colourless and contained in various preparations mercury /5.7 - 8.9 %/ and copper/3.0 - 4.5 %/. The estimates of molecular weight varied in various series from 7.500 to 9.500. Ion exchange chromatography on DEAE-cellulose gave poor protein recovery. When separated by electrophoresis on polyacrylamide gel, it yielded four fractions /Fig. 5/ of which all contained mercury and copper as well as detectable SH groups /staining in gel/ /21, 22/.

The electrophoretic pattern could not have been reproduced by in vitro saturation with mercury of horse kidney metallothionein.In the latter case mercury appeared to be bound in a labile way allowing for dissociation of mercury from the protein during electrophoresis.

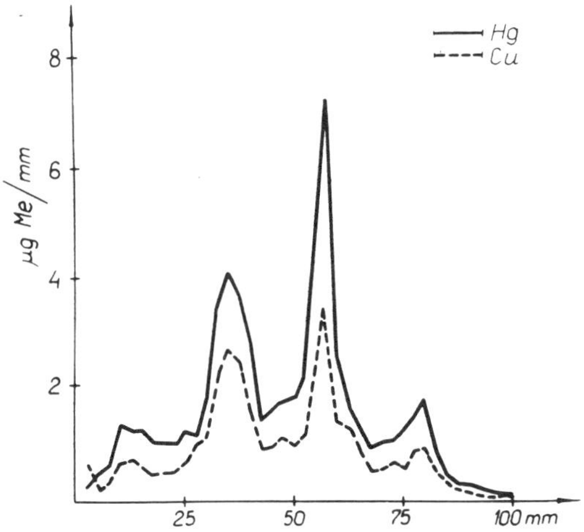

Fig. 5. Mercury and copper distribution following electrophoresis of the mercury-binding proteins on polyacrylamide gel.

DISCUSSION

Among the three metals discussed in this report, the first evidence of interaction with a low molecular weight renal protein was obtained for mercury as early as 1970 /23/. The early identification studies /24/ pointed to identity of the said protein with metallothionein. In these experiments, however, partial binding of mercury on the cadmium-induced metallothionein could not have been excluded. Subsequent attempts to characterize the protein more closely /25, 26/, performed in other laboratories, did not bring unequivocal conclusions. Assuming that the mercury binding protein was identical with metallothionein it has been difficult to understand, why mercury and cadmium do not bind in the same organ, and why mercury does not induce metallothionein synthesis in the liver where the induction exerted by cadmium is enormously high /Table 2/.

Our subsequent studies revealed another metal, bismuth, which behaved in a much similar way as mercury. In this case, however, the metal binding protein isolated from the kidneys displayed potentially significant differences as compared with metallothionein. Our special attention has been called by the relatively high content of copper, exceeding by one order of magnitude that reported for metallothionein /27/.

The third metal which exhibited evident interaction with the system of renal binding proteins, gold, was shown to induce synthesis of proteins containing also high levels of copper.

The above facts have prompted us to reexamine the nature of the mercury-binding renal protein. Evidence gathered till now strongly suggests that the protein in question belongs to the same family as those binding bismuth and possibly gold. The relation of these proteins to that induced in rat kidneys by cadmium which also appeared to be a copper-protein /28, 29/ is under investigation.

Rat kidneys contain copper in concentrations ranging between 5 and 20 μg per g of tissue. In rats non exposed to metals part of copper is bound by low molecular weight proteins. This fraction isolated by Sephadex filtration, displayed a molecular weight of approximately 7.500 /19/. Following electrophoresis on polyacrylamide gel it separates into two major and at least one minor fraction. The electrophoretic mobility of the major fractions resembles that of the proteins induced by bismuth and mercury.

The low molecular weight renal copper proteins have not been yet thoroughly investigated. Nevertheless a physiological role of these proteins in the copper homeostasis seems probable. Since these proteins appear in tissues of non exposed rats in several fractions it seems probable that the interaction with heavy metals occurs with some preference of a given metal /bismuth, mercury, gold/ to certain proteins of this family. Such hypothesis would offer explanation to the differences of RMBP induced by these three metals, especially to their variable Cu-content. Verification of this hypothesis requires further studies.

REFERENCES

1. Piotrowski, J.K., Trojanowska, B., and Sapota, A. /1974/ Arch. Toxicol. 32, 351 - 360.
2. Trojanowska, B., and Piotrowski, J.K. /1974/ Bromat. Chem. Toksykol. 7, 137 - 142.
3. Mogilnicka, E.M., Piotrowski, J.K., and Trojanowska, B. /1975/ Med. Pracy 26, 147 - 155.
4. Piotrowski, J.K., and Szymańska, J.A. /1976/ J. Toxicol. Environ. Health 1, 991 - 1002.
5. Mogilnicka, E.M., Piotrowski, J.K., and Tomaszewski, A. /1976/ Bromat. Chem. Toksykol. 9, 357 - 359.
6. Piotrowski, J.K., Bolanowska, W., and Sapota, A. /1973/ Acta Bioch. Polon. 20, 207 - 215.
7. Szymańska, J.A., and Żelazowski, A.J. /in preparation/.

8. Mogilnicka, E.M. /1978/ PhD-thesis, Medical Academy of Łódź.
9. Piotrowski, J.K., Trojanowska, B., Wiśniewska-Knypl, J.M., and Bolanowska, W. /1974/ Toxicol. Appl. Pharmacol. 27, 11 - 19.
10. Squibb, K.S., and Cousins, R.J. /1974/ Environ. Physiol.4, 24 - 30.
11. Squibb, K.S., Cousins, R.J., and Feldman, S.L. /1977/ Biochem. J. 164, 223 - 228.
12. Trojanowska, B. /1970/ PhD-thesis, Medical Academy of Poznań.
13. Szymańska, J.A., Zychowicz, M., Żelazowski, A.J., and Piotrowski, J.K. /1978/ Arch. Toxicol. 40, 131 - 141.
14. Mogilnicka, E.M., and Piotrowski, J.K. /1977/ Biochem. Pharmacol. 26, 1819 - 1820.
15. Komsta-Szumska, E., and Chmielnicka, J. /1977/ Arch. Toxicol. 38, 217 - 228.
16. Szymańska, J.A., and Zychowicz, M. /1977/ Chem. Anal. 22, 985 - 988.
17. MacNulty, B.J., and Woollard, L.D. /1955/ Anal. Chim. Acta 13, 154 - 158.
18. Mejbaum-Katzenellenbogen, W. /1955/ Acta Biochim. Polon. 2, 279 - 296.
19. Szymańska, J.A. /1977/ PhD-thesis, Medical Academy of Łódź.
20. Benesh, R.E., Lardy, H.A., and Benesh, R. /1955/ J.Biol. Chem. 216, 663 - 670.
21. Żelazowski, A.J., and Piotrowski, J.K. /1978/ XVI Meeting of Polish Bioch. Soc., Łódź.
22. Żelazowski, A.J. /1978/ XVI Meeting of Polish Bioch. Soc., Łódź.
23. Jakubowski, M., Piotrowski, J., and Trojanowska, B. /1970/ Toxicol. Appl. Pharmacol. 16, 743 - 753.
24. Wiśniewska, J.M., Trojanowska, B., Piotrowski, J., and Jakubowski, M. /1970/ Toxicol. Appl. Pharmacol. 16, 754 - 763.
25. Nordberg, M., Trojanowska, B., and Nordberg, G.F. /1974/ Environ. Physiol. Biochem. 4, 149 - 158.
26. Cherian, M.G., and Clarkson, T.W. /1976/ Chem. Biol. Interactions 12, 109 - 120.
27. Kägi, J.H.R., and Vallee, B.L. /1960/ J. Biol. Chem. 235, 3460 - 3465.
28. Suzuki, T.K., Kubota, K., and Takenaka, S. /1977/ Chem. Pharm. Bull. 25, 2794 - 2796.
29. Szymańska, J.A., and Żelazowski, A.J. /1978/ Environ. Res. /submitted/.

AUTHOR INDEX

SUBJECT INDEX

Numbers in italics refer to pages in the Report.
For a detailed subject index of the Report see pp. 122-124.

A

B

C

Abbreviation: MT, metallothionein

Abbreviation: MT, metallothionein

Abbreviation: MT, metallothionein

Abbreviation: MT, metallothionein